Lithium-Ion Batteries and Applications

A Practical and Comprehensive Guide to Lithium-Ion Batteries and Arrays, from Toys to Towns

Volume 2

Applications

For the *Artech House Power Engineering Library*
go to the back of this book.

Lithium-Ion Batteries and Applications

A Practical and Comprehensive Guide to Lithium-Ion Batteries and Arrays, from Toys to Towns

Volume 2

Applications

Davide Andrea

ARTECH HOUSE

BOSTON | LONDON
artechhouse.com

Library of Congress Cataloging-in-Publication Data

A catalog record for this book is available from the U.S. Library of Congress.

British Library Cataloguing in Publication Data

A catalog record for this book is available from the British Library.

ISBN-13: 978-1-63081-769-5

Cover design by John Gomes

© 2020 Artech House
685 Canton Street
Norwood, MA 02062

To Ann

CONTENTS

CHAPTER 2
LARGE, LOW-VOLTAGE BATTERIES 45

CHAPTER 3
TRACTION BATTERIES   161

CHAPTER 4
HIGH-VOLTAGE STATIONARY BATTERIES 279

CHAPTER 5
ACCIDENTS 347

APPENDIX A
BATTERIES 357

APPENDIX B
APPLICATIONS 399

APPENDIX C
RESOURCES 429

PREFACE

This book is the second volume of a two-volume set. This book is incomplete without Volume 1. Please refer to the Preface in Volume 1 for an introduction to both books.

Volume 1 discusses batteries (Figure P.1(a)):

- Basic concepts, common misunderstandings, and introduction of new concepts and terms (Chapter 1);
- Lithium-ion cells (Chapter 2) and their arrangement (Chapter 3);
- Battery management systems (BMS) (Chapter 4);
- All this is put together into a battery (Chapter 5) or multiple batteries (Chapter 6) and assembled (Chapter 7);
- Preventing and addressing dysfunctions and accidents (Chapter 8).

Volume 2 discusses four classes of applications that constitute the majority of Li-ion battery usage (Figure P.1(b)):

- Small batteries (Chapter 1);
- Low-voltage batteries and battery arrays (telecom, residential, house power) (Chapter 2);
- Traction battery packs for vehicles (unmanned aerial vehicles (UAVs), passenger electric vehicles (EVs), marine, industrial) (Chapter 3);
- High-voltage, stationary batteries, on-grid and, off-grid (Chapter 4);
- Case studies of accidents (Chapter 5).

Thank you for reading.

FIGURE P.1
Book flow: (a) Volume 1—
batteries, and
(b) Volume 2—applications.

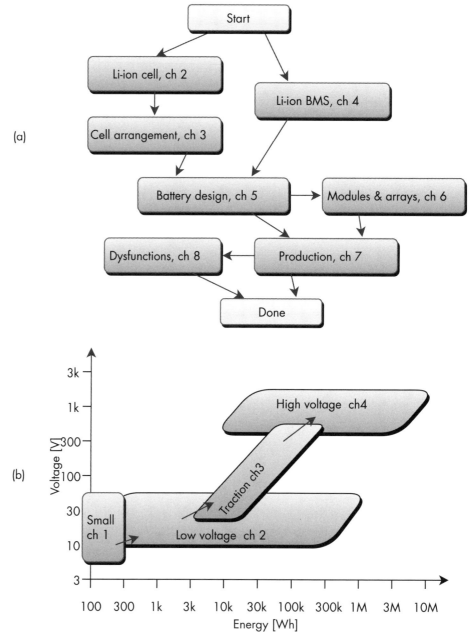

CHAPTER 1

SMALL BATTERIES

1.1 INTRODUCTION

Small batteries make up the vast majority of Li-ion batteries in the world. This book defines small batteries as ranging from 3V to 48V and from 0.03 to 300 Wh. This chapter explores a range of solutions for small batteries to help you select the best one for your battery design.

1.1.1 Tidbits

Some interesting items in this chapter include:

- Your phone's charger is actually a power supply; the actual charger is inside the phone (Section 1.2.1).
- A battery with no output voltage is probably not dead; it merely needs to be woken up (Section 1.4.4).
- A charger with no output voltage probably needs to see a battery voltage before it will start charging (Section 1.2.1.3).
- A power bank with no output probably shut down simply because the load current is too low (Section 1.5.1.1).
- Sometimes the BMS is not in the battery; it's in the product (Section 1.4.2.1).
- Some products work only with the original battery; those batteries work only with the original product (Section 1.4.1.3).
- Chinese PCMs have a poor reputation; unfortunately, the problem may be the user (Section 1.6.2.1).

1.1.2 Orientation

This chapter starts with a discussion of chargers and alternating current (AC) adapters used with small batteries. It then discusses internal (inside the product) and external (removable) batteries, with one or more cells; batteries for power tools and laptops are used as examples. Although there are thousands of applications for small batteries, this chapter only looks at a few of them. Finally, it discusses small battery design and their battery management systems (BMSs).

The focus of this chapter is the battery, specifically the BMS[1]. It is not a complete guide for designing small batteries. Appendix C recommends books that specialize in the design of small batteries.

1.1.3 Applications Classification

This chapter divides the most common applications for small batteries based on the number of cells in series and the placement of the battery within a product:

- *Internal:* installed inside a product (Figure 1.1(a)): single-cell (Figure 1.2(a)) and multicell (Figure 1.2(b));
- *External:* on the outer surface of a product, removable by the user (Figure 1.1(b)): single-cell (Figure 1.2(c)) and multicell (Figure 1.2(d)).

Of course by "single-cell," I also mean a block of cells in parallel.

FIGURE 1.1
(a) Internal battery, and
(b) external battery.

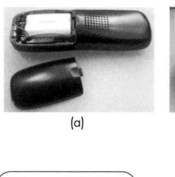
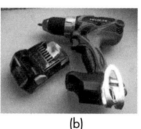

FIGURE 1.2
Small battery classification:
(a) single-cell, internal,
(b) multicell, internal,
(c) single-cell, external,
and (d) multicell, external.

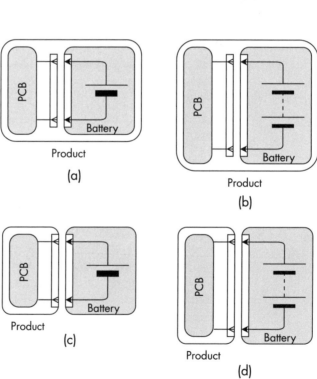

---

1.   This book refers to particular sections in Volume I, first edition; the section numbering of future editions may differ. See the "Resources" section in the Appendix C

2.   Answer: "c." (a) NiMH cells charger. (b) A dinner plate on a charger (a decorative base setting upon which the serving dinnerware is placed for formal occasions). (c) A USB AC power adapter (not a charger). (d) A Dodge Charger muscle car

1.2 DEVICES

Applications that use small batteries may use some of the following devices.

1.2.1 AC Adapters and Chargers

Can you tell a charger when you see one (Figure 1.3(b))?

A common misunderstanding is that the device that powers a laptop or a phone is a charger (Figure 1.4(a)). It isn't; it's a power supply (also known as an AC adapter). The charger itself is inside the laptop or phone (Figure 1.4(c)). The typical AC/DC consumer product can be powered by AC or it can be run on a battery. Such a product uses both an external power supply and an internal charger.

Conversely, a stand–alone battery charger truly is a charger (Figure 1.4(b)).

1.2.1.1   AC/DC Circuits

For example, the power supply for a laptop computer (Figure 1.5(a)) may produce 18V. The charger inside the laptop drops this voltage down to the battery voltage (e.g., 10V to 16.8V, depending on the battery's state of charge (SoC)). Finally, a DC-DC converter changes this variable battery voltage to a fixed voltage (e.g., 5V) for the actual electronics.

A cell phone is similar, except that the voltages are different (Figures 1.4(c) and 1.5(b)): the USB input is at 5V, the single-cell battery ranges from 2.5V to 4.2V, and the load is powered at 3.3V.

A simpler device may use only an AC-powered charger and operate directly at the battery voltage (Figure 1.5(c)). When looking at an AC adapter, you can't tell whether it's a plain power supply (constant voltage output only) or a charger (constant current/constant voltage (CCCV) output) (see Volume 1, Section 1.8.3).

FIGURE 1.3
One of these things is not like the others[2].

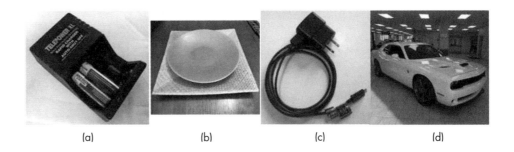

(a)　　　　　　(b)　　　　　　(c)　　　　　　(d)

FIGURE 1.4
Devices: (a) manufacturer adds to the confusion, (b) stand-alone charger, and (c) AC adapter and charger in a cell phone.

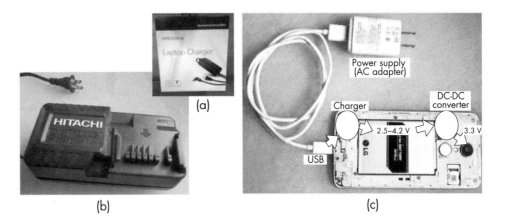

(a)

(b)　　　　　　(c)

FIGURE 1.5
Power supplies and
chargers: (a) laptop
computer, (b) cell
phone, (c) simpler
consumer product, and
(d) battery charger.

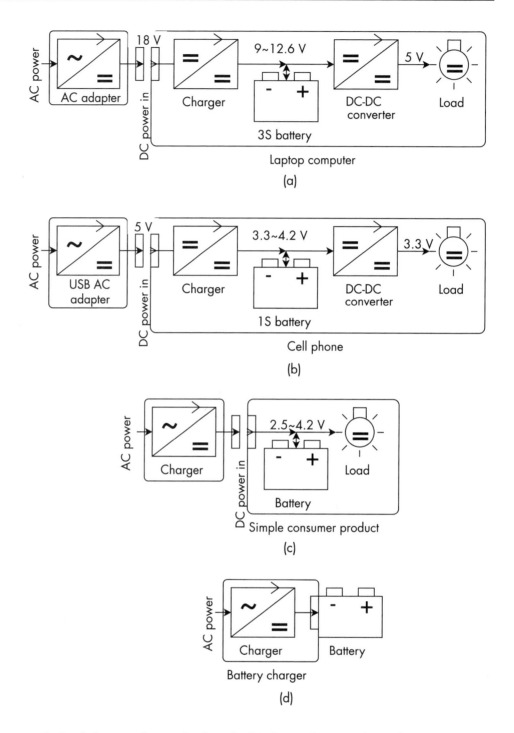

FIGURE 1.5
Power supplies and chargers: (a) laptop computer, (b) cell phone, (c) simpler consumer product, and (d) battery charger.

A simple battery charger (such as the kind you plug into the wall and then insert cells in it) truly is a charger (Figures 1.4(b) and 1.5(d)).

A power supply for a product that uses a small battery may be a wall wart (a power supply mounted directly on an AC outlet), or a brick (it has an AC power cord going to an AC outlet).

1.2.1.2 Laptop Computer Power Management

The circuit in the previous example (Figure 1.5(a)) is simple (see Section 1.7.1.1), although it has a few disadvantages:

- The load is powered through two converters; the charger and the DC–DC converter; the total efficiency is worse than if the DC–DC converter were powered directly from the DC input.

- The current into the battery depends on the load current; if the load is light, more current is available for charging the battery, which may be more than the battery can handle.

A laptop computer uses a more sophisticated power management circuit to overcome these limitations. A switch selects the source that powers the DC-DC converter (Figure 1.6) (see Section 1.7.1.4):

- *AC power present:* the switch selects power directly from the DC power input.

- *AC power absent:* the switch selects power from the battery; the charging process is better managed because the current from the charger only feeds the battery.

1.2.1.3 Chargers That Require a Battery Before Starting

Before they start charging, some chargers ensure that they are connected to a battery by sensing the voltage on their output. This is done for safety, and to avoid charging an overly discharged battery. While this is fine for a lead–acid battery (Figure 1.7(a)), it doesn't work for a Li–ion battery when its protection switch is open (Figure 1.7(b)).

If a Li–ion battery has disabled discharging (i.e., because its SoC is too low), there is no voltage on its terminals. Yes, charging is enabled, but no current is allowed to flow out of the battery to power the battery terminals. As the charger does not detect a battery, it does not start charging.

An ugly workaround is to prime the charger by temporarily connecting a smaller charger or another battery to its output (Figure 1.7(c)). A cleaner solution is to allow a tiny bit of discharge current (Figure 1.7(d)). A high-resistance resistor bypasses the discharge metal oxide semiconductor field effect transistor (MOSFET), placing a voltage across the battery output even when discharge is disabled. That current is too small to power a load connected to the battery; it is sufficient to tell a charger that a battery is connected. The capacitor stores enough energy to feed the charger until the charger starts charging the battery; without the capacitor, the load in the charger's output may drag the battery's output voltage too low.

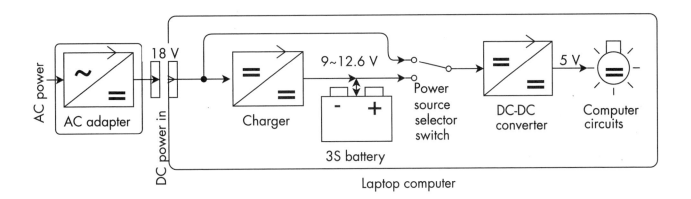

FIGURE 1.6 Laptop power management.

FIGURE 1.7
Charger that needs to see
a battery before starting:
(a) does not see a Li-
ion battery that disabled
discharge, (b) because the
discharging MOSFET
is off, (c) priming with
a 12V lead-acid battery,
and (d) self-priming
circuit inside the battery.

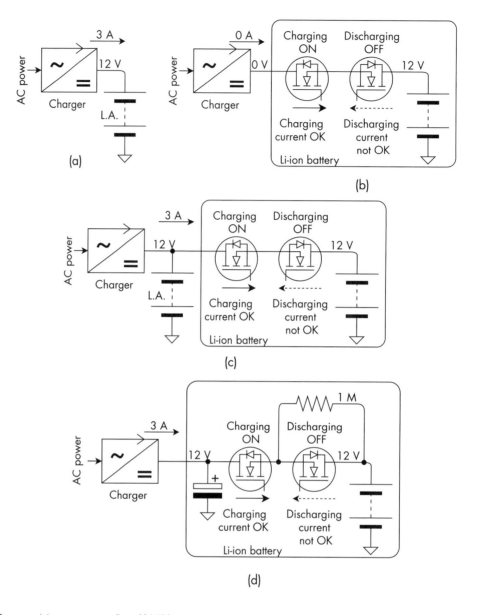

1.2.2   System Management Bus (SMB)

SMB is the standard communication bus [1] used by PCMs for small external batteries. The bus is typically used for[3]

- Configuring the PCM (at the factory);
- Reporting the status of the battery to the product;
- Letting the product control any optional inputs and outputs on the PCM.

This bus is based on the I^2C bus standard, which was designed for communication between integrated circuits within a single printed circuit board (PCB). It is a master/slave, multipoint bus. The host (i.e., the product) is the master[4]. It is only appropriate for short distances (a few centimeters) and low data rates, up to 100 kbps. It uses two

---

3.   See the specification sheets for protector BMS ICs, from Maxim, TI, AD, and others.
4.   The battery may become a master if it needs to transmit a warning.

lines: clock and data. The master generates the clock and initiates communications. Data can flow in either direction, depending on the message. Either device may pull these lines down to ground; otherwise, resistors in the master pull the lines back up to the logic supply voltage (typically 5V).

1.3 INTERNAL BATTERY

An internal battery is integrated into a product. The battery is not intended to be removable by the user, although a service technician may access it after opening the product.

1.3.1 Technical Considerations

An internal battery is simpler, as its product provides it with mechanical protection.

1.3.1.1 Options

An internal battery may use many options, resulting in many permutations:

- May be single-cell or multicell (from two to 16 cells in series);
- May or may not have a PCM or may have only a portion of a BMS; if with a PCM:
  - With or without SMB communications;
  - Single-port or dual-port.
- For each power terminal it may use one or more contacts;
- May use one or two connectors; if two connectors:
  - A high-power connector;
  - A low-power balance connector.
- May or may not include a thermistor to report the battery temperature to the product;
- May or may not include an enable line.

1.3.1.2 High-Power Battery Connector

For the high-power connector, RC[5] batteries use connectors of uncertain parentage and wildly speculative specifications. They are known by their generic names (unrelated to brands): RCY (also known as BEC), XT30, XT60, Deans, EC3, and Tamiya.

Alternatively, batteries may use individual terminals on each wire, such as bullet or HXT terminals.

1.3.1.3 Low-Power Battery Connector

While the general public refers to the low-power connector in an internal battery as a JST connector[6], technically it is a single-row, female rectangular connector[7]. Its pitch[8]

---

5. Radio Control miniature vehicles.
6. Many manufacturers make these connectors; yet vendors catering to hobbyists call these "JST connectors," even though this is technically incorrect, as JST is a particular manufacturer, not a type of connector.
7. The full description for its housing is "rectangular connector, housing, single-row, non-latching, for crimped female contacts."
8. Center-to-center spacing between adjacent contacts. The best way to measure the pitch of an N-pin connector is to measure between the first and the last terminal and divide by N-1. For example, if the center-to-center distance between the first and the last pin in a 7-pin connector is 12 mm, the pitch is 12 mm / (7 − 1) = 2 mm.

ranges from 0.8 to 2.5 mm, depending on the battery current and physical size. Table 1.1 lists some possibilities.

The small-battery industry has tried to standardize the power connector and its pin-out, with partial success. The XH connector (by JST) seems to be winning right now [2].

Notes

- The connector on the wires consists of a housing and matching female crimp contacts (i.e., the sockets) that are bought separately.
- The mating connector is a PCB mount male header, either through–hole or SMD, either straight or right angle.
- Pin 1 is marked with a triangle or with the manufacturer's name; when in doubt, consult the specification sheet.
- Watch out for polarity! Some manufacturers place B– on pin 1, and others place B+ on this pin.

1.3.1.4   Mechanical Design

The product's enclosure cradles the cells to keep them from rattling inside the case. For pouch cells, the enclosure should provide retention against cell expansion.

1.3.1.5   Assembly

Battery assembly may include the following steps:

- Interconnect cells as appropriate (see Volume 1, Sections 5.4 and 5.6):
  - Weld cylindrical cells to nickel strips.
  - Weld or clamp the tabs of pouch cells; soldering may be possible, if done with care.
- If using a thermal fuse, place it in series with the cells, at the end of the string.
- For multicell batteries, prepare a balance connector with wire tails; connect the balance connector wires to the cells.
- Prepare a power connector with two– wires and possibly a thermistor.
- Connect the power connector wires to the two end terminals.
- Install the thermistor, if any.
- Wrap the battery in large heat shrink tubing and shrink.

TABLE 1.1
Low-Power Connectors
for Internal Batteries

Pitch	Manufacturer
1 mm	Harwin: M40; JST: SH; Molex: Picoclasp; TE: HPI; Wurth: WR-WTB
1.25 mm	Hirose: DF13, DF14; Molex: PanelMate, Picoblade; TE: HPI
1.5 mm	JAE: IL-Y; JST: ZE, ZH; Molex: Picospox; TE: HPI, Mini CT; Wurth: WR-WTB
2 mm	Hirose: DF3; HR-connector: A2006 (1); JST: PA (2), PH; Molex: Microblade, Micro-Latch; Sullins: SWH201; TE: CT, HPI
2.5 mm	JAE: IL-G; JST: EH, XH; Molex: Mini-lock, Mini-spox; TE: EI, HPI, MIS; Yeonho: SMH250 (3)
Notes: (1) Used by Thunder Power (code TP); (2) Used by CellPro (code CP); (3) Used by Hyperion (Polyquest) (code HP). Common connectors are in bold.	

1.3.2   Single-Cell Internal Battery

A single-cell battery (see Volume 1, Section 5.1.3) may be integrated directly into a product. This battery may use a single pouch cell, or, less likely, a single small cylindrical cell.

1.3.2.1   Topologies

A few topologies are possible:

- A permanently connected cell (Figures 1.8(a) and 1.9(a))
- An unprotected battery (Figures 1.8(b) and 1.9(b))
- Only the protector switch in the battery (Figure 1.9(c))
- A stand-alone (protected) battery (Figures 1.8(c), 1.8(d), 1.9(d))
- A stand-alone SMB battery (Figure 1.9(e))

If a thermistor is included in the battery and made available on the connector, the product may read the battery's temperature to provide redundant protection. If the cells are permanently connected to the PCB, so is the thermistor (Figure 1.9(a)). Otherwise, the battery connector includes an extra pin for the thermistor (Figures 1.8(d) and 1.9(b–e)). Each of these topologies is described in more detail below.

1.3.2.2   Permanently Connected Cell

Small cheap products, such as toys that do not expect the cell to be replaced, may use a single cell permanently soldered to the product's main PCB (Figure 1.9(a)). There is no connector. This battery is likely to consist of a pouch cell whose tabs are soldered directly to the PCB or a small cylindrical cell with welded tabs that are then soldered to the PCB (Figure 1.8(a)).

The BMS is in the product. It may consist of only a CCCV charger and a low-voltage cutoff (see Section 1.6.2.2). This solution is straightforward, as it saves interconnections.

1.3.2.3   Unprotected Battery

Other, not-so-small consumer products may use an unprotected battery (Figure 1.9(b)). Once more, the BMS is in the product and may consist of only a CCCV charger and a low voltage cutoff. The battery consists of a single cell (likely to be a

FIGURE 1.8
Common single-cell internal batteries: (a) 18650 cells soldered to PCB, (b) unprotected, single cell, (c) stand-alone battery with two-position connector, and (d) stand-alone battery with thermistor and three-position connector.

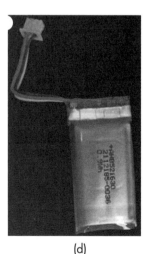

(a)   (b)   (c)   (d)

FIGURE 1.9
Topologies for an internal,
single-cell battery:
(a) permanently connected
cell, (b) unprotected
battery, (c) only switch in
battery, (d) stand-alone
battery, and (e) stand-
alone SMB battery.

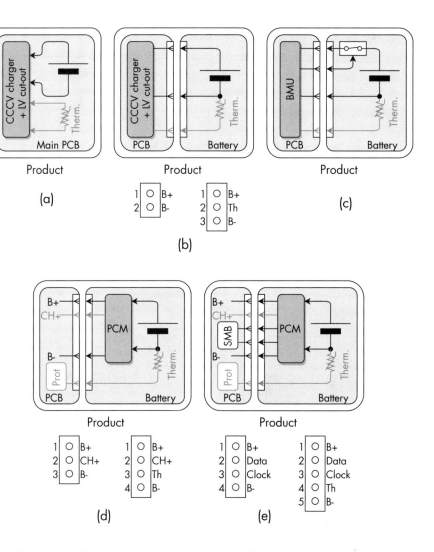

pouch cell), optionally a thermistor against the cell, and a wire tail terminated by a connector.

Without a thermistor, the battery uses a 2-position connector (Figure 1.8(c)). Although there is no standard for the pin-out, this pin-out is most common:

1. Positive;
2. Negative.

With a thermistor, the battery typically uses a three-position connector (Figure 1.8(d)):

1. Positive;
2. Thermistor;
3. Negative.

Be aware that some manufacturers use the opposite polarity.

1.3.2.4   Only the Switch in the Battery

With these batteries, the BMS is divided into two parts (Figure 1.9(c)):

- A BMU on the main PCB in the product;
- A protection switch in the battery.

As a result, the connector has no voltage when the battery is disconnected. A three-position or four-position connector has the two power terminals and the switch control. This topology is unlikely in a single-cell internal battery as it offers little advantage because, as the battery is only handled by a qualified technician, having battery voltage always present on the battery connector contacts is acceptable.

1.3.2.5 Stand-alone Battery

A stand-alone battery is safe because it doesn't need to rely on the product for its protection. The battery includes a PCM (Figure 1.9(d)). A two-position connector has the power terminals (Figure 1.8(c)).

This battery may be as simple as a protected 18650, a small cylindrical cell that includes a tiny PCM (See Volume 1, Section 4.3.4). Or it may be a pouch cell with a PCM soldered to its tabs and a wire tail with a connector (Figure 1.8(c)). A thermistor may be placed on the body of the cell, so the PCM may turn off the battery in case of overtemperature. A three-position connector is used in a battery that includes a thermistor that reports the temperature to the product (Figure 1.9(d)). The circuit marked "Protection" in the diagram reads the thermistor value and shuts down the battery if it is too warm.

Typically, a stand-alone battery uses a single-port PCM. Alternatively, it may use a dual-port PCM, one port for charging and one for discharging. If so, the connector has an additional contact, marked CH+ in the diagram. The connector has two to four positions, depending on which optional functions are included. Although there is no standard, typically, the positive terminal is in position 1, the negative terminal is at the opposite end of the connector, and any other functions use intermediate positions.

1.3.2.6 Stand-alone SMB Battery

A stand-alone SMB battery is the same as the previous one, except that the PCM communicates with the product to report the battery's state using an SMB digital link (Figure 1.9(e)). A four-position connector has the two power terminals plus two digital link lines: data and clock. If a thermistor is made available to the product, it uses a five-position connector: one more pin for the thermistor.

This topology is unlikely for an internal battery as it is not advantageous: an internal battery would not normally include communications because, normally, the battery remains in the product, so there's nothing that the battery's PCM would know that the product would not already know.

1.3.2.7 Topology Recommendations

Which topology should one use? I recommend

- Cheap products: unprotected cell (Figure 1.9(b)); protect the cell by using the CV setting of the charger and a low-voltage cutoff in the load; consider soldering the cell directly to the main PCB (Figure 1.9(a)).
- Low-volume applications: ready-made stand-alone battery (Figure 1.9(d)); ideally, one with a thermistor, supported in the product; if designing a new battery, use a pouch cell and a single-port PCM.

1.3.3 Multicell Internal Battery

A multicell internal battery has two or more cells in series and is intended to be integrated into a product. The battery may consist of a stack of pouch cells (Figure 1.10(a)) or a bundle of small cylindrical cells (Figure 1.10(b)).

1.3.3.1 Topologies

Some of the topologies used in a single-cell battery are possible in a multicell battery:

- Unprotected battery (Figure 1.11(a));
- Stand-alone battery (Figure 1.11(b));
- Stand-alone SMB battery (Figure 1.11(c)).

The other two topologies are not likely:

- Permanently mounted: multiple cells are not likely to be permanently connected to a PCB.
- Only the switch in battery: the connector has no B+ voltage when the battery is disconnected, yet the tap voltages are still present, so there is no advantage.

The previous section lists details for each of these topologies. Additional details specific to multicell batteries are listed below.

1.3.3.2 Unprotected Battery

An unprotected battery has a string of cells and no BMS (Figure 1.11(a)). The BMS is inside the product (see Section 1.4.2.2). Typically, it would be used in a larger consumer product.

The battery includes a balance connector whose primary function is to let the BMS monitor each and every cell and balance them. For N cells in series, the connector has N+1 positions: N cell sense taps plus the B- terminal. Generally, pin 1 is connected to B-[9], although, in some batteries, this order is reversed. When connecting

FIGURE 1.10
Common multicell internal batteries: (a) stack of pouch cells, (b) bundle of small cylindrical cells in shrink wrap, and (c) bundle of 18650 cells with PCM.

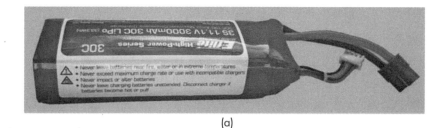

(a)

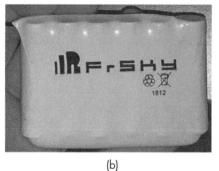

(b)

(c)

---

9.   Note that this is opposite from single-cell batteries, in which pin 1 is often B+.

FIGURE 1.11
Topologies of an internal,
multicell battery:
(a) unprotected battery,
(b) stand-alone battery,
and (c) stand-alone
SMB battery.

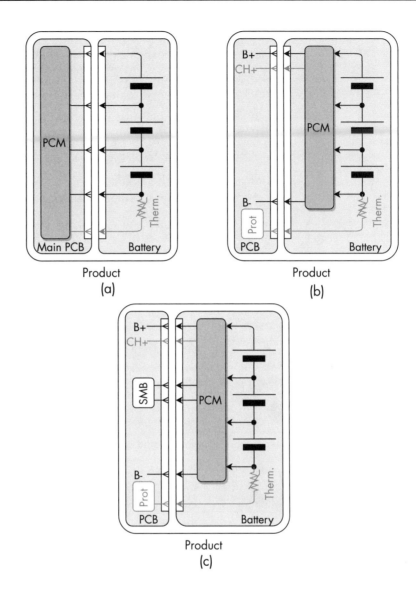

a battery to a BMS, do double-check the polarity, as there's no guarantee that they both use the same pin-out. If the battery has a large number of cells in series, it may use two or more balance connectors to handle all the wires. The connectors should differ in some manner so they may not be interchanged. A high-power battery also has a separate two-conductor connector for the power terminals: B+ and B-. If the battery has a thermistor, its line could be included in any of these connectors, or it could have its own, two-position connector.

Shops that sell to RC and drone hobbyists offer unprotected stacks of pouch cells for power applications (Figure 1.10(a)). They may advertise a 2-Ah battery with a C-rating of 60 C (see Volume 1, Section 1.2.2.8). I must caution you:

- A 60 C discharge of such a battery is 120A; that level of current, if even possible, would discharge the battery in 1 minute, with dire consequences to the life of the cells.

- At that current, the efficiency is likely to be, at best, 50%, meaning that half the power goes into heating the cells; the voltage sags badly.

- Neither the connector nor the wires are sized to carry that much current.

All too often, inexperienced designers decide to build a large battery by connecting a number of these pouch cell stacks in parallel and series. That is impractical:

- If using the series-first arrangement, that requires a BMS that can support it, which is hard to find, and quite expensive.
- If using the parallel-first arrangement (connecting cells directly in parallel through the balance wires), an inrush of current between adjacent cells (due to differences in cell resistance) would melt the balance wires or connectors.

Instead of forcing these cell stacks into a service they are not designed for, design a battery around individual cells, especially cells that already have the desired capacity.

1.3.3.3   Stand-alone Battery

A stand-alone battery includes a PCM (Figure 1.11(b)). This battery may consist of a stack of pouch cells with a PCM soldered to its tabs or a bundle of 18650 cells with a PCM mounted on the top (Figure 1.10(c)). It uses a two-position connector for B+ and B-, or a three-position connector with a third wire for a thermistor.

1.3.3.4   Stand-alone SMB Battery

A stand-alone SMB battery is the same as the previous one, except that the PCM uses an SMB link (see Section 1.2.2) to report its state to the product (Figure 1.11(c)).

1.3.3.5   Topology Recommendations

Which topology should one use? I recommend

- High-volume applications: use an unprotected battery and design a PCM on the main PCB (Figure 1.11(a));
- Low-volume applications: buy a ready-made stand-alone battery (Figure 1.11(b)) with a thermistor; support the thermistor in your product.

1.3.4   Procurement

Ready-made internal batteries are readily available for sale, mainly from specialty hobby stores and from AliExpress. These include

- Protected 18650 batteries;
- Single pouch cells with a two- or three-position connector, with or without a PCM;
- Bundles of 18650 small cylindrical cells, either in a hex pattern or in a line inside a tube
- Stacks of pouch cells wrapped in heat-shrink tubing.

If a pouch battery has a thermistor wire, it is an input; you need to supply a thermistor and connect it to this wire. It is not a thermistor output that reports the battery temperature.

1.4   EXTERNAL BATTERY

An external battery is in its own, hard enclosure. The user may remove it from the product to replace a degraded battery with a new one or a discharged battery with a full one. Typical uses are for cell phones (Figure 1.12(a)), power tools [3] (Figure 1.12(b)), or laptop computers (Figure 1.12(c)).

1.4.1   Technical Considerations

An external battery is more complex, as it requires an enclosure.

1.4.1.1   Options

A removable battery may use various options, resulting in many permutations:

- May be single-cell or multicell (from two to 16 cells in series);
- May or may not have a PCM, or may have only a portion of the BMS; if with a PCM:
  - With or without SMB communications;
  - Single-port or dual-port.
- May use more than one contact for each power terminal;
- May or may not include a thermistor whose resistance can be measured by the product;
- May or may not include an enable line;
- May use a male or female connector;
- May or may not recess the contacts to protect against accidental shorts.

1.4.1.2   Battery Connector

For ease of mating, these batteries use blind mate[10] connectors, using one of two approaches:

1. Spring-loaded pins on the product (Figure 1.13(b)) that land on flat pads on the battery (Figure 1.13(a));

FIGURE 1.12
Typical removable batteries: (a) cell phone, (b) power tool, and (c) laptop computer.

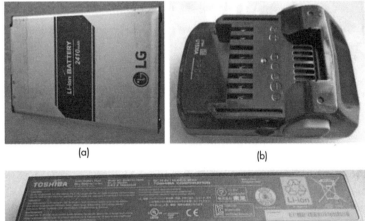

(a)   (b)

(c)

---

10. "Blind" is in the sense that the user does not need to watch the connector while mating the battery to the product.

FIGURE 1.13
External battery connector:
(a) cell-phone battery,
exposed pads, (b) cell
phone, spring loaded pins,
(c) laptop battery,
recessed sockets, and
(d) laptop, male blades.

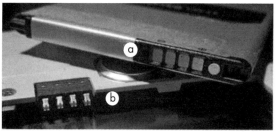

2. Leaf sockets in the battery (Figure 1.13(c)) and blades in the product (Figure 1.13(d)).

When the battery is not in the product, its terminals are exposed to an accidental short circuit should a conductor (e.g., a paper clip, a metal plate) come into contact with live terminals.

Protection against such accidents may be implemented using either or both of two methods:

- *Electrical:* The battery may include an electronic protector switch (regardless of whether the BMS function is implemented in the battery or the product) that is turned off while the battery is not in the product; therefore, there is no voltage on the terminals, and a short circuit cannot occur (Figure 1.13(a)).

- *Mechanical:* The battery terminals are recessed into the connector, making the chance of contact with a conductor unlikely (Figure 1.13(c)); ideally, the battery connector should be female, since male pins would be more prone to an accidental short circuit; yet many batteries use male blades, sometimes for backward compatibility to older, non-Li-ion batteries

1.4.1.3   Authentication

Some products work only with their original battery. Ostensibly, they do so for safety and performance reasons. The product is designed to work optimally with the original battery; it may not work as expected with a replacement battery from another manufacturer. In reality, authentication has more to do with economics than technology; the manufacturer wants to sell replacement batteries at a nice profit and doesn't want the consumer to buy a cheaper, knock-off battery.

The product authenticates the battery through a digital communication link (usually SMB). The product sends a challenge message to the battery and checks its response message. As only a battery made by the same manufacturer knows the proper response, the product will not work with an aftermarket battery [4].

A second effect is that the original battery will not work with any other device; this frustrates the hobbyist who wants to reuse a consumer battery for some other purpose.

1.4.1.4 *Mechanical Design*

The product enclosure is designed to contain the cells in various ways:

- The enclosure cradles a bundle of 18650 cells to prevent relative movement of the cells and to avoid damage, abrasion of insulation, and tearing of welds and other interconnects including for thermistors.
- The enclosure contains the expansion of a stack of pouch cells.

1.4.2 Single-Cell External Battery

Single-cell batteries are ubiquitous.

1.4.2.1 *Topologies*

Single-cell external batteries could use any of these seven topologies:

1. Unprotected battery (Figure 1.14(a));
2. Battery with only a switch (Figure 1.14(b));
3. Battery with only a BMU (Figure 1.14(c));
4. Stand-alone battery (Figure 1.14(d));
5. Stand-alone SMB battery (Figure 1.14(e));
6. Dual-port (See Volume 1, Sections 4.3.1 and 5.12.2), stand-alone battery (Figure 1.14(f));
7. Dual-port stand-alone SMB battery (Figure 1.14(g)).

1.4.2.2 *Unprotected Battery*

Unprotected single-cell batteries (Figure 1.14(a)) are used in cheap toys, consumer products, and small computing products.

The BMS is in the product, not in the battery. The battery includes a thermistor that is made available to the BMS inside the product. The battery uses a three-position (or more) connector. The unprotected cell voltage appears on the connector, although mechanically recessed contacts mitigate this problem.

Placing the BMS in the product shifts part of the cost from the battery to the product. This approach may reduce the overall cost by removing the PCB assembly inside the battery. Since there is already a PCB assembly in the product, adding the BMS function to this assembly doesn't increase the cost by much. This also reduces the cost of replacement batteries, which is an advantage to the consumer. Moving the BMS function to the product may result in improved performance because a computer in the product may implement a more advanced BMS function than a standard PCM in the battery.

A contact in the connector may be used to identify the battery (the ID line shown in gray in the diagrams). Inside the battery, a resistor encodes the type of battery. The product detects the battery type by checking the value of this resistor. For example, a 10-kΩ resistor to B- may indicate a 3-Ah Li-ion battery, 0Ω to B- may indicate a NiMH battery, and a 1-kΩ resistor to B+ may indicate a 6-Ah battery. Each manufacturer uses a different coding scheme.

The product monitors a thermistor in the battery to detect if the battery is overheating. It may also do so to prevent charging below freezing. Knowing the battery temperature may be used to improve its management and to optimize its performance.

FIGURE 1.14
Topologies for a
removable, single-cell
battery: (a) unprotected
battery, (b) battery with
only a switch, (c) battery
with only a BMU,
(d) stand-alone battery,
(e) stand-alone SMB
battery, (f) dual-port
stand-alone battery,
and (g) dual-port SMB
stand-alone battery.

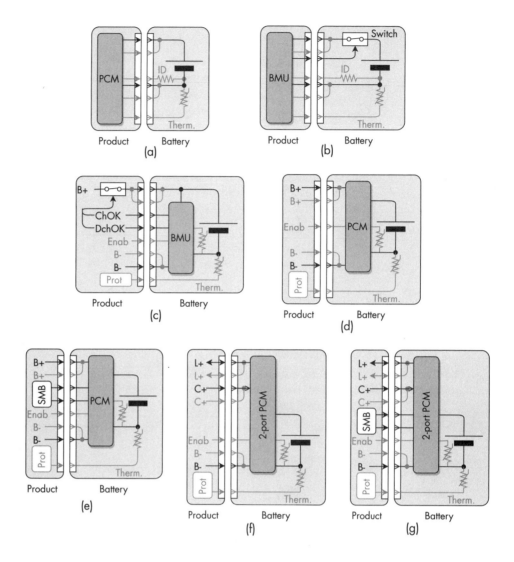

An unprotected battery is dumb and cannot remember its state. When the user installs a new battery, the BMS must quickly learn its state, especially its SoC. TI specializes in ways of doing so, using the cell's impedance[11].

1.4.2.3 Battery with Only a Switch

A battery with only a protection switch may be used for a small power tool. Having the switch in the battery removes the risk of an accidental short against the contacts when the battery is not in the product. The BMS is divided into two parts (Figure 1.14(b)):

- The BMU is in the product.
- The protection switch is in the battery.

This solution does require a PCB assembly in the battery, so it doesn't much reduce the cost compared to placing a complete PCM in the battery. Still, as explained above, it's better if a powerful computer in the product performs the BMS function, rather than a dumb PCM in the battery.

---

11. "Impedance Track" technology (trademark of TI).

The connector uses at least four contacts: two or four contacts for power, one for a thermistor, and one or two to control the switch.

1.4.2.4 Battery with only a BMU

This topology could be used in products that have a motor, such as an electric toothbrush or a power drill. In this topology, the BMS is also divided into two parts, but the other way around (Figure 1.14(c)):

- The BMU is in the battery.
- The protection switch is in the product.

This approach is only advantageous in products that already have a switched load: instead of using two high-power switches, a protection switch for the BMS and a control switch in the power circuits, the latter switch can perform both functions (see Volume 1, Section 1.6.2.3).

1.4.2.5 Stand-alone Battery

A stand-alone battery (Figure 1.14(d)) is inherently protected and does not have to rely on the product. In particular, the battery is still protected even if a user repurposes it for another application.

The battery may include a thermistor for use by the product, in which case it uses a three-position connector[12]. Otherwise, it uses a two-position connector with only the power terminals[13]. Regardless, the battery includes a thermistor for the internal BMS.

1.4.2.6 Stand-alone SMB Battery

A stand-alone SMB battery is used in communications and computing devices (cell phones and laptop computers). Not only is this battery (see Section 1.2.2) protected, but it also reports its state to the product (Figure 1.14(e)). The product only needs to concern itself in displaying the state of charge to the user, although it may also use this information to adjust its functionality (e.g., reduce consumption by dimming the lights when the battery is nearly empty).

The connector has between four and eight contacts:

- B+ (one or two each): battery positive;
- B– (one or two each): battery negative;
- SMB data, bidirectional;
- SMB clock input[14];
- Thermistor output (optional);
- Enable input (optional).

1.4.2.7 Dual-Port Stand-alone Battery

A dual-port stand-alone battery is the same as the previous battery, except that it has separate ports for charging and discharging (Figure 1.14(f)) (see Volume 1, Section

---

12. For example, the 18V Ridgid R84008 and the 19.2V Craftsman 315 for power tools.
13. For example, the 18V DeWalt DC9180 battery for power tools.
14. May change to an output if it needs to transmit a warning.

4.3.1). This improves the efficiency of the protector switch because current flows through only one transistor. However, the BMS is unable to protect the battery from any attempt to discharge through the charging port. It is also unable to protect the battery from any attempt to charge through the discharging port (see Volume 1, Section 5.12.2.3).

If the charging and discharging currents are significantly different (i.e., slow charging and fast discharge, or fast charging and slow discharge), a dual-port battery may save a bit of the cost of the transistors, since only one of the transistors in the protector switch is high power. However, the connector has an extra contact, which cuts into the savings.

1.4.2.8 Dual-Port Stand-alone SMB Battery

A dual-port stand-alone SMB battery is the same as above, except that it includes an SMB communication link (see Section 1.2.2) (Figure 1.14(g)).

If the protector switch is on the positive side, the connector has between four and eight contacts:

- L+: positive discharge to the load, one or two each;
- CH+: positive charge from the charger, one or two each;
- B-: battery negative, common for charging and discharging, one or two each;
- SMB data (bidirectional);
- SMB clock;
- Thermistor output (optional);
- Enable input (optional).

If the protector switch is on the negative side, the power terminals are different:

- B+: battery positive, common for charging and discharging, one or two each;
- L-: negative discharge from the load, one or two each;
- CH-: negative charge to the charger, one or two each.

1.4.2.9 Topology Recommendations

Which topology should one use? I recommend

- Low-power, low-cost applications: unprotected battery (Figure 1.14(a)) with recessed contacts to minimize accidental shorts when the battery is not in the product.
- Better performance: stand-alone battery (Figure 1.14(d)) because its BMS can remember the battery state between uses; a dual-port battery may be a bit more efficient (Figure 1.14(f)).
- High-power applications whose load may be shut-off with a low-power signal (see Section 1.6.2.3): battery with only a BMU (Figure 1.14(c)).
- Small computing devices (e.g., smartphones, tablets): stand-alone SMB battery (Figure 1.14(e)); for higher efficiency, consider a dual-port battery (Figure 1.14(g)).

1.4.3 Multicell External Battery

Multicell batteries are common in larger consumer products.

1.4.3.1 Topologies

Multicell external batteries may be divided into eight classes:

1. Unprotected battery (Figure 1.15(a));
2. Unprotected battery with a balancer (Figure 1.15(b));
3. Unprotected battery with a fuel gauge (Figure 1.15(c));
4. Unprotected battery with a BMU (Figure 1.15(d));
5. Stand-alone battery (Figure 1.15(e));

FIGURE 1.15
Topologies for a removable, multicell battery: (a) unprotected battery, (b) unprotected battery with balancer, (c) unprotected battery with fuel gauge, (d) unprotected battery with BMU, (e) stand-alone battery, (f) stand-alone SMB battery, (g) dual-port stand-alone battery, and (h) dual-port stand-alone SMB battery.

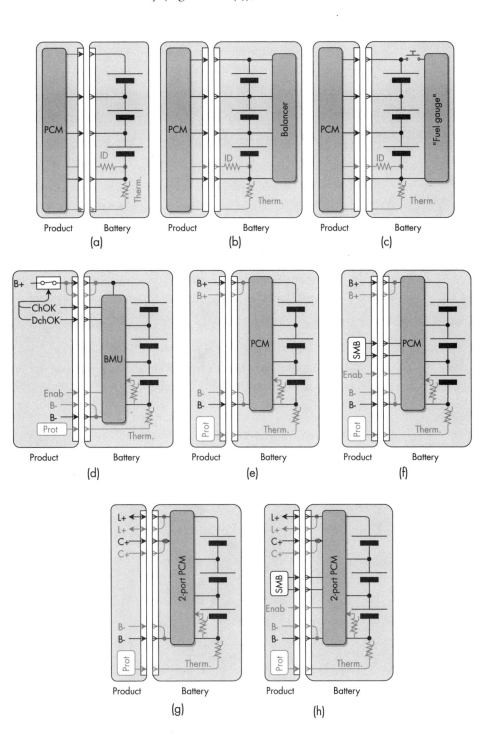

6. Stand-alone SMB battery (Figure 1.15(f));
7. Dual-port (See Volume 1, Sections 5.12.2 and 4.3.1), stand-alone battery (Figure 1.15(g))
8. Dual-port, stand-alone SMB battery (Figure 1.15(h)).

These are the same topologies as for single-cell batteries, with the addition of unprotected batteries with auxiliary electronics. I excluded from this list batteries with only a protection switch because it doesn't make sense for a multicell battery; even though the B+ voltage is disconnected when the battery is not in the product, all the tap voltages are still exposed in the connector.

The same points covered for single-cell batteries apply to multicell batteries, with few variations, as noted below.

1.4.3.2   Unprotected Battery

Unprotected multicell batteries (Figure 1.15(a)) are used in cheap power tools. For example, the 12V Ridgid R86048 battery uses this topology (Figure 1.16(a)).

The connector includes contacts to sense the cell voltages. While not having a PCM in the battery may save money, the additional contacts in the battery connector reduce this cost advantage.

1.4.3.3   Unprotected Battery with Balancer

This unprotected battery contains only the balancing portion of the BMS (Figure 1.15(b)). The rest of the BMS is in the product. This has the same advantages as an unprotected battery, plus the battery can continue to balance after it's removed from the charger and until it's used in a tool. This solution is effective because that's precisely when the cell voltages are the highest, and therefore the easiest to top-balance. The DeWalt DCB200 20V batteries for power tools use this topology (Figure 1.16(b)).

1.4.3.4   Unprotected Battery with Fuel Gauge

Batteries may use a simple voltmeter to estimate the SoC of the battery and display it with three or more LEDs (Figure 1.15(c)). The user activates the circuit by pressing

FIGURE 1.16
Inside unprotected, multicell, removable batteries: (a) 12V battery, (b) with balancer, and (c) with fuel gauge. (All courtesy of Russel Graves, Syonyk's Project Blog.)

(a)

(b)

(c)

a push-button switch. This meter is vaguely accurate. The assumption is that the user would not be checking the state of charge while using the power tool at a time when the battery voltage sags. The DeWalt DCB200 battery for power tools uses this topology (Figure 1.16(c)).

1.4.3.5 Unprotected Battery with BMU

This battery contains only the BMU. The protector switch is in the product (Figure 1.15(d)). This topology has the same consideration as the single-cell version (see Section 1.4.2.4). For example, this battery could be used in a power drill.

1.4.3.6 Batteries with PCM

All four of these multicell topologies (Figures 1.15(e–h)) have the same advantages and disadvantages as the equivalent single-cell topologies (Figures 1.14(d–g)), so there is nothing more to add. Two examples of such batteries are:

- Single-port, 18V Ridgid battery for a power tool (Figure 1.17(a)); its connector has four contacts, for B+, B– and two externally accessible thermistors.
- Dual-port, 18V laptop battery with SMB communications (see Section 1.2) (Figure 1.17(b)); the connector has seven contacts: two each for B+ and B–, two for SMB data, and one for a thermistor.

1.4.3.7 BMS without Voltage Taps

Surprisingly, there are commercial products that violate the cardinal rule that the BMS must monitor the voltage of each and every cell[15]. One supposes that careful control of the manufacturing process ensures that all the cells in a given battery are nearly identical and that the battery is perfectly balanced at the time of manufacture. The hope would be that the cell voltages track each other extremely closely during a cycle

FIGURE 1.17
Protected multicell batteries. (a) Power tool. (Courtesy of Russel Graves, Syonyk's Project Blog.) (b) Laptop, with SMB communications. (Courtesy of karosium.com.)

(a)

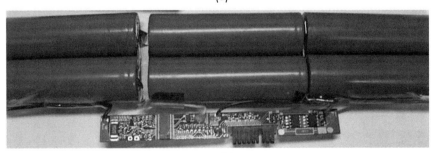

(b)

---

15. This includes some older batteries for Makita power tools. See Volume 1, Section 3.2.10.

and that the reduction in capacity due to imbalance is masked by degradation. By the time the battery becomes severely imbalanced, it is also so degraded that it's time for a replacement anyway. Doing so may shave $0.50 off the cost of the product, which is only worthwhile if the average cost of in-warranty replacement is $0.25[16].

1.4.3.8   Topology Recommendations

Which topology should one use? I recommend:

- High-volume applications: the need for cost reductions may point to the un-protected topology (Figure 1.15(a)):
- Backward compatibility to a non-Li-ion battery: any of the stand-alone batteries without SMB (Figure 1.15(e, g)); a single-port PCM must be included in the battery;
- Computing device: a battery with SMB (Figure 1.15(f, h));
- General use: single-port protected battery (Figure 1.15(e)) because it is straightforward and safe and because it uses a simple connector.

1.4.4   Troubleshooting

Often, a small, sealed Li-ion battery is considered "dead" because there is no voltage on its contacts. Rather than being dead, it's more likely that the battery is OK and simply needs to be woken up.

The battery may have no voltage due to one of several reasons:

- The battery is empty, and its BMS disabled discharging; all it needs is to be charged.
- The battery shuts itself down after a long period of inactivity; charging it for a bit may wake it up.
- There's an enable input on the connector that needs to be connected to B-.
- The battery uses authentication to ensure that it works only with the original product; if so, repurposing it elsewhere is not practical.

In particular, you should know how to distinguish an 18650 cell that is completely discharged from a protected 18650 battery that has disabled discharging (see Volume 1, Section 4.3.4).

1.4.5   Laptop Battery Repair

If your time is free, you may consider replacing the cells in a laptop battery, rather than buying a new battery. Before you do, be aware that working with Li-ion batteries involves a risk of fire, burns, and sparks. Reread the cautions in the assembly section (see Volume 1, Section 7.2). The hardest part is opening the battery because the case is sealed and ultrasonically welded. The biggest mistake is soldering the cells. The second biggest mistake is to replace only some cells. The third mistake is to use mismatched cells, such as cells at different SoC levels.

To repair a laptop battery, follow these procedures:

---

16. In the case of Makita, that may have been a bad assumption because later versions of that same product use a new battery design that does include voltage taps.

- With the battery still in the laptop computer, charge it fully, so that the BMS remembers that the SoC is 100%.
- Remove the battery.
- Open the case as well as you can.
- Note the number of cells; confirm that they are 18650 cells.
- Buy the same number of cells:
  - 18650 shape, LCO chemistry[17] (3.6V nominal, 4.2V maximum).
  - Important! Get them with tabs already welded.
  - Higher capacity is OK.
  - Protected 18650 batteries are not OK.
  - Buy from reputable vendors of Li-ion products[18], not from generic companies[19].
- Buy a single-cell Li-ion charger with a 4.2V final voltage.
- Fully charge each new cell individually with the single-cell Li-ion charger.
- Take pictures of the battery, so you can refer to them when connecting the new cells and placing the thermistor.
- Disconnect the thermistor from the original cells.
- Break off the nickel strips from the original cells, as close as possible to the cell, so that the remaining stubs are as long as possible; be very careful with any metal tools, as you may short out the cells and cause a loud spark.
- With a meter, confirm that all the new cells are at about 4.1V.
- Make doubly sure that the cells are oriented with the correct polarity; confirm by looking at the pictures you took.
- Solder the tabs of the new cells to the original stubs; do not solder directly to the new cells; again, be very careful not to cause shorts.
- Place the thermistor back where it was; confirm by looking at the pictures you took.
- Close the battery temporarily.
- Place the battery back into the laptop computer.
- Check that the computer reports that there is a battery and that it is full.
- Disconnect the AC adapter and check that the battery powers the computer.
- Run the computer on battery power for a while to discharge the battery.
- Plug the AC adapter to the computer.
- Check that the computer reports that the battery is charging.
- Disconnect the battery.
- Close the battery case permanently.

1.4.6 Reverse Engineering

You may wish to reverse engineer the connector of a sealed, production battery. Because there are so many possibilities, it takes some work to deduce the function of

---

17. If they are labeled just "Li-ion," they are probably LCO
18. Sparkfun, BatterySpace, Hobby King.
19. eBay, Amazon, AliBaba, AliExpress.

each contact. A DMM is required[20]. Knowing the number of cells in series helps as well.

Start by probing the contacts, to see if there are any voltages; proceed with one of the following three sections based on what voltages you see:

- No voltage;
- Only the battery voltage;
- Battery and cell voltages.

1.4.6.1   No Voltage

If there is no voltage, it may be for one of the following reasons:

- The battery is empty and a BMS disabled discharge (Figure 1.18(a)).
- There's a BMS that requires communications, or it has an enable line that needs to be connected (probably to B-), or it requires authentication, which it can only get from the original product (Figure 1.18(b)).
- It's a single-cell battery with a protection switch, and it is disabling discharge (Figure 1.18(c, d)).

FIGURE 1.18
Reverse engineering a sealed battery with no voltage: (a) empty battery, (b) BMS requires communications or enabling, (c) switch requires activation, and (d) two switches require activation.

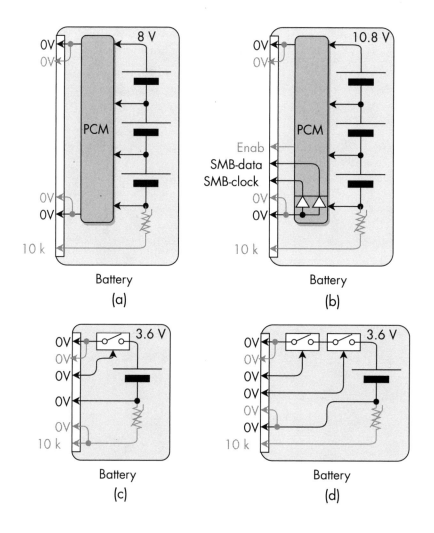

---

20.   Digital multimeter.

First, look for the contacts that are likely to be the power terminals (B+ and B-):

- Larger contacts;
- Contacts that are connected: with the meter in the OHMS range look for 0Ω between two contacts; typically, two large contacts are next to each other for B- at one end of the connector, and another two large contacts are next to each other for B+ at the other end.

Next, look for a thermistor:

- With the meter in the ohms range, look for about 10 kΩ[21] between two contacts.
- One of these contacts is B- (usually the larger one).

Then look for SMB lines:

- With the meter in the DIODE mode, look for a diode's cathode between two contacts; you should find two such diodes (Figure 1.18(b)), one for the SMB-data line and one for the SMB-clock line, sharing the anode connection; this shared connection is the B- contact.
- Reversing the probes shows an open circuit.

If you found an SMB port, you'll need to talk to the BMS [5]. Chances are that the protector BMS uses a Texas Instrument IC, in which case you may be able to talk to it using a TI communication board[22].

Maybe there's a BMS that disabled discharging while charging is still enabled. If so, you can try to wake up the battery by charging it. Without knowing which contacts are the power terminals, this has to be done carefully:

1. Get a power supply whose voltage exceeds the expected voltage of the battery; to limit the current, place a 1-kΩ resistor in series with the positive output.

2. Connect a voltmeter to the far side of the resistor to check the voltage (Figure 1.19); now you have two probes with which to look for the B+ and B- contacts.

3. Connect the probes to the first pair of contacts (Figure 1.19(a)); repeat with all other possible combination of two contacts (Figure 1.19(b–f)); try exchanging the polarity (Figure 1.19(g–l)); you're looking for a voltage that is between zero and the power supply voltage; if found, that is the voltage of the string of cells inside the battery.

4. If you found a thermistor, then one of the two leads connected to the thermistor is likely to be B-; don't look for a battery voltage between the two thermistor contacts: we already know there's a thermistor there, so they can't be B+ and B-; if you found an SMB port, then don't test those contacts either.

5. When the voltmeter sees 0V, that means that the two contacts that are connected (Figure 1.19(l)); they are likely to be either B- or B+.

6. When the voltmeter sees a lower voltage, but still greater than 0, it probably is the battery voltage; increase the power supply voltage; if the voltage on the voltmeter stays the same, it is indeed the battery voltage.

---

21. The colder the battery, the higher the resistance.
22. The Texas Instruments EV2300 SMB to USB adapter.

FIGURE 1.19
Reverse-engineering
a sealed battery with
no voltage looking
for B+ and B–: (i)
found the two power
contact; and (l) found
B– or B+: two contacts
connected together.

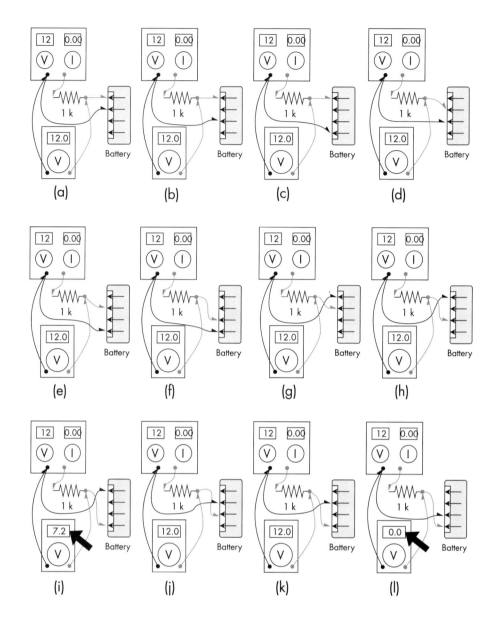

7. Now that you found B+ and B–, remove the 1-kΩ resistor, reduce the power supply voltage to a voltage slightly above the measured battery voltage, and charge the battery directly from the power supply for a while.

8. After some time, turn off the supply and see if the voltage remains; if so, the battery is ready and you can start using it.

Table 1.2 lists possible functions for the contacts, given the number of contacts and whether you found a thermistor or an SMB port.

1.4.6.2 Only the Battery Voltage

If, when you first check, you see voltage on only two pins, those are B+ and B–. The battery may be

- A single-cell unprotected battery (Figure 1.20(a));
- A protected battery that has not disabled discharge (Figure 1.20(b));

	No SMB port		With SMB port	
	No thermistor	With thermistor	No thermistor	With thermistor
2 pins	B+, B-			
3 pins	B+, B-, switch control B+, B-, enable line B+, B-, battery ID resistor	B+, B-, thermistor		
4 pins	B+ (2 each), B- (2 each) B+, B-, charge switch, discharge switch B+, B-, enable line, battery ID resistor	B+, B-, thermistor (2 pins, isolated) B+, B-, thermistor, switch control or enable line or battery ID	B+, B-, SMB-clock, SMB-data	
5 pins	B+ (2 each), B- (2 each), switch control B+ (2 each), B- (2 each), enable line B+ (2 each), B- (2 each), battery ID resistor	B+ (2 each), B- (2 each), thermistor B+, B-, charge switch, discharge switch, thermistor B+, B-, enable line, battery ID resistor, thermistor	B+, B-, SMB-clock, SMB-data, enable line	B+, B- SMB-clock, SMB-data, thermistor
6 pins	B+ (2 each), B- (2 each), charge and discharge switch B+ (2 each), B- (2 each), enable line, battery ID resistor	B+ (2 each), B- (2 each), switch control, thermistor B+ (2 each), B- (2 each), enable line, thermistor	B+ (2 each), B- (2 each), SMB-clock, SMB-data	B+, B- SMB-clock, SMB-data, thermistor (2 pins, isolated)
7 pins			B+ (2 each), B- (2 each), SMB-clock, SMB-data, enable line	B+ (2 each), B- (2 each), SMB-clock, SMB-data, thermistor

TABLE 1.2   Reverse-Engineering a Sealed Battery with No Voltage: Possible Contact Functions

- A protected SMB battery which doesn't require communications to operate (Figure 1.20(c)).

If the voltage is 2.5~4.2V, then it's quite likely that it's a single-cell unprotected battery. (It's also possible that it's a multicell unprotected battery that is severely overdischarged and should be discarded.)

Table 1.2 lists possible functions, given the number of contacts and whether you found a thermistor or an SMB port (Table 1.2). Look for a thermistor and SMB contacts, as described in the previous section.

1.4.6.3   Battery and Cell Voltages

If the contacts have multiple voltages, this is a multicell, unprotected battery (Figure 1.20(d)). It has no PCM, and the cell taps are fed to the connector.

Table 1.3 lists possible functions for the contacts given the cell voltages and the number of contacts.

1.5   APPLICATIONS

There are thousands of applications for small batteries. I touched on power tools, phones, and laptop computers in previous sections. In this section, I mention two more:

1. Power banks: because of hobbyists' strong desire to build one;

FIGURE 1.20
Reverse engineering a
sealed battery with voltage:
(a) single-cell, unprotected,
(b) protected battery,
(c) SMB battery, and
(d) multicell, unprotected.

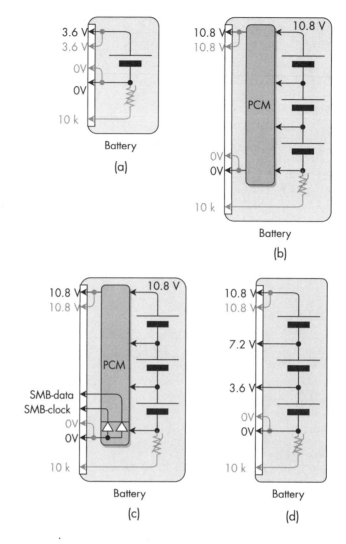

TABLE 1.3
Possible Contact Functions
in an Unprotected Multicell
Battery

		2 cells	3 cells	4 cells	5 cells
LFP →		3.2V and 6.4V	3.2V, 6.4V, and 9.6V	3.2V, 6.4V, 9.6V, and 12.8V	3.2V ... 16V
Other →		3.6V and 7.2V	3.6V, 7.2V, and 10.8V	3.6V, 7.2V, 10.8V, and 14.4V	3.6V ...18V
3 pins		B+, B-, one center tap			
4 pins		B+, B-, one center tap, thermistor	B+, B-, 2 taps		
5 pins		2 × B+, 2 × B-, center tap	B+, B-, 2 taps, thermistor	B+, B-, 3 taps	
6 pins		2 × B+, 2 × B-, center tap, thermistor	2 × B+, 2 × B-, 2 taps	B+, B-, 3 taps, thermistor	B+, B-, 4 taps
7 pins			2 × B+, 2 × B-, 2 taps, thermistor	2 × B+, 2 × B-, 3 taps	B+, B-, 4 taps, thermistor
8 pins				2 × B+, 2 × B-, 3 taps, thermistor	2 × B+, 2 × B-, 4 taps
9 pins					2 × B+, 2 × B-, 4 taps, thermistor

2. Li-ion AA cell: because I find it to be an interesting application.

1.5.1 Power Banks

Power banks address many needs for portable power such as to extend the operating time of laptop computers, to start cars with a dead battery (Figure 1.21(b)), to go camping, for location movie shoots and news reporting, to charge phones.

Hobbyists have a keen interest in making power banks out of cells harvested from laptop computers.

When asked how to make a power bank, my standard answer has been: "Don't! Just buy one!" because, compared to anything you could possibly come up with, a commercial power bank is cheaper, safer, more reliable, guaranteed to work, properly enclosed, and meets regulatory standards.

Normally, the reply is: "But I want to learn and want to build something." My answer is: "Great! How about building something safer, something that won't risk burning your house down if you built it wrong?" followed by a project suggestion.

This has been my answer because the knowledge required to build a safe and effective power bank is too vast to be presented in a single comment in social media. However, my answer to you is different because, presumably, you have read this book up to this point, and you already know the nuances and safety concerns of battery design.

1.5.1.1 Operation

A power bank (Figure 1.22) may be charged from one or more sources, such as AC power, 12 Vdc from a car or solar panel, 5 Vdc from a USB computer port, or a charger. It may provide one or more outputs such as 18 Vdc to power a laptop, 14 V to recharge a car battery, 5 V for USB devices, and AC power for small appliances. Most power banks would not implement all these inputs and outputs. They are likely to limit themselves to only one such input and one such output.

On the input side, a power bank requires one or more chargers (with a CCCV output). The input may be AC, USB (5V), 10~16 Vdc (from a car's starter battery), or a solar panel (using a solar charge controller). On the output side, a power bank requires one or more DC-DC converters. For an AC output, it requires an off-grid inverter (see Section 2.2.4).

To avoid needlessly discharging the battery, a power bank without a power switch turns off these converters when not in use. If the load current is below a certain level,

FIGURE 1.21
Two devices: (a) Texas Instruments EV2300 SMB to USB adapter, and (b) power bank/ car battery charger.

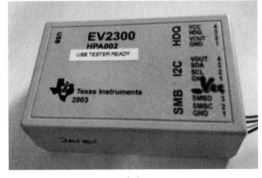

(a)   (b)

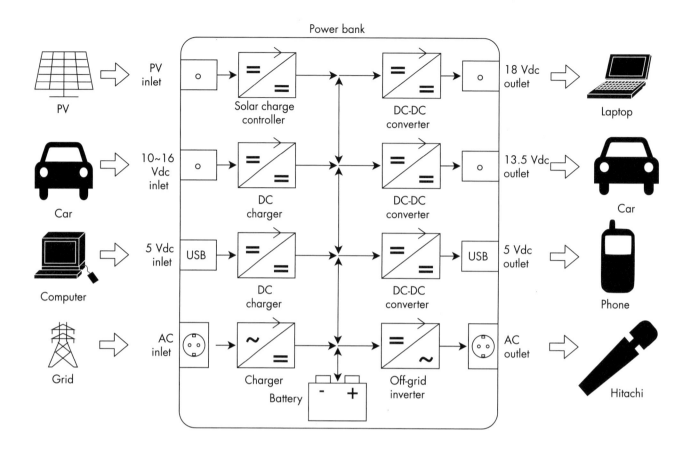

FIGURE 1.22 Power bank with maximum extent of AC and DC inputs and outputs.

the power bank goes to sleep. When powering a small load, a power bank shuts down prematurely. As a workaround, users resort to adding a higher-power load to keep the power bank from going to sleep.

Some power banks prevent the user from connecting to a charger and a load at the same time; this is not because you can't charge and discharge a battery at the same time, as some people may assume (see Volume 1, Section 1.3.1). This is simply because the unit can only handle the heat from one converter at the time.

1.5.1.2 Characteristics

Power banks are often rated by Ah capacity, which is silly because the battery voltage is unknown; a 10-Ah power bank with a 12V battery holds 3.3 times as much energy as a 10-Ah power bank with a 3.6V battery. Even if a power bank is rated by energy, it is of limited use because the energy that can be extracted from a power bank depends on how fast it is done. Regardless, the rating of a power bank is generally moot because many power bank vendors grossly exaggerate the ratings (see Volume 1, Section 2.8.1).

Users ask how to calculate how long a power bank will power a product given the capacity of the power bank and the current rating of the product. This question cannot be answered because:

- The battery is at one (unknown) voltage, and the output is at another voltage.
- The current rating of the product is the maximum, not the actual current, which depends on the use.

On the same token, users also ask why their USB power meters measure a different value for the total charge into a power bank than the stated capacity of the bank. This is because the USB port is at a different voltage than the battery: there's a DC-DC converter (a charger) between them (Figure 1.23). This device converts the 5V from the USB port down to the 3.6V nominal cell voltage, while at the same time raising the current due to conservation of power. A 2-Ah charge out of the USB port increased to about a 2.9-Ah charge at the battery.

Not all the energy (see Volume 1, Section 1.4.4.1) that goes into a power bank's input is available out of its output due to the inefficiencies of the charger, the cells, and the DC-DC converter (see Volume 1, Section 1.4.5.4).

1.5.1.3 Design

Most power banks use an internal, single-cell battery (see Section 1.3.2) or a block of cells in parallel.

A standard protector BMS would be appropriate for a power bank. In practice, a power bank with a single-cell battery may not use an explicit BMS and instead rely on a CCCV charger and a low-voltage cutoff in the DC-DC converter to perform rudimentary BMS functions (see Section 1.6.2.2).

Most people who decide to build a power bank from recycled cells are unaware of the technical challenges ahead:

- Safe handling of AC power coming into the unit, and going out of the unit;
- Thermal design, to keep the heat from converters from degrading the cells;
- Deciding which converter may be turned on at a given time, to limit heat generation;
- Containing radio emissions from the converters;
- Negotiating with USB devices;
- Designing a shutdown timer and low-voltage cutoffs in the DC-DC converters;
- Protecting against an excessive voltage, a negative voltage, and surges on an input;
- Protecting against overcurrent on an output;
- Protecting against surges and voltages forced into an output;
- Reliable design to prevent accidental internal shorts and loose wires;
- Sturdy mechanical design to minimize the effect of external impact and drop;
- Dealing with mismatched cells taken from old laptop batteries;
- If for sale, performing regulatory testing and obtaining required certificates;

Given these challenges, I am going back to my original advice: just buy one.

FIGURE 1.23
Charging a power
bank from USB.

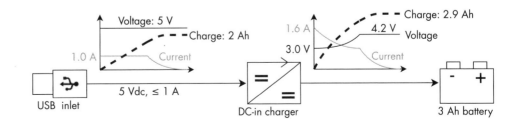

1.5.2   Rechargeable AA Battery

Typically, people use NiMH cells as a rechargeable replacement for alkaline cells; they don't use Li-ion cells due to the voltage difference: 3.2V or 3.7V versus 1.5V. Two AA cells in series may be replaced with a real LFP cell in an AA format, plus a dummy cell that is nothing but a jumper. The total voltage is 3.2V, which is a good match for the total voltage of two AA alkaline cells in series.

1.5.2.1   Li-Ion AA battery

A clever solution is to place a small Li-ion cell and a powerful DC-DC converter inside an AA case (Figure 1.24(a))[23].

The DC-DC converter drops the 3.7V from the Li-ion cell to a constant 1.5V output, regardless of SoC (Figure 1.24(b)). When the cell is almost empty, the DC-DC converter slowly drops the output voltage to allow the device to recognize a low voltage and trigger a low battery voltage warning and then finally shuts down completely (Figure 1.24(c)). The converter also shuts off in case of overcurrent: 2A maximum.

The battery is recharged in a special charger (not a NiMH charger!). A diode across the DC-DC converter allows the current to flow from the battery positive terminal back into the Li-ion cell.

FIGURE 1.24
Li-ion AA battery, rechargeable with special charger: (a) internal structure, (b) circuit, and (c) discharge curves.

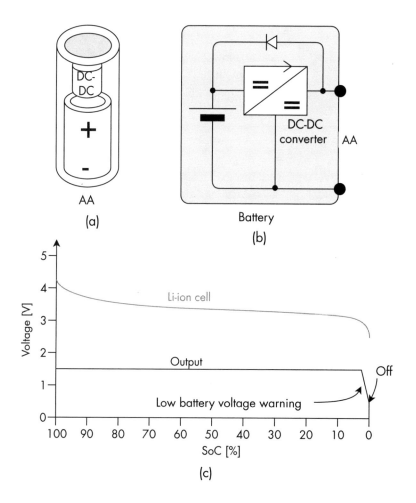

---

23.   Kentli PH5 rechargeable Li-ion AA 1.5V battery.

The DC-DC converter goes to a low-power sleep mode when the cell is not in use, to save the charge.

Surprisingly, the battery delivers less energy at low current levels than at maximum current due to the lower efficiency of the DC-DC converter at low power. This is the opposite of what we expect with just a cell, which is more efficient at low current.

Is a Li-ion AA cell worthwhile? Not really; its total charge is still less than the charge in today's NiMH cells, and its price is significantly higher [6]. The only advantage is that it has a longer lifetime than a NiMH cell.

1.5.2.2 USB AA Battery

Some batteries incorporate a USB connector and a 5V Li-ion charger (Figure 1.25(a)). The USB connector is accessed by removing a cap at one end of the AA battery (Figure 1.25(b)). The battery includes two DC-DC converters (Figure 1.25(c)):

1. A charger from the USB 5V down to the voltage of the Li-ion cell;

2. A stepdown converter from the Li-ion cell voltage down to 1.5V.

Again, the efficiency of this battery increases as higher currents, unlike the efficiency of a naked Li-ion cell (Figure 1.25(d)); regardless, the efficiency of this battery is always less than the efficiency of a Li-ion cell.

FIGURE 1.25
USB rechargeable AA cell:
(a) in use, (b) cap removed
to reveal USB charging
connector;
(c) circuit, and
(d) efficiency
versus current.

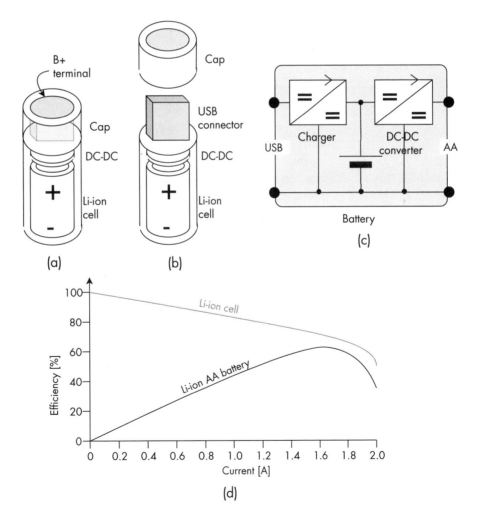

1.6   BATTERY DESIGN

This section discusses some design aspects applicable to small batteries.

1.6.1   Cell Selection

The best cell for a small battery depends on the application.

1.6.1.1   Single-Cell Battery

In general, the pouch format is best for a single-cell small battery because its dimensions can be matched to the available space; a pouch cell can be quite thin. For permanent mount, the best solution may be a 18650 cell with welded tabs that are then soldered to the main PCB.

1.6.1.2   Multicell Battery

In general, an 18650 cell is the best cell format for a small multicell battery because it is reliable, widely available, most economical, robust, and easily replaced with a cell from a different manufacturer.

Cells are available optimized for power, energy, or both (see Volume 1, Section 2.2.3). To a certain extent, cells may be physically arranged to fit the shape of the available space.

Pouch cells are best for weight-sensitive applications, when a particularly thin battery is required, or when the available space is oddly shaped. These cells require a complex design; yet, in significant volumes, the extra effort is worthwhile.

1.6.2   BMS

A PCM (see Volume 1, Section 4.3) protects the battery in small products. The PCM may be in the battery, in the product, or shared between the two. The PCM can be single-port (the battery is charged and discharged through the same wire) or dual-port (the battery is charged through an input, and discharged through a separate output) (see Volume 1, Section 5.12.2). Two essential points about two-port PCMs bear repeating:

1. You cannot convert a dual-port PCM to a single-port one by connecting the charging and discharging terminals; doing so places the charging and discharging transistors in parallel (they should be in series), and the protector won't be able to control either charging nor discharging.

2. The battery can be overdischarged through the charge port because a dual-port PCM cannot stop this current; similarly, the battery can be overcharged through the discharge port because a dual-port PCM cannot stop this current.

In a single-cell battery, its functions may be performed by the charger and the load (see Section 1.6.2.2). A BMU (without protector switch) may be used instead of a PCM if the BMS can turn off the load with a low-power signal (see Section 1.6.2.3).

1.6.2.1   PCM Selection

When selecting a protector BMS for your application, check the following:

- Cell nominal voltage: 3.2V for LFP or 3.6V otherwise; alternatively, check the maximum cell voltage—3.6V and 4.2V, respectively;

- Number of cells in series, typically 1 to 16;
- Maximum current, often the same for charging and discharging;
- Whether a thermistor is supported;
- Single-port or dual-port.

Note the following:

- The protection limits are fixed in hardware; to change them, the PCM company must make a custom version for you.
- Most of these PCMs are analog (no communication, no information of what a problem might be); some PCM companies are starting to offer sophisticated digital BMSs with an elegant user interface for your smartphone[24].
- In general, these BMSs do not prevent charging at temperatures below freezing; yet doing so is likely to damage the cells through dendrite growth.
- Double-check the voltage sense harnesses before connecting it to the PCM; incorrect wiring would damage the PCM.
- The PCM must have a minuscule current drain (see Volume 1, Section 5.8.1); as the cells power the PCM, if the battery is left unused for a long time, the PCM will overdischarge the cells.

Many ever-changing vendors market PCMs. A few Chinese companies manufacture practically all PCMs, including the ones listed in Table 1.4.

Some notes about buying Chinese PCMs:

TABLE 1.4
List of Chinese
Manufacturers of PCMs
for Small Batteries

Company	URL	Maximum Cells	Notes
A123RC	a123rc.com	12	For LFP
All Battery	all-battery.com	13	
Apollo	apollogroup.com.hk	4	
AYAA	ayaatech.com/pcm/	32	Sold through Alibaba
BesTech Power	bestechpower.com	32	Also digital BMSs with fuel gauge and communications: CAN, Bluetooth
BMS Battery	bmsbattery.com	30	May be nothing more than a reseller
EVPST	evpst.com	16	
Green Bike	greenbikekit.com/ bms-pcm.html	24	May be nothing more than a reseller
Qwawin	quawin.com	32	
Shenzhen Leadyo Technology	leadyo-battery.com	35	
Shenzhen Li-ion Bodyguard Tech	lws-pcm.com/en/ product/	20	
Shenzhen Smartec	(none)	16	Through Alibaba: szsmartec. en.alibaba.com; with Bluetooth
SU Power Battery	batterysupports.com	24	Direct sales
XJR Technology	(none)	29	Through AliExpress: aliexpress. com/store/2344164

24. BesTech Power.

- The part numbers are, shall we say, "fluid"; a given part number may apply to widely different PCMs (such as a different number of cells supported); when ordering a second time, you may receive a significantly different PCM; therefore, when ordering, specify both a part number and all the PCM specifications.

- Chinese BMSs suffer from a poor reputation, which I believe is undeserved; in reality, they are usually assembled well, and problems are more likely due to misuse or misapplication by the user; it's easier to blame the "cheap Chinese parts" rather than yourself for having miswired the cells or having specified a BMS that is a poor match to your application.

- These products have jaw-dropping low prices: you must buy 100 units, but this costs less than a single unit of a U.S.-made BMS!

- Work directly with your Chinese vendor, who is quite eager to make sure you're happy; be aware of cultural differences and do not be offended by their directness.

1.6.2.2   Single-Cell, No Explicit BMS

With a single-cell battery, a rudimentary BMS function may be implemented by devices that don't look anything like a BMS:

- A CCCV charger limits the maximum cell voltage.
- A low-voltage cutout on the load limits the minimum cell voltage.
- A thermal fuse blows in case of overcurrent or overtemperature.

Although rudimentary, this solution does protect the cell to a great extent.

1.6.2.3   Discharge Disable Through a Speed Controller

Products such as power tools may include power electronics that control the speed of a motor (Figure 1.26(a)). Two high-power transistors control the battery current:

- In the PCM: the transistor that controls the discharge inside the battery;
- In the tool: the transistor that controls the motor and its speed.

If the motor controller can be turned off through a low-power signal, a protected PCM may be replaced by an unprotected BMU that disables discharging by instructing the motor controller to shut down (Figure 1.26(b)). Using only one transistor to

FIGURE 1.26
Discharge disable:
(a) protected PCM, and
(b) unprotected BMU.

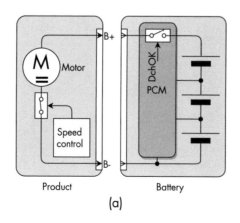
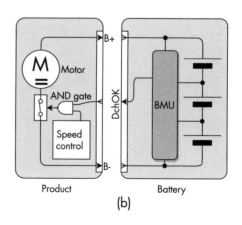

(a)   (b)

control the speed and to disable discharge reduces the cost and the size of the product, while also increasing its efficiency. There still needs to be a separate switch to control charging; usually, this uses a smaller transistor because, typically, the charging current is lower than the peak discharging current.

If the battery is removable, there is no PCM to protect its terminals. Therefore, a mechanical solution is required to prevent accidental shorts across the battery terminals (see Section 1.4.1.2).

1.6.3 Thermal Design

In general, most applications for small batteries are low power, so the battery generates little heat. If this is not the case, at least the enclosure should be designed to handle high temperature, while at the same time protecting the user from getting hurt from excessive temperatures. In extreme cases, the product should circulate air through the battery to cool it.

1.6.3.1 Cooling

If a battery requires cooling, it is using the wrong cells: energy cells instead of power cells. Even if power cells have a lower capacity, the battery is more efficient and may actually power a product longer (see Volume 1, Section 2.6.2.2).

1.6.3.2 Thermistor

Any decent small battery places at least one thermistor on the cells to monitor their temperature (Figure 1.27(a)). The BMS uses the thermistor to shut down the battery in case of overtemperature. It may also use it to prevent charging below freezing. The battery may make the thermistor available to the product so that the product may independently shut off the battery in case of overtemperature.

There can be a problem if the thermistor becomes disconnected (open) because it is interpreted as indicating a frigid temperature. The typical PCM doesn't shut down the battery if the temperature is low. Therefore, this battery loses overtemperature protection.

1.6.3.3 Thermal Fuse

A small battery may also include a thermal fuse, which blows in case of overtemperature or overcurrent. Like a regular fuse, a thermal fuse does not reset. Therefore, it incapacitates the battery permanently.

In a non-Li-ion battery, a thermal fuse is placed between two adjacent cells, in place of an interconnection bar (Figure 1.27(b)). This placement doesn't work as well in a Li-ion battery because, if the fuse opened under load, the BMS would be damaged (see Volume 1, Section 8.3.2.4). Therefore, the thermal fuse must be placed at one end of the string.

You can recognize a thermal fuse by noting that it has two thick wires and is placed in series with the battery current. Conversely, a thermistor does not carry battery currents and has two thin wires.

1.7 SYSTEM INTEGRATION

This section discusses the interface between the small battery and the product.

FIGURE 1.27
(a) Thermistor, and
(b) thermal fuse.

(a)

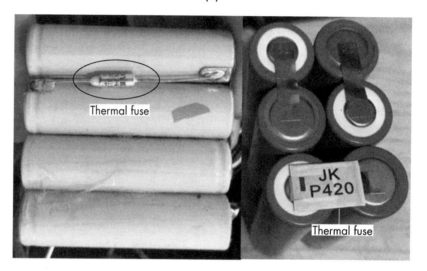

(b)

1.7.1 AC/DC Switchover

AC/DC products can be powered from an AC outlet or a battery. The switchover between the two must be seamless. There are two general approaches: (1) connect everything in parallel, and (2) use a switchover circuit.

1.7.1.1 Battery-Fed Circuit

The battery-fed circuit is quite simple: the charger, battery, and load are all in parallel[25]. In most cases, this circuit works well: the current naturally flows from where it's available to where it's needed. There is no electronic circuit that decides where the current should flow and routes it accordingly.

If there is no AC power:

- If the battery is not empty, it powers the load (Figure 1.28(a)).
- If the battery is empty, everything is off (Figure 1.28(b)).

---

25. This is the circuit used in telecom installations. See Section 2.4.

FIGURE 1.28
Parallel circuit in the
absence of AC power:
(a) battery OK, and
(b) battery empty.

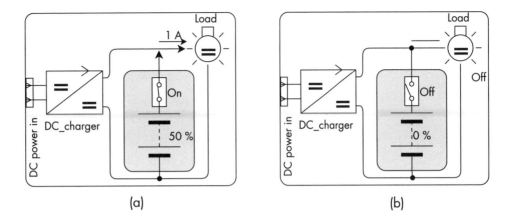

If there's AC power:

- If the battery is not full, the battery current is the difference between the charger current and the load current:

  - If the load current is low, the extra current from the charger is available to charge the battery (Figure 1.29(a)).

  - If the load current is the same as the maximum current that the charger can provide, there's no current left for the battery (Figure 1.29(b)).

  - For a short time, if the load current is higher than the charger can provide, the charger provides as much as it can, and the battery provides the rest (Figure 1.29(c)).

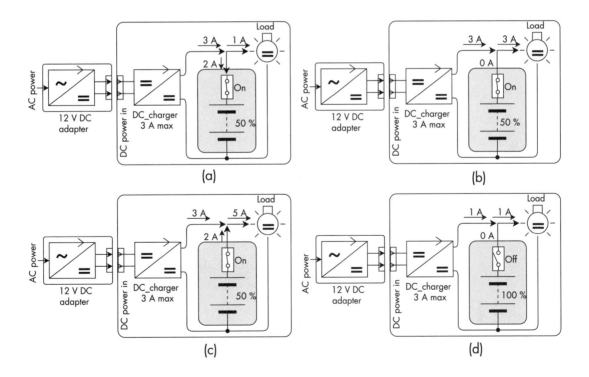

FIGURE 1.29   Parallel circuit in the presence of AC power: (a) charging, (b) load takes all the current from the charger, (c) load takes more current than the charger can provide, and (d) battery is full.

· Eventually, the battery will be charged (as long as the load's average power does not exceed the charger's peak power).

· If the battery is full, no current goes into it, and the load takes from the charger precisely the amount of current it needs (Figure 1.29(d)).

If the AC power goes away, the battery starts powering the load instantaneously because discharging is enabled, even is charging isn't.

This circuit does have two disadvantages: maximum voltage and excessive charging current.

The first limitation is that it requires a way to prevent the cells from being kept at their maximum voltage. If the charger CV setting is configured based on the maximum cell voltage (e.g., 4.2V), once the battery is charged, the charger keeps the cells at this maximum voltage, degrading them. The solution is to use a BMS that disables charge at a lower cell voltage than the rated maximum (e.g., 4.0V). Once the cells reach 4.0V, the BMS disables further charging, so that the cells are not kept at the maximum voltage. The disadvantage of this solution is that the battery is never fully charged, which is greatly offset by the noticeable increase in battery life.

The second limitation of this circuit is that it may charge the battery too fast. A high-power load requires a high-power charger, one that can provide more current than the battery can handle. If the load is turned off, the entire current from the charger goes into the battery, degrading it. The solution is to let the BMS control the charger to tell it to reduce the current if necessary. The problem is finding a PCM that has a CCL output and a charger with a CC control input.

1.7.1.2 Power Path

A switchover circuit (known as power path) solves the disadvantages of the battery-fed circuit: the battery is charged fully and at the ideal current; yet, it is not exposed to a constant charge at full voltage. However, this circuit is more complicated.

This circuit includes a switch that selects how the load is powered (Figure 1.30):

· AC power present: from the DC input;

· AC power absent: from the battery.

The switchover must be quite fast so that the load is always powered.
This circuit works as follows:

· If there's no AC power:
  · If the battery is not empty, it powers the load (Figure 1.30(a)).
  · If the battery is empty, everything is off (Figure 1.30(b)).
· If there's AC power:
  · If the battery is not full, the charger charges it at the ideal current (Figure 1.30(c)).
  · Once the battery is full, either the charger shuts off, or the BMS does (Figure 1.30(d)).

This circuit does have a disadvantage: when AC power is present, the battery is not available to help power the load at times of peak demand. This means that the AC adapter must be larger and able to handle the maximum power of the load, rather than just the average power.

FIGURE 1.30
Switchover circuit:
(a) powered by the battery,
(b) empty battery,
(c) powered by the AC,
charging the battery,
and (d) powered by
the AC, full battery.

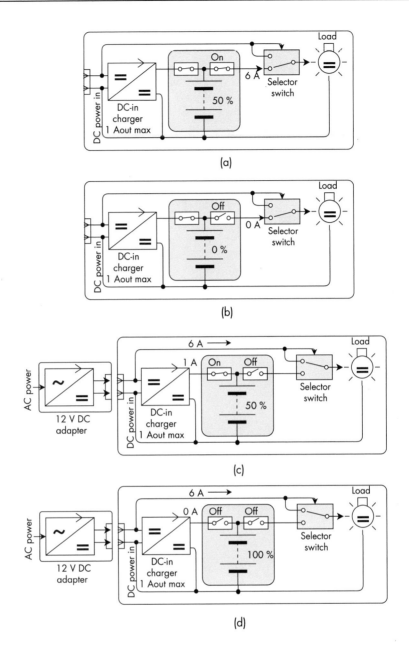

REFERENCES

[1]   Duracell and Intel, "Smart Battery Data Specification" sbs-forum.org/specs/sbdata10_a. pdf.

[2]   "Popular Connectors for LiPo Batteries," www.lipolbattery.com/li-po-battery-connectors.html.

[3]   Graves, R., "Syonyk's Project Blog," syonyk.blogspot.com.

[4]   Barsukov, Y., and J. Qian, *Battery Power Management for Portable Devices*, Norwood, MA: Artech House, 2013, Section 6.6.

[5]   Barsukov, Y., and J. Qian, *Battery Power Management for Portable Devices*, Norwood, MA: Artech House, 2013, Chapter 7.

[6]   "Performance Analysis/Review of Kentli PH5 Li-Ion 1.5V AA Battery," Jason's electronics blog-thingy, ripitapart.com/2015/06/17/.

CHAPTER 2

LARGE, LOW-VOLTAGE BATTERIES

2.1 INTRODUCTION

Large, low-voltage batteries are used to power on-grid or off-grid stationary systems. They also provide the house power for certain vehicles[1].

This book defines large, low-voltage batteries as ranging from 12V to 48V, and from about 300 Wh to 1 MWh. In some cases, a stationary battery array may fill many rooms.

Applications include telecom sites, residential backup power, solar array installations, yachts, trucks, and airplanes. Low voltage is preferred in these applications for safety[2], for historical reasons[3], because all the components in the system use low-voltage DC, or because widely available power devices are designed to use a low voltage battery.

Lead-acid batteries have been used traditionally in these applications, although Li-ion batteries have penetrated this market, using either ready-made, complete Li-ion batteries or custom-made batteries with individual Li-ion cells and a BMS.

This chapter focuses on the battery and specifically on the BMS. It is not a guide on the complete applications that use large, low-voltage batteries.

2.1.1 Tidbits

Some interesting items in this chapter include

- The best battery for a stationary system is not a battery; it's the grid (Section 2.5.2.1).
- A battery saves fuel for a generator (Section 2.4.2.2).
- A floating battery is the safest; yet safety regulations require that batteries be grounded (Section 2.6.3.1).
- Having two batteries on a boat can start a fire (Section 2.6.3.11).
- Car alternators can burn up while charging the 12V battery (Section 2.2.1.1).
- Maintaining LFP batteries fully charged degrades them very fast (Section 2.7.2.4).

---

1. They are not used to propel a vehicle. For those batteries, see Chapter 3.
2. A system with a nominal voltage of 48V is generally considered low voltage and, therefore, safe. Various standards define low voltage. In the United States, the National Electric Code variously specifies up to 50 Vdc or <60 Vdc while UL specifies 30 Vac (42V peak). In Europe, IEC specifies an extra-low voltage (supply system) of up to 12 Vdc. IEC60950-1 defines SELV (Safety Extra Low Voltage) as <60 VDC. Above these voltages, safety rules become strict, and handling methods are regulated, as are spacing requirements and insulation levels.
3. Telephone central offices generally use 48V batteries.

- Invergers and solar charge controllers are mostly incompatible with Li-ion batteries (Section 2.11).
- Some inverters are leaders, some are followers, and some are chameleons (Section 2.2.4)
- A docked boat can kill swimmers (Section 2.6.3.2).
- A battery won't normally power your home during a blackout (Section 2.5.2.1).
- Airplanes can crash after a jump-start (Section 2.8.3.2).

2.1.2   Orientation

This chapter starts with a list of some of the devices typically used in these applications. It lists possible system topologies. It discusses applications, listing issues that arise when using Li-ion batteries, giving possible solutions and suggested circuits. It explores ways to overcome the intrinsic incompatibility of today's invergers and solar charge controllers with Li-ion batteries. Finally, it discusses battery technology and system integration.

2.1.3   Applications Classification

One may assume that all low-voltage battery applications are essentially the same. In reality, they differ because the equipment designed for each application evolved independently, each with its set of challenges requiring mutually incompatible solutions. For example, the power supply in a grid-tied telecom system continues operating if a battery is disconnected (e.g., when a Li-ion battery is full). Yet a DC-coupled solar charge controller and an inverger are unable to start if the battery is disconnected. Therefore, a residential system requires a solution to this problem, while a telecom system does not.

This chapter divides the most common applications for large, low-voltage batteries as follows:

- Stationary:
  - Telecom: for example, a cell phone tower;
  - Residential: batteries for on-grid or off-grid houses.
- Mobile:
  - Marine: house power (not for traction);
  - Recreational: power for a caravan independent of the truck's battery;
  - Auxiliary power units: for long-haul trucks, utility trucks, airplanes;
- Other: UPS, microgrid, engine starter, battery modules.

2.1.4   Major Technical Challenges

Large, low-voltage, Li-ion batteries are burdened by three significant technical challenges that are nearly insurmountable:

- Some devices require the presence of a battery to turn on, yet Li-ion batteries may be off to protect the cells.
- When a battery is connected to a capacitive load, there is a large inrush of current.

· When two batteries at different voltages are connected in parallel, high current flows between them.

This chapter offers partial solutions to these challenges.

Lead-acid batteries are inherently always connected, while Li-ion batteries may be turned off to protect their cells. Power products designed for lead-acid batteries do not work well with Li-ion batteries. This chapter offers workarounds that are required until the day when power products natively compatible with Li-ion batteries will become widely available.

2.2 DEVICES

To discuss applications that use large, low-voltage batteries, we need to understand some of the associated devices (Figure 2.1).

2.2.1 Power Sources

The power source for a system may be either AC[4] or DC[5], provided by the following:

- *Grid:* AC power from the electric utility:
  - Permanent, metered connection to the grid;
  - Shore power for vehicles (vessels, trucks, recreational vehicles) when stationary[6]
- *AC generator, also known as AC genset:* a fuel-powered engine driving a high-voltage AC generator;
- *DC generator:* same as above, but with a low-voltage DC output:
  - A DC genset;
  - An alternator (driven by the engine in a vehicle) with a voltage regulator.
- Renewable resources[7]

FIGURE 2.1
Devices used in large, low-voltage batteries:
(a) TS22432 inverger;
(b) TurboCharger solar charge controller, and
(c) large LFP battery. (All courtesy of Apollo Solar.)

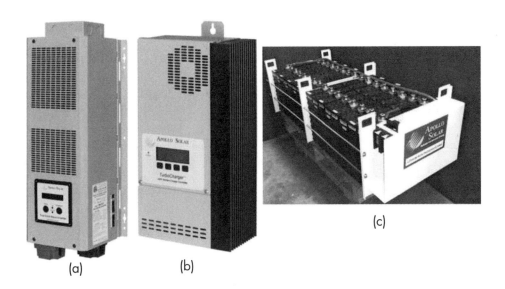

(a)   (b)   (c)

---

4. Alternating current; in this context, 110, 240, 480 Vac, 50 or 60 Hz.
5. Direct current; in this context, 12, 24, 36, or 48 Vdc.
6. For safety, the outlet is protected by a ground fault interrupter (GFI) also known as ground fault circuit interrupter (GFCI).
7. All together commonly referred to as distributed generation (DG).

· *Solar array:* also known as PV, photo-voltaic, solar panels;

· *Wind generators:* also known as wind turbines, wind energy converters;

· Hydro generators:

     · *Stationary:* powered by a stream;

     · *On vessels:* powered by the vessel's motion through the water.

2.2.1.1   Alternator, Voltage Regulator

In vehicles propelled by an engine (e.g., cars, trucks, and vessels), the low-voltage DC electrical system is powered by a battery. The engine drives an alternator with a voltage regulator that charges the battery (Figure 2.2(a)).

In this context, an alternator is a DC generator[8], consisting of

· A three-phase AC generator whose magnetic field is generated by a field coil;

· A three-phase rectifier.

The output voltage of the alternator on its own depends on its speed and load. A voltage regulator stabilizes this voltage. It does so by driving the alternator's field coil with a current that is inversely proportionally to the DC bus voltage (Figure 2.2(b)). Together, the alternator and the voltage regulator form a servo loop to maintain a constant voltage on the DC bus. If the DC voltage is too high, the regulator reduces the current to the alternator's field coil, which results in a lower output voltage; this restores the desired level on the DC bus.

FIGURE 2.2
Alternator and regulator:
(a) simplified circuit,
and (b) regulator
transfer function.

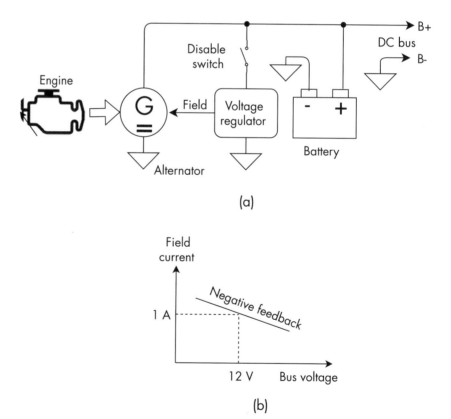

---

8.   In other contexts, such as in a power plant, alternator means an AC generator. It produces AC because it does not include a rectifier.

The alternator is shut down by removing power to the voltage regulator. Doing so removes power to the field coil, eliminating the magnetic field in the alternator and disabling the generator.

Often, the alternator's output current is not limited. This may be suitable for lead-acid batteries because they are more forgiving and have high internal resistance. It is not suitable for Li–ion batteries, as their low internal resistance does not limit the current as well, and because they degrade when charged at excessive current. When charging Li–ion batteries, the voltage regulator must also limit the output current and the alternator's temperature. This is not only for the sake of the Li–ion cells but also for the sake of the alternator, which is not designed to operate continuously at high current while charging a Li–ion battery, unfortunately. Few regulators include current limiting[9].

2.2.2 Loads

Loads include

- *AC loads:* equipment typically operating at 120 or 240 Vac:
  - Heating and air conditioning (also known as HVAC);
  - Lighting and appliances;
  - Grid: it may be possible to back-feed power to the grid.
- *DC loads:* equipment typically operating at 12, 24, or 48 Vdc:
  - Telecommunication equipment;
  - Vehicle lighting.

2.2.3 Power Converters

Power converters convert between or withi+n AC and DC (see Volume 1, Section 1.8.1). The following power converters are used with large, low-voltage batteries.

2.2.3.1 *Current Limited Power Supply*

A current limited power supply (also known as charger, SMPS[10], rectifier[11]) is AC-powered. Its output is CCCV[12] and is suitable to charge a battery.

Specialized chargers use a profile, varying the CV setting over time, for fast yet safe charging, through three or more stages, for example, for a 12V lead-acid battery:

- *Bulk:* CC mode at 0.2 C current;
- *Topping (absorption):* CV mode at 14.5V, for a preset time;
- *Float:* CV mode at 13.5V, indefinitely.

Optimally, the CV setting is temperature compensated to match the -3 mV/°C thermal coefficient of the voltage of lead-acid cells.

---

9.   The Balmar MC-614 does.
10. Switch mode power supply: uses a high-frequency electronics and a smaller transformer.
11. That is confusing to an electrical engineer: a "rectifier" is a just high-power diode, incapable of changing voltage and of limiting current, which is what a charger does. Yes, a charger uses rectifiers, but calling a charger a "rectifier" because it uses four rectifiers is like calling a car a "piston" because it uses four pistons. This is what the lead-acid battery industry has called them historically, and this term remains in use.
12. Constant current/constant voltage. See Volume 1, Section 1.2.1.

2.2.3.2 Charger with DC Input

This charger is the same as above but with DC input.

2.2.3.3 Inverter

In these applications, an inverter generates AC power at a fixed frequency of 50 or 60 Hz. It is unidirecional.

On the DC input, an inverter may be connected to

· A low voltage DC bus: often called battery-based;
· Low-voltage solar panels or other renewable resources;
· High-voltage solar panels: often called a string inverter (see Section 2.5.2.1);
· Wind generator: weirdly, also called a string inverter[13].

On the AC output, an inverter may be connected to

· AC loads (other than variable speed motors);
· Back-feed to the grid.

2.2.3.4 Invergers

An inverger[14] is placed between a DC bus and an AC bus. It is bidirectional: it operates either as a charger or as an inverter, as required at a given moment. While inverters may use a variety of DC sources, in these applications, invergers are always battery-based: connected to a battery.

Invergers used for grid applications (Figure 2.3(a)) and residential use (Figure 2.3(b)) are slightly different:

FIGURE 2.3
Types of inverger:
(a) for grid applications,
and (b) for residential use.

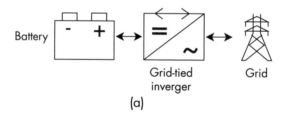

(a)

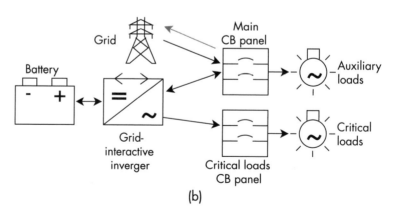

(b)

---

13. The unfortunate term "string inverter" is used for an inverter for a high-voltage DC input (as in "a string of solar panels in series"), even if the source is not a string of solar panels.
14. Charger/inverter or combi. See Volume 1, Section 1.2.4.1.

- *Grid applications:* A grid-tied inverger transfers power from the grid to a battery, or vice versa.
- *Residential:* The inverger may or may not be grid-tied; it has multiple AC ports (see Section 2.2.5.1).

The control circuits of these invergers are powered by the DC bus (not by the AC side), which means that they require a battery on the DC side to start operating. If a Li–ion battery has disabled discharging to protect its cells, it cannot recover when the AC power returns. I urge any designers of invergers who may be reading this to remedy this oversight and power the inverger from the battery or the AC power, whichever one is available.

2.2.3.5 Transverter

In this context, a transverter is a universal power converter. It combines and converts power from various sources (solar panels, batteries, the grid, generators) to power both DC and AC loads. Specifically, a transverter reduces the number of power devices, simplifying a system by integrating three functions: charger, inverter, and solar conversion (Figure 2.4). The quintessential tranverter for these applications is the Heart Transverter, by Heart Akerson[15].

2.2.4 Off-Grid, Grid-Tied, Grid-Interactive

Inverters, invergers, and transverters can either operate on-grid, off-grid, or automatically switch between the two modes:

FIGURE 2.4
A transverter converting from any source to any load.

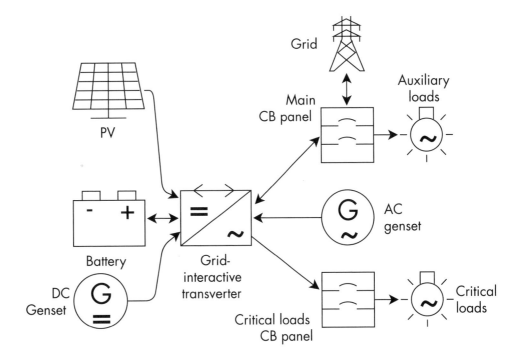

---

15. Heart Akerson is the grandfather of solar inverters. He designed the original Trace inverters, later Xantrex, now part of Schneider Electric. He now lives in Costa Rica, where he manufactures the Heart Transverter and other flexible products for micro-grid applications.

- *Off-grid*[16] (Figure 2.5(a)): generates the AC sinusoidal voltage waveform on its own; it is an independent voltage source; it is a leader, and as such it must not be connected to other independent AC sources (such as the grid, and AC genset, another off-grid inverter).

- *Grid-tied* (Figure 2.5(b)): the AC bus is connected to the grid or some other independent AC source, such as a genset; the inverter pushes additional power onto the AC bus; that independent AC source is the leader and this inverter follows its frequency and phase; without that independent AC source, this inverter shuts down.

- *Grid-interactive*[17]:

  - AC power present: When the grid or another source of AC power is present, it operates in grid-tied mode (Figure 2.5(b)); it may power the loads and it may back-feed to the grid.

  - AC power absent: Switches to the off-grid mode, to power critical loads[18]; it does not power the grid (Figure 2.5(a)).

This distinction is crucial and often overlooked by the system designer. Not understanding this distinction causes disappointment: a system that doesn't provide backup power during a power outage. It may even result in damage if two leaders conflict with each other.

2.2.4.1   Single Leader

Various AC sources may provide power to an AC bus at various times. However, at any given time, there can only be one and only one independent AC source such as the grid (Figure 2.6(a)), an AC genset (Figure 2.6(b)), an off-grid inverter (Figure 2.6(c)). Two such AC sources connected simultaneously to a single AC bus would fight each other, with dire consequences.

This single AC source provides the AC reference voltage to any grid-tied inverter on the same AC bus.

An analogy may help us understand this distinction:

- An off-grid inverter and/or other independent AC power source is like a bicyclist, who both pedals and steers.

FIGURE 2.5
Inverter types or modes: (a) off-grid or grid-interactive in off-grid mode, and (b) grid-tied or grid interactive in grid-tied mode.

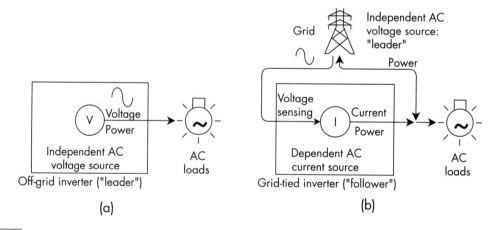

(a)   (b)

---

16.   Also known as stand-alone.
17.   Also known as bimodal.
18.   Also known as no-break loads.

FIGURE 2.6
Only one independent AC source, a leader, may be connected to an AC bus: (a) grid, (b) AC genset, and (c) off-grid inverter.

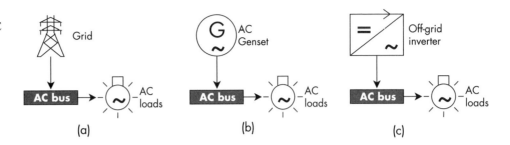

- A grid-tied inverter is like the person in the back of a tandem bike, who provides extra torque but does not determine the route; should the person in the front faint, the person in the back cannot control the bike; the bike falls.

- A grid-interactive inverter is like the bicyclist who normally pedals in the back of the tandem but who can jump on the front seat if the partner decides to stop at a microbrewery.

- Connecting an off-grid inverter to the grid is like two bikes tied by a rope; one rider heads North, and the other rider heads East; both bikers fall.

2.2.4.2 Grid-Interactive Inverter

A grid-interactive inverter is used in systems that require that certain specific loads remain powered in case of failure of the grid. The system includes two circuit breaker panels: one for critical loads, and one for noncritical ones. Each panel is connected to a separate port on the grid-interactive inverter.

- *Grid power present* (Figure 2.7(a)): The inverter operates in the grid-tied mode, powering the critical loads; if back-feeding to the grid is allowed, it may send any extra power to the grid.

- *Grid power absent* (Figure 2.7(b)): For safety, the inverter disconnects itself from the main AC panel; then it switches to the off-grid mode to continue powering the critical loads while receiving power from the DC bus.

- *Grid power returns* (Figure 2.7(a)): The inverter reconnects itself to the main AC panel, and switches back to the grid-tied mode.

Note how, at any given time, there is always one and only one leader: if the grid power is present, it is the leader. If the grid power is absent, the inverter is the leader. At no time are there two leaders.

Grid-interactive inverters include an automatic transfer switch to select either the grid or the local power source. This keeps the inverger from powering the grid. Without it, the inverger would attempt to power nearby houses and endanger workers who may be working on the power lines (see Section 2.5.2). When the grid is present, the switch connects the inverter to the grid (Figure 2.7(a)). When absent, the switch disconnects the inverter from the grid (Figure 2.7(b)).

2.2.4.3 Multiple Inverters in a System

For increased power, one or more grid-tied inverters may be added to an AC bus. The leader may be the grid (Figure 2.8(a)), an AC genset (Figure 2.8(b)), or an off-grid inverter (Figure 2.8(c)).

FIGURE 2.7
Operating modes of a grid-
interactive inverter: (a) grid
power present, grid-tie
mode, and (b) grid power
absent, off-grid mode.

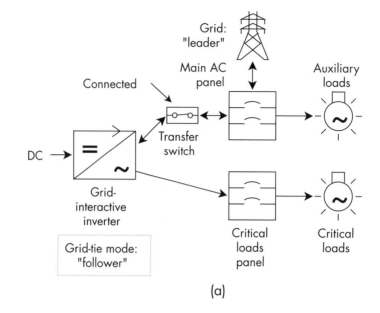

(a)

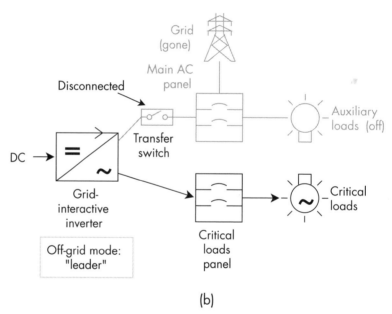

(b)

FIGURE 2.8
Grid-tied inverters
used with: (a) grid,
(b) AC genset, and (c)
off-grid inverter.

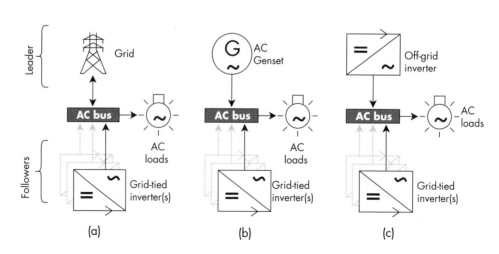

Note that, in each example, there is a single leader (top row in the figure) that provides the AC reference voltage, and any number of followers (bottom row) that follow the reference from the leader.

Grid-tied inverters may also be used with a grid-interactive inverter:

- Grid power present (Figure 2.9(a)): The grid is the leader; the grid-interactive inverter operates in the grid-tied mode, following the grid; all the grid-tied inverters follow the grid as well; if back-feeding to the grid is allowed, any extra available power is sent to the grid.

- Grid power is absent (Figure 2.9(b)): The grid-interactive inverter operates in the off-grid mode and becomes the leader; all the grid-tied inverters follow the grid-interactive inverter.

FIGURE 2.9
Grid-interactive inverter with multiple inverters: (a) grid-tie mode, and (b) off-grid mode.

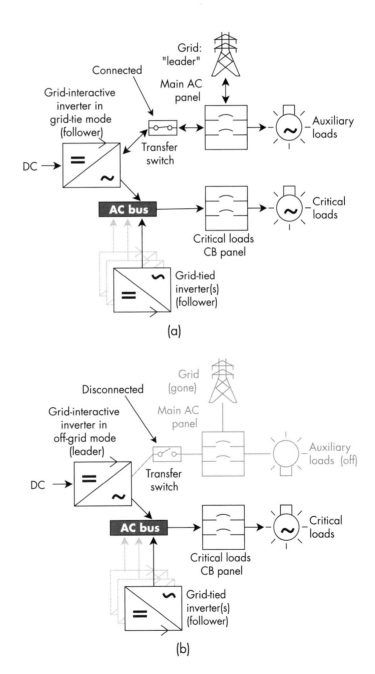

2.2.5 Solar

Power from PV solar panels is converted to either high-voltage AC or low-voltage DC[19]. One of two topologies may be used:

1. C-coupled (Figure 2.10(a)): Uses a solar inverter that produces AC and an inverger to convert back to DC; when charging the battery, power goes from DC to AC and then back to DC.

2. DC-coupled (Figure 2.10(b)): Uses a solar charge controller that produces low-voltage DC; when charging the battery, power remains as DC.

Each approach has advantages and disadvantages (Table 2.1).

2.2.5.1 Solar Inverter

A solar inverter is a specialized grid-tied inverter. It is designed to handle the wide range of input voltage from the solar panels and to maximize the generated power through MPPT[20] technology.

Just like other inverters, solar inverters may be

- Off-grid;
- Grid-tie;
- Grid-interactive (rare).

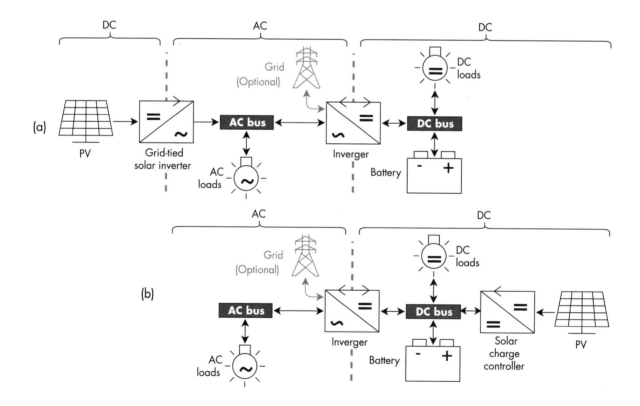

FIGURE 2.10   Solar coupling topologies: (a) AC-coupled and (b) DC-coupled.

---

19. Or high-voltage DC, but this is a subject covered in Chapter 4.
20. Maximum power point tracking; maximizes the charging power by adjusting the operating point of its DC-DC converter.

TABLE 2.1
Comparison of AC-Coupled and DC-Coupled Solar

	AC-Coupled	DC-Coupled
Efficiency	Higher efficiency when powering AC loads (about 95%) due to having a single converter between the PV and the AC loads. Lower efficiency (about 92%) when charging or powering DC loads, due to the two-step conversion between the PV and the battery.	Higher efficiency when charging (90%–99%) and powering DC loads due to having a single converter between the PV and the battery. Lower efficiency (about 90%) when powering AC loads, due to the two-step conversion between the PV and the AC loads.
Cost	Cheaper for systems above 3 kW. Better when retrofitting a system to add batteries because the grid-tied inverter is reused.	Cheaper for systems below 3 kW. Ideal for micro system with one or two panels.
Li-ion	The solar inverter does not need to be compatible with Li-ion (although the inverger may not be).	Both the solar charge controller and the inverger may not be compatible with Li-ion.
Complexity	The solar inverter and the inverger must be synchronized to use the same AC frequency and phase.	The solar charge controller and the inverger do not need to be synchronized.
Regulatory	Fewer regulations concerning net metering.	More limitations when using net metering, yet allows unlimited solar power even if the power utility limits the AC power that can be back-fed to the grid.
Large PV array	Compatible with any solar inverter including micro inverters. High AC power transmission is more efficient than low voltage DC.	Each PV set requires its own solar charge controller; low-voltage DC power transmission is inefficient and requires large cables.

Based on the type of connection to the PV panels, the industry also divides solar inverters as

- Micro: Low-voltage, low-power PV panels; one grid-tie inverter on each PV panel; the AC outputs of many micro-inverters are connected in parallel.
- Standard: Low voltage, medium-power PV panels.
- String: High-voltage, medium-power PV panels.
- Central: High-voltage, high-power PV panels.

2.2.5.2 Solar Charge Controller

A solar charge controller is like a solar inverter except that it has a DC output with CCCV regulation, suitable to charge a battery. Using a solar charge controller with a Li-ion battery, which at times may be turned off, presents significant challenges:

- Their control circuits are powered by the output DC bus; therefore, if a Li-ion battery is turned off, they can't recover.
- The power electronics may not be robust enough to survive being disconnected from the DC bus while there's sunshine[21].

The loads may be unpowered even if the sun is out, and the battery is full; this is how:

---

21. Because they rely on the battery to absorb the impulses of energy from their powerful DC-DC converters.

- Initially, it's daytime; the solar panels are charging the battery and powering the load.

- The battery is unbalanced. As soon as any one cell in the battery is fully charged, the BMS disconnects the battery from the DC bus[22], to keep this cell from being overcharged; the solar charge controller continues powering the loads as well as itself.

- A cloud arrives; the solar charge controller stops powering the DC bus; the loads are unpowered.

- The cloud leaves; there is still no voltage on the DC bus, so the solar charge controller cannot restart.

The load receives no power, neither from the battery (it's shut down because a cell is full) nor from the charge controller (it's shut down because it sees no battery). The user now experiences the frustrating situation of having no power to the loads even though the battery is full, and there's plenty of sunshine.

Note that the line between the DC bus and the solar charge controller is bidirectional (Figure 2.10(b)) because current may flow in either direction:

- If there is sunshine and the solar charge controller has started, current flows from the solar charge controller to the DC bus.

- Otherwise, a small current flows from the DC bus to the solar charge controller to power its electronics and to allow it to resume delivering power when the sun returns.

2.2.5.3   Tomorrow's Solar Charge Controller

Because the control circuits of today's solar charge controllers are powered only by the DC bus, they require a battery to start operating. If a Li-ion battery has disabled discharging to protect its cells, it cannot recover when the sun returns. I urge designers of solar charge controllers to remedy this oversight[23] and power their product from one of these two sources:

1. Entirely by the sun (Figure 2.11(a));

2. Through a separate DC supply input (Figure 2.11(b)).

In the first case, recovery would occur as soon as the sun returned, even if the battery is empty. In the second case, the battery could provide an independent power source (that does not power the loads), through a small switch that remains turned on until the battery is completely depleted.

2.2.5.4   Solar Heating of Batteries

Li-ion batteries should not be charged below freezing. If it's too cold for charging, electrical power from the solar panels may be diverted to battery heaters; once the battery temperature is sufficient, the solar charge controller may start charging the battery.

---

22. We're assuming a single-port, single-switch battery; once disconnected, it disables both charging and discharging.
23. I can imagine the manufacturers' argument: "Hey, our present designs have worked quite well for decades with lead-acid batteries, why change it? Let the Li-ion battery people solve the problem they created."

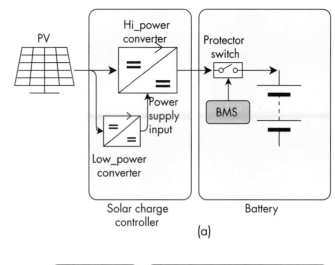

FIGURE 2.11
Ideal solar charge
controller: (a) entirely
solar-powered, and
(b) with separate
power supply input.

(a)

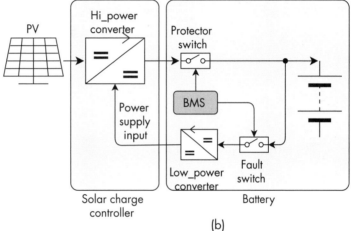

(b)

2.2.6 Battery Monitor

A battery monitor[24] measures and displays the battery voltage and current (measured through a shunt) to report them and to calculate the battery's SoC[25]. They work independently of a BMS. A classic one is the E-meter[26] (Figure 2.12).

2.3 SYSTEM TOPOLOGIES

One would think that applications that use a large, low voltage battery could be divided into only two or three topologies. For example, telecom applications are normally divided into only two topologies: off-grid and on-grid. In reality, if one considers all these applications, and all the possible arrangements of its components, many topologies are possible:

- There are five options for power sources:
  - Unidirectional: power flows only in one direction:
    - DC: such as solar;

---

24. Also known as a Coulomb counter or Ah counter.
25. Xantrex LinkPRO, Victron BMV 700.
26. Nothing to do with the Church of Scientology.

FIGURE 2.12
Battery monitor.

• AC: grid or genset;

• DC + AC: both of the above.

• Bidirectional: power flows in either direction (i.e., back-feed to the grid):

• AC: back-feed to the grid;

• DC + AC: local DC power generation plus back-feed to the grid.

• There are two options for the main loads (which may be powered by the battery):

• DC loads only (e.g., only 48-Vdc loads);

• DC plus inverter and AC loads: any load.

This results in a total of 10 permutations (Figure 2.13).

A total of 10 topologies seems excessive to those who are aware of only two or three topologies in their particular industry. The fact is that some applications use a small subset of these 10 topologies, while other applications use another small subset; by the time you add them all up, most of these 10 topologies are used in one application or another.

Specifically:

• Telecom applications are divided into two topologies (Figure 2.14(a)):

• Off-grid: may use topologies a, A, b, B, c, or C;

• Grid-tie: may use topologies b, B, c, or C.

• Solar applications are divided into three topologies (Figure 2.14(b)):

• Batteryless: the most common topology for solar systems doesn't use a battery and is therefore not one of the 10 topologies.

• Off-grid: may use topologies a, A, b, B, c, or C.

• Grid-tie: may use topologies b, B, c, C, d, or e.

• Other applications use a variety of topologies (Figure 2.14(c)):

• A car uses topology A.

• An auxiliary power unit uses topology A.

• A UPS uses topology B.

• A yacht uses topology C.

• A recreational vehicle may use topologies B or C.

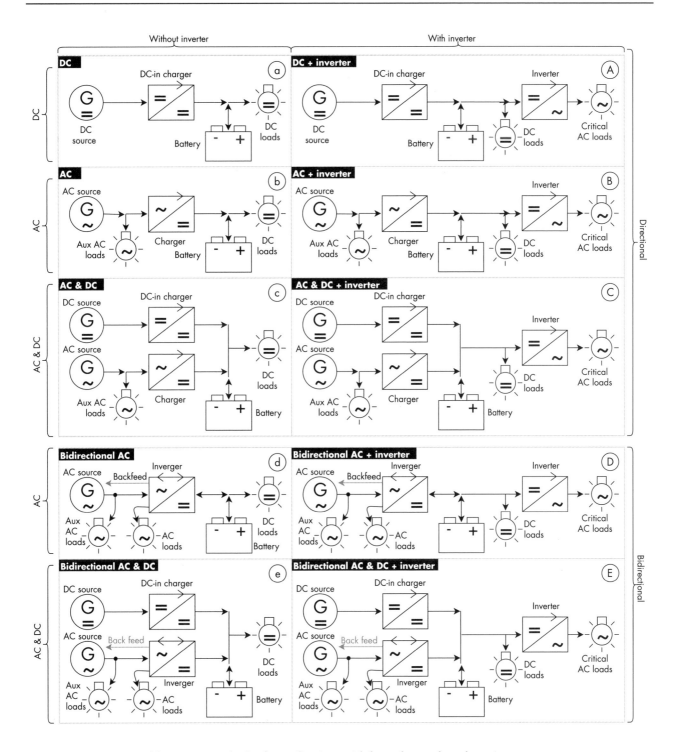

FIGURE 2.13   Ten possible system topologies for applications with large, low-voltage batteries.

· A microgrid designer may consider topology a, d, or e.

When you consider every application that uses a large, low-voltage battery, you see that most of these 10 topologies are employed. Table 2.2 lists the characteristics of each topology in detail.

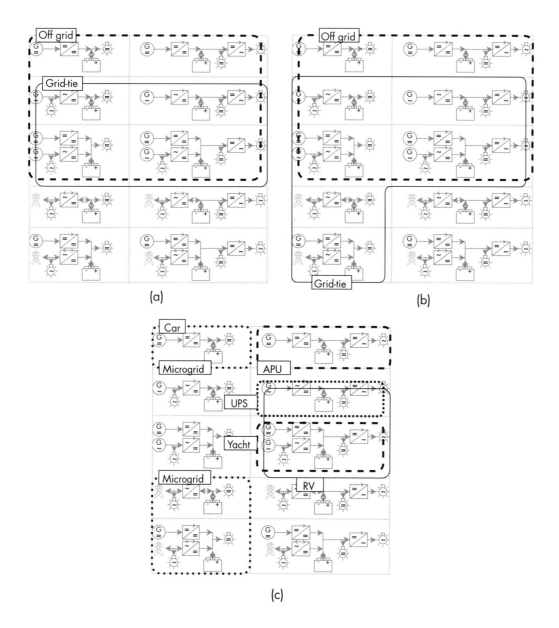

FIGURE 2.14   Topologies used by given applications: (a) telecom; (b) solar; and (c) other.

In the following sections, the discussion of each application points out which of these topologies it uses.

2.4   TELECOM APPLICATIONS

This section discusses telecom sites[27]. These sites may be on-grid or off-grid and provide wireless communication services, such as cell-phone service, two-way radios for law enforcement and emergency services, microwave links for phone companies, radio and TV, and paging. The typical site consists of a secure hut or cabinet housing communication equipment, backup power, and climate control.

27.   Many thanks to John Pfeifer of Apollo Solar.

	Topology		AC In (1)	DC In (2)	Solar (3)	Inverter (4)	Main Loads (5)	Auxiliary Loads (6)	Notes	Applications
Unidirectional	a	DC		✓	DC		DC	—	Off-grid systems that consist entirely of low voltage DC devices	Off-grid telecom sites
	A	DC + inverter		✓	DC	✓	AC and DC	—	Off-grid sites with AC loads but without AC generation	Utility trucks (to power a cherry-picker arm or power tools), tiny houses and long-haul trucks
	b	AC	✓		AC		DC	AC		Telecom site with AC power, either on-grid or off-grid
	B	AC + inverter	✓		AC	✓	AC and DC	AC	Two separate AC buses, one on the input and one on the output; DC loads are optional	Uninterruptible power supplies, simple recreational vehicles
	c	AC and DC			AC or DC (7)		DC	AC	Systems with both AC and DC power sources	Telecom sites
	C	AC and DC + inverter	✓	✓	AC or DC (7)	✓	AC and DC	AC	For critical AC loads that must remain powered at all times	Telecom sites, yachts, recreational vehicles
Bidirectional	d	AC	✓		AC		DC	AC	Uses a grid-interactive inverger that powers the critical AC loads from the battery when there is no AC power; may back-feed to the grid; otherwise, some AC loads must be included to receive power from the inverger	Residential systems with AC-coupled solar, micro-grids, and arbitrage
	D	AC + inverter	✓		AC	✓	AC and DC	AC		Not many; some advantages (4)
	e	AC and DC			AC or DC (7)		DC	AC		Grid-tied solar home with DC-coupled solar
	E	AC and DC + inverter	✓	✓	AC or DC (7)	✓	AC and DC	AC		Not many; some advantages (4) (8)

Notes: 1. AC sources: grid, AC-coupled solar (uses a solar inverter), genset, AC wind generator. 2. DC sources: car alternator, DC genset, PV (uses a solar charge controller), DC wind generator. 3. Solar coupling: AC or DC coupled. 4. Inverter: powers main AC loads with clean AC and with no disruption. 5. Critical loads: remain powered if the power source goes away. 6. Auxiliary loads: go off if the power source goes away. 7. Typically you would not have both AC and DC-coupled solar in a given system, but nothing precludes it. 8. A transverter that integrates all three functions (charger, inverter, and solar) would simplify the system.

TABLE 2.2   Characteristics of Each Topology

2.4.1   Telecom Devices

The telecom industry uses the following devices in addition to the ones described at the start of this chapter (see Section 2.2).

2.4.1.1   Base Transceiver Station (BTS)

This communication equipment operates at low voltage (12, 24, or 48 Vdc) and includes radio receivers and transmitters. Supposedly, these devices include a low-voltage cutoff to shut off if the bus voltage is too low, so as not to overdischarge a lead–acid battery. Whether they actually do is hard to say, as it is not specified.

2.4.1.2   System Controller

Except for the most basic sites, a system controller manages the site; this may be a PLC[28] or a personal computer. At a minimum, the system controller turns on a genset

---

28. Programmable logic controller.

when the battery voltage drops below a given threshold. In sites with other power sources, the system controller turns off the genset after charging the battery just a bit, just enough to wait until a cheaper power source becomes available. In sites with no other power source, the system controller leaves the genset running until the battery is full.

The system controller may extend the battery on-time by turning off noncritical loads before the battery is depleted. The system controller may also perform other functions, such as controlling a transfer switch or a power supply.

2.4.1.3   Automatic Transfer Switch

An automatic transfer switch automatically selects the primary AC power source when available, such as the grid. Else, it selects an alternate AC power source, such as from a genset.

2.4.2   Telecom Technical Considerations

A telecom site has specific requirements.

2.4.2.1   Up-Time Requirement

Telecom applications categorically expect communication equipment to have an up-time of 100%, no matter what[29]. For this reason, a telecom site is typically designed to have enough backup capacity to power the site for 3 to 5 days. Of course, if there is no charging source, eventually the battery will be depleted, and the site will go down.

A telecom site is powered if at least one of these two conditions is true:

1. A charging source (other than solar) is present.
2. The battery's discharge is enabled.

Solar is a special case because the DC bus powers the solar charge controller. Even if the sun is shining, it won't start generating power if a Li-ion battery has disabled discharging[30]. There are ways to minimize this risk (see Section 2.11).

The good news for telecom applications is that the chargers used in this industry do not require the presence of a battery, unlike the ones used in residential applications.

2.4.2.2   Strategies

Different battery management strategies are used depending on the local circumstances. In developed countries, the grid is reliable, and fuel is cheap. As power failures are rare and short, the goal is to minimize the cost of the battery. A small, float-type battery is selected since it is nearly always kept full. It only needs to power the load until the genset comes on. Having solar power is environmentally desirable but economically unjustifiable. If there is a genset, it is turned on when the battery voltage is low. It runs until the battery is full or the AC power returns (Figure 2.15(a)).

In developing countries, the grid is unreliable. As fuel is expensive, the goal is to minimize its usage. Solar power is an unaffordable luxury. The long duration and frequency of power outages make the upfront cost of large batteries worthwhile. The high cost of fuel justifies the cost of large batteries; lead-acid and Li-ion batteries may be used sequentially in a smart way to reduce costs (see Volume 1, Section 6.6).

---

29. The penalty that the site operator suffers for each minute off-air is much higher than the operating cost of the backup system.
30. This is not a problem with a lead-acid battery, which does not shut down.

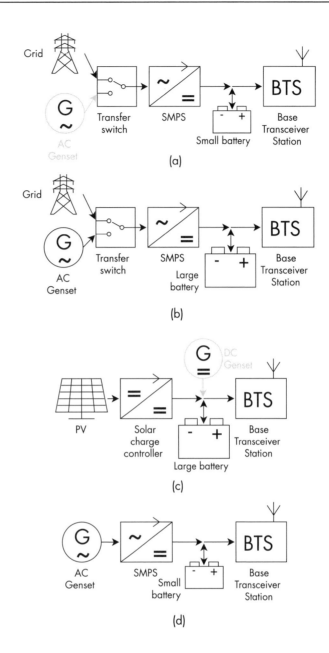

FIGURE 2.15
Telecom site strategies:
(a) developed country,
(b) developing country,
on-grid, (c) off-grid solar;
and (d) off-grid, genset.

The strategy is to turn on the generator when the SoC of the battery drops to ~10% and run it until the SoC reaches ~25% or grid power returns, whichever occurs first (Figure 2.15(b)).

In remote areas, there is no grid, and transportation costs make fuel prohibitively expensive. Solar is the preferred source of power. The goal is to keep the site running for up to 5 cloudy days and to avoid using fuel. Therefore, a large battery is used, and a backup DC generator remains mostly unused. The strategy is to run from solar power and batteries, and to turn on the DC generator only as a last resort if the SoC drops below 10% (Figure 2.15(c)).

In remote areas with just a generator and no solar power, a battery reduces operating costs: without a battery, the genset is inefficient because it runs continuously and inefficiently, as it is designed to generate the peak power requirement of the load. The addition of a small battery allows the genset to run part-time (to recharge the battery every hour or so) and at full power (for maximum efficiency) (Figure

2.15(d)). This results in significant fuel savings[31], quickly justifying the upfront cost of the battery. In theory, the battery would allow the use of a smaller genset, running continuously, to provide just the average power requirement. The small battery would act just as a power buffer. In reality, the load in a telecom site is pretty constant, so the peak power is only slightly above the average power, meaning that using the battery as a power buffer does not result in any noticeable savings.

2.4.2.3   Minimum Battery Capacity

Typically, the batteries in a telecom site are sized to power the site without charging for 3 to 5 days. This works out to a discharge rate of about 0.001 C. The charge rate is much higher, although still well within the safe limits for the battery.

A site may use a smaller battery, although its minimum capacity depends on the maximum current that the power supply can deliver. The capacity must be large enough so that the battery is not charged too fast when the loads are turned off, and all the current from the power supply goes into charging the battery. For example, assuming that the cells in the battery are designed for a maximum charging current of 0.5 C, a power supply capable of 100A requires a battery of at least 200 Ah.

If the battery is smaller than this, a system controller must monitor the load and charging currents and instruct the power supply to lower its output current to prevent charging a Li-ion battery too fast. In reality, few sites have a system controller. In any case, the typical power supply does not provide a way to control the maximum current. Therefore, the system designer must match the size of the power supply and the batteries. Any individual battery must be large enough to handle the full charging current provided by the power supply, as shown in Table 2.3. The row for lead acid says "total" while the rows for Li-ion say "each battery"; this is because lead-acid batteries are connected permanently to the DC bus, while in an array of Li-ion batteries all but one may be disconnected. This single battery would receive the full current from the charger.

2.4.2.4   Typical Sizing for Battery, PV Panels, Genset

As a rule of thumb, for every kilowatt of DC load, a telecom site requires

- Solar panels with a peak power of 8~11 kW;
- A 5-kW genset;
- 20-kWh worth of Li-ion batteries or 60-kWh worth of lead-acid batteries.

This results in specific currents of about C/2 to C/20 for Li-ion and C/6 to C/60 for lead acid.

TABLE 2.3
Minimum Battery
Capacity

Type of Battery	Maximum Charging Specific Current	Minimum Capacity Relative to Charger Current	Example for a 100-A SMPS
Lead-acid	0.2 C	5 Ah/A_{SMPS}, total	500 Ah total
Standard Li-ion	0.5 C	2 Ah/A_{SMPS}, for each battery	200 Ah each battery in an array
High power Li-ion	2 C	0.5 Ah/A_{SMPS}, for each battery	50 Ah each battery in an array

31.   Orun Energy, in Nigeria, reports a 50% reduction in fuel use.

The capacity of the lead-acid battery is three times as high because only about 30% of its capacity is used to extend its life. A Li-ion battery can be completely charged during the 4 hours or so of peak sunshine at a current of 0.25 C. A lead-acid battery cannot be charged at this high current. Increasing the capacity by a factor of 3 reduces the specific current by a factor of 3, from 0.25 C down to 0.08 C, which is a safe level for lead acid.

2.4.2.5   End of Battery Charge

If, despite all precautions, the battery charge is exhausted, the site goes down. This is bad. What's worse, though, is if the site cannot recover automatically once a power source returns.

This is an issue with a site with DC-coupled solar: once the sun returns, if the battery can no longer discharge, it can't power the solar charge controller, and the site does not recover. The standard solution to this problem is to use loads with a low-voltage cutoff that is higher than the minimum operating voltage of the solar charge controller.

Assuming a 48V battery with 16 LFP cells in series, the threshold voltages in Table 2.4 may be used. Because the battery voltage is not divided equally among the cells, some voltages are in parentheses:

- *System controller actions*: bases its thresholds on the battery voltage; cell voltages are in parentheses.

- *BMS actions*: bases its thresholds on the cell voltages; battery voltage is in parentheses.

Given these voltage thresholds, toward the end of a battery charge, a series of events unfolds (Figure 2.16):

1. When the battery voltage drops below 48V, the system controller tries to turn on a genset. If there is one and it does turn on, it recharges the battery, until the battery voltage reaches 52.8V; otherwise, the battery keeps on discharging.

2. When the battery voltage drops below 46.4V, the loads turn off due to their low-voltage cutoff function, preventing further discharge; the site shuts down; the battery (which still has some charge left) keeps on powering the solar charge controller; the battery voltage drops much more slowly because the current is so much lower.

3. If the sunshine returns, the charge controller starts recharging the battery, and the system recovers.

TABLE 2.4
Threshold Voltages at the End of Discharge for a 48V Li-Ion System

Battery Voltage [V]	Cell Voltage [V]	System Controller Action	Loads Action	BMS Action	Solar Charge Controller
>52.8	(>3.3)	Turn off the generator	On	Discharge OK	Powered
<48.0	(<3.0)				
<46.4	(<2.9)	Turn on the generator	Low-voltage cutoff		
(<45.6)	<2.85		Off	Shuts off battery and itself; the system is bricked	Not powered

FIGURE 2.16
Operation of a telecom site
toward the end of charge.

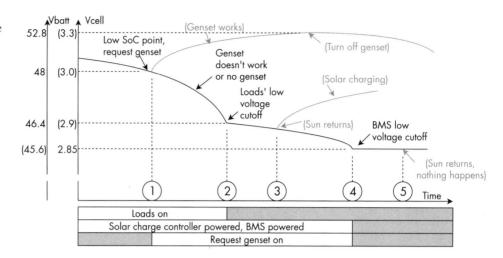

4. If the sun does not return, when the lowest cell drops down to 2.85V, the BMS shuts down the battery; this removes power to the BMS itself and the solar charge controller; the system is bricked.

5. Even if the sun returns, the site does not recover because there is no power to the solar charge controller.

At this point, the site is no longer able to recover automatically, and manual intervention is required to recharge the battery somehow to jump-start the system. This is conundrum is not easily solved (see Section 2.11.3).

2.4.2.6   Grounding

Typically, the positive terminal of low-voltage stationary batteries for telecom applications is grounded to earth to minimize electrolysis in buried cables. Even though this may no longer be an issue, this is still done for compatibility with existing equipment (see Section 2.12.2).

2.4.2.7   Arrays

Rather than being implemented as a single large battery, a telecom storage system may consist of multiple batteries connected to a shared parallel bus (see Volume 1, Chapter 6).

2.4.2.8   RF Exposure

In a telecom site, the tower may be shared with radio or TV broadcast stations. Their high-power radio frequencies (RF) penetrate the equipment hut. The telecom electronics, including the BMS inside the battery, must be immune to such EMI[32].

2.4.2.9   Lightning and Surge Protection

Telecom towers attract lightning. Most of the lightning's current flows straight to the earth because the tower includes lightning protection: air terminals, bonding, and direct earthing at the base of the tower. However, the coaxial cables from the antennas conduct a portion of the impulse into the equipment hut. Nearby lightning strikes

---

32.   Electromagnetic interference. See Volume 1, Sections 4.12.3, and 5.16.

generate surges that enter the hut through external AC power lines (such as the grid) and network lines (for phone and internet).

Measures are implemented to reduce the effects of a lightning strike:

- Lighting protection: on the tower, to conduct the current to earth ground;
- Surge protection: in the equipment hut, to bypass the surge to earth ground and attenuate any remaining energy before it reaches the electronic equipment.

Various surge protection measures should be implemented inside the hut:

- Metal hut: acts as a Faraday cage;
- Single point ground for everything in the hut onto a grounding point to the metal of the hut, which is then earthed just outside the hut: avoids ground loops;
- Lightning protectors on the RF lines: divert the surge directly to ground;
- Parallel mode surge suppressors[33] on AC power lines and network lines: divert the surge directly to ground;
- Series mode surge suppressors: convert part of the energy in the surge in an AC power line into heat; these are in effect large lowpass filters that let 50~60 Hz through and attenuate anything at a higher frequency, including surges.

Old-timers tell us that a simple precaution is to twist each coaxial cable into a complete loop immediately after it enters the hut. The resulting inductance, although small, is supposed to be sufficient to provide some series impedance, to allow the other lightning protection measures to be more effective. Others say that this is just a myth.

These measures help to protect the battery from surges:

- Float the battery itself: do not connect a terminal to the metal case (even if some other equipment grounds the B+).
- Run the battery output through a DC suppressor[34].
- If the communication bus in the BMS is not isolated, use a data isolator on this line.
- Add a loop on the battery cable and each communication line, just in case this trick does work.

2.4.3 Off-Grid Site, DC Powered

Let's now look at specific examples of telecom sites, starting from a basic off-grid site that uses only DC components (Figure 2.17). The site uses the DC topology (Figure 2.13(a)) and consists simply of

- DC sources to charge the battery, such as
  - DC genset with remote start and a current limited power supply;
  - DC-coupled solar: PV panels and a solar charge controller.
- The battery:
  - DC loads:
    - Communication equipment (BTS);

---

33. Surge protection device (SPD) or transient voltage suppressor (TVS).
34. Polyphaser IS-48VDC-30A-FG Impulse suppressor.

DC

FIGURE 2.17
Off-grid telecom solar
site, DC powered.

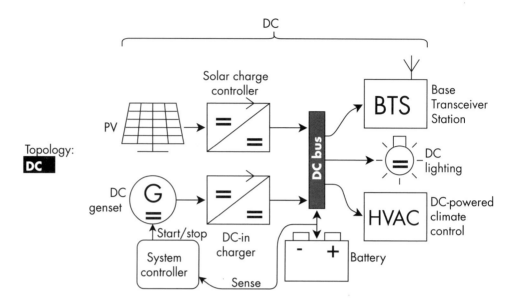

- Low-voltage lighting;
- DC powered climate control (HVAC).
- A system controller.

Connecting the DC genset directly to the DC bus would damage the battery because it is a constant voltage (CV) device and would fight the battery, which is also a CV device at a different voltage. Instead, a CCCV device, placed between the DC genset and the DC bus, limits the current and allows the battery to set the voltage of the DC bus.

For such a simple system, the only job of the system controller is to turn on the genset when the bus voltage drops below a given threshold and turn it off when the voltage goes above another threshold. This keeps the battery minimally charged until it is fully charged by the sun. If the system uses a Li-ion battery, the system controller may use the SoC value from the BMS instead. A BMS designed specifically for telecom applications may include a line to control the genset, in which case a system controller is not needed.

2.4.4   Off-Grid Site, DC Powered with Inverter

This site is like the previous one, except that it includes an inverter to power AC loads such as lighting, AC-powered climate control (which is more common than DC powered), and AC outlets (Figure 2.18).

This site uses the DC + inverter topology (Figure 2.13(A)).

2.4.5   AC Powered Site

This site (Figure 2.19) is powered only by AC sources, while the primary loads are DC-powered. Auxiliary AC loads turn off in the absence of any AC power source.

This site uses the AC topology (Figure 2.13(b)) and typically includes:

- AC power sources, such as
  - AC-coupled solar[35];

---

35. In this discussion, we consider AC-coupled solar to be an AC source, even though the PV panels are DC.

FIGURE 2.18
Off-grid telecom solar
site, DC powered,
with inverter.

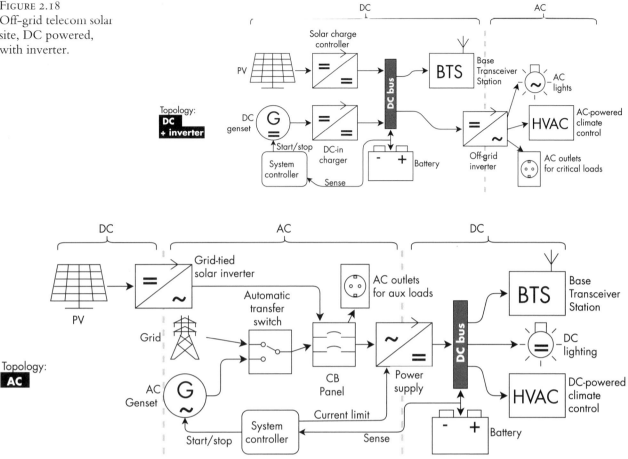

FIGURE 2.19   AC-powered telecom solar site.

- Grid;
- AC genset with remote start.
- Auxiliary AC outlets: powered by the selected AC source when present;
- An automatic transfer switch to select either the grid or the genset;
- A current limited power supply[36]: that charges the battery and powers the loads;
- The battery;
- DC loads:
  - Telecommunication equipment (BTS);
  - DC lighting;
  - DC powered climate control.
- A system controller.

As the power supply operates even if the BMS has disabled discharging of the battery[37], this system can recover on its own when AC power returns.

This system operates as follows:

---

36. In the United States, it would be called a "rectifier," while elsewhere it would be called a switch mode power supply (SMPS).
37. Unlike the solar charge controller.

- If the grid is present, the automatic transfer switch selects its voltage; otherwise, it selects the AC genset.

- The grid-tied solar inverter synchronizes with the selected source of AC power; if neither is present, then the solar inverter is disabled and cannot power the system even if there is sun.

- If the battery needs to be charged, the power supply operates in constant current (CC) mode; the battery sets the voltage of the DC bus.

- When the battery voltage reaches the CV setting, the power supply switches to the CV mode and generates a constant voltage. This voltage powers the loads, yet does not overcharge the battery:

  - With a lead-acid battery: the power supply is configured for the float voltage of the battery; the battery is charged fully.

  - With a Li-ion battery: the power supply is configured for somewhat lower voltage to reduce cell degradation; for example, for a battery using 16 LFP cells in series, the top voltage should be set at 3.4V × 16 cells = 54.4V[38].

- Charging continues at constant voltage:

  - If the Li-ion battery is balanced, the charging current naturally and gradually drops to zero; the BMS doesn't need to open the battery's power switch.

  - If the Li-ion battery is unbalanced, when any one cell is full, the BMS disables charging switch; the BMS continues to balance he battery. A sophisticated power supply may allow remote control of the output current; if so, the BMS may tell the power supply to reduce the output current down to the balance current instead of disabling charging (see Volume 1, Section 4.7.5.2).

A system controller is used to turn the genset on and off (unless the BMS can do so).

The type of solar inverter used affects the operation:

- *Grid-tied:* As it needs an AC voltage reference to operate, it would produce power only if the grid is present or the AC genset is running.

- *Off-grid:* A three-way transfer switch would be required, to select it, the grid, or the genset; at a given time, only one source can power the system.

- *Grid-interactive:* The solar inverter would operate in grid-tied mode when the grid is up, or the genset is running; it would operate in off-grid mode otherwise; this would be ideal but, unfortunately, the typical solar inverter is grid-tied[39].

2.4.6   AC Powered Site with Inverter

This site is like the previous one, except that there is an inverter to power the AC loads such as lighting and climate control (Figure 2.20).

This site uses the AC + inverter topology (Figure 2.13(B)): an inverter powers critical AC loads. When the battery is nearly empty, the system controller could turn off the inverter, or at least the climate control, to reduce the load and delay shut down as long as possible.

---

38. LFP cells would rapidly degrade if held at 3.6V. For 16 cells in series, that corresponds to a battery voltage of 57.6V.
39. Outback makes grid-interactive inverters.

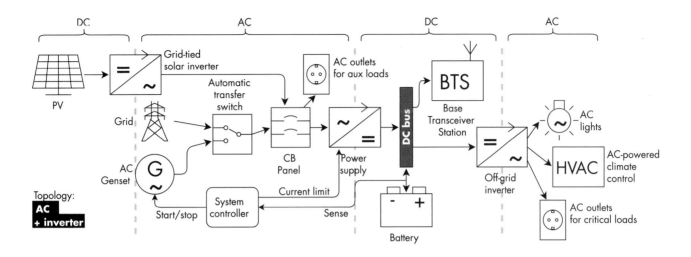

FIGURE 2.20 AC-powered telecom solar site, with inverter.

The selected AC power source powers the auxiliary outlets. They can power light and tools while repairing the site, and the inverter is off. Optionally, a transfer switch (not shown) could select what powers the primary AC loads: the input AC power or the inverter's output.

2.4.7 AC and DC-Powered Site

This site is powered by a variety of AC and DC power sources for maximum reliability and lowest operating cost (Figure 2.21).

This site uses the DC and AC topology (Figure 2.13(c)). Normally, it includes

- AC side:
  - AC power sources, such as the grid, an AC genset with remote start, AC-coupled solar;

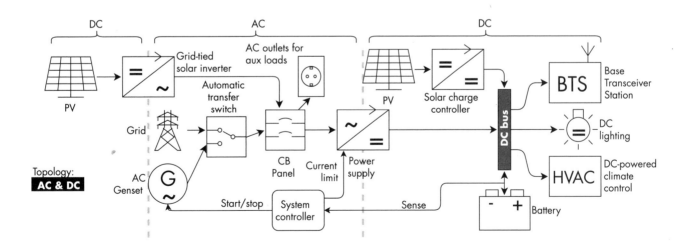

FIGURE 2.21 AC and DC-powered telecom solar site.

- An automatic transfer switch and a circuit breaker panel;
- Auxiliary AC outlets, powered by the selected AC source when present.
- A current limited power supply:
  - DC side:
    - The battery;
    - DC-coupled solar;
    - DC loads.
- A system controller.

Although the illustration shows both AC-coupled and DC-coupled solar, an installation would use only one. Each has advantages:

- DC-coupled solar:
  - More efficient charging;
  - The whole issue of grid-tied/off-grid goes away: solar charging is available regardless of the state of the AC power sources.
- AC-coupled solar:
  - With a grid-interactive inverter, the system can recover even if the battery is discharged (the solar charge controller requires the presence of a charged battery).

2.4.8 AC and DC-Powered Site with Inverter

This site is like the previous one, except that there is an inverter to power AC loads such as lighting and climate control (Figure 2.22). This site uses the DC and AC + inverter topology (Figure 2.13(C)).

2.5 RESIDENTIAL APPLICATIONS

Large, low-voltage batteries are included in distributed energy systems in residences in growing numbers. The three most common applications for residential distributed energy are:

- Batteryless;
- Grid-tie;

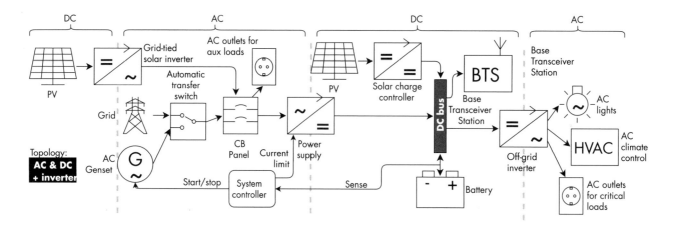

FIGURE 2.22   AC and DC-powered telecom solar site, with inverter.

- Off-grid.

This chapter[40] examines examples of those applications, particularly those that include a low-voltage battery.

2.5.1   Residential System Devices

Residential applications use the following devices, in addition to the ones described at the start of this chapter (see Section 2.2).

2.5.1.1   Inverger for Residential Use

An inverger for residential use (Figure 2.23) may have three AC ports:

- Grid: wired to the main breaker panel that feeds noncritical loads such as a hot tub;
- Genset: for an AC generator;
- Critical loads: wired to the critical loads panel that feeds appliances that must remain powered in case of grid failure, such as safety lighting; a solar inverter, if present, is also connected to this port.

The inverger switches very rapidly and automatically between charging and discharging, as needed at the moment.

2.5.2   Technical Considerations

Many of the same issues and proposed solutions for telecom applications also apply to residential applications. Differences between residential and telecom include

FIGURE 2.23
Application of
residential inverger.

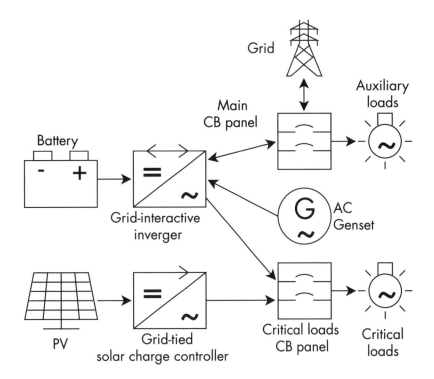

---

40. Many thanks to Timothy Schoechle of Colorado State University, Rhodes Carter of Schneider Electric.

- Keeping the loads powered continuously in a house is not as crucial as it is in a telecom site.
- Houses use invergers and may back-feed to the grid; telecom sites don't.
- Telecom sites use a system controller to manage power sources; houses don't need one: the residents can manage those resources.

2.5.2.1   Benefits of a Battery in an On-Grid System

The obvious benefit of having a battery in a grid-connected system is to provide back-up power if the grid goes down. Less understood is that having a battery brings even more benefits when the grid is up:

- Time-of-day pricing: Electrical bills are reduced by buying electricity at times when the price is low to charge the battery and using energy from the battery when prices are high (see Section 4.3.1.1).
- Demand charges: Limit the power drawn from the grid[41] to a maximum, by taking more energy from the battery when the house load is high; this is more an issue with businesses and factories that are charged based on peak demand (see Section 4.5.3) but can also be of use in residential applications.
- Peak shaving: Conversely, it is possible for the power company to request energy from the house's battery when needed because of peak energy demand from other users (see Section 3.9.1.8); this may be sufficient to avoid starting a peaker power plant; as this could be an older and dirtier power plant, not turning it on would have obvious environmental benefits.
- Reduced wear and tear on the battery:
  - Keep the battery charged to power the house in case of a blackout.
  - Use the grid as the "battery": sell extra power to the power company when available and buy it back when needed.
- Extra income: by selling extra power to the power company.

When employing one or more of these strategies, the battery SoC is kept above a certain level so that the house is not exposed to the risk of running out of battery charge should the grid go down. It is possible to use different SoC goals for the battery's SoC based on time of day and expected amount of solar power available[42].

2.5.2.2   Back-Feeding to the Grid and Islanding

Invergers in grid-tied residential applications power the very same AC circuits that are fed by the grid. This may result in

- *Back feeding:* Purposely sending extra power to the power company when the grid is up;
- *Islanding:* Inadvertently doing so when the grid is down. Islanding is categorically prohibited.

---

41. Either during times of peak demand on the grid or when the local load needs to draw peak current above a set threshold.
42. The BMS tells the inverger what the limits are, but the inverger takes or gives as much or as little current as it wants. Therefore, the inverger or a system controller manages this process, not the BMS.

Back-feeding may be allowed with a contract[43]; islanding is categorically prohibited. Islanding occurs when the grid goes down, and a grid-interactive inverger powers not just the house but also the neighborhood. This creates an "island of lights" in an otherwise dark region. Islanding is unacceptable, especially for the safety of workers repairing the local power lines who do not expect them to be powered. The inverger must include an anti-islanding function to detect when the grid goes down[44]; in this case, the inverger shuts down, or it continues to power the house after disconnecting it from the grid[45]. Specific laws regulate anti-islanding[46].

2.5.2.3 Low-Voltage DC Versus High-Voltage DC

Today, most residential systems use a low-voltage battery for historical reasons (lead acid is not ideal for high-voltage batteries) and due to the simplicity, safety, and lower cost of low voltage batteries. As a result, many power products that use a low voltage battery have been developed.

Yet high-voltage batteries for residential use[47] and power products compatible with them are being introduced (see Section 4.6.2). Using a high-voltage DC bus is more efficient[48], and power products compatible with it are simpler and, therefore, less expensive. However, a high-voltage battery is more expensive because its BMS requires more voltage taps. There isn't a clear answer on whether low-voltage or high-voltage battery is better for residential use. A careful analysis of the advantages and disadvantages of a system for a given application is required.

2.5.2.4 Invergers and Solar Charge Controllers with Li-Ion Batteries

As mentioned, today's invergers and solar charge controllers expect a lead-acid battery. When used with a Li-ion battery (which may be turned off to protect its cells), these products may unexpectedly shut down the house power, even if the grid or the sun is up. This problem will remain unsolved until products that are designed to work with Li-ion batteries become widely available. Until then, the system designer must take some measures that are only a partial solution (see Section 2.11).

2.5.2.5 Invergers That Disobey

Regretfully, some invergers do not obey messages from the BMS. In particular, under certain conditions, they may charge the battery when the BMS instructs them not to do so[49]. Therefore, the battery must prevent charging on its own with a protector switch in case the inverger disobeys. The single-port/dual-switch topology is best to achieve this (see Volume 1, Section 5.12.2.2).

---

43. Contractually, the utility accepts back-fed energy under feed-in tariff, net metering, or power purchase agreement.
44. Detecting that the grid is down is not easy. Induction motors keep on spinning, powered by the inverger. In so doing, they generate an AC voltage, which, in turn, the inverger uses as a reference, in a vicious cycle. The trick is to detect the slight reduction in frequency when the grid goes down by noting that the induction motors slow down (they go from a negative slip to a positive one). See Section B.3.1.3.
45. The better invergers can do so in 10 ms or so, which is unnoticed by most residential loads.
46. Regulated in the United States by IEEE-1547 and UL-1741.
47. Most notably, Tesla's Powerwall.
48. The power efficiency of a DC-DC converter is best when the input and output voltage levels are close; additionally, connections in the battery of a given size do not waste as much power in heat at the lower currents typical of a high-voltage system.
49. See Section B.6.2 for details and a rebuttal from SMA.

2.5.2.6 *System Controller*

While not required, a system controller may provide some benefits, such as load management (shutting down particular AC loads when the battery is close to empty) or reporting the house's status over the Internet.

2.5.3 Batteryless Generation

Batteryless systems convert a local resource (typically solar, possibly wind) to generate electricity for local use or to sell to the power utility. This is by far the most common application of solar power[50].

Batteryless generation could be

- *Off-grid* (Figure 2.24(a)): When the resource is not available (e.g., at night), there is no power.

- *Grid-tied distributed generation*: The vast majority of solar systems in developed countries are of this type; if the grid is down, and the resource is not available, there is no power:

  - *Without back-feed* (Figure 2.24(b)): Does not sell back to the utility; it simply offsets the electrical bill by reducing the demand from the grid.

  - *With back-feed* (Figure 2.24(c)): Sells excess energy to the utility, in effect using the grid as a battery: it provides power to the house when needed, and takes excess power from the house when available.

FIGURE 2.24
Typical batteryless systems: (a) off-grid, (b) grid-tied, and (c) back-feed.

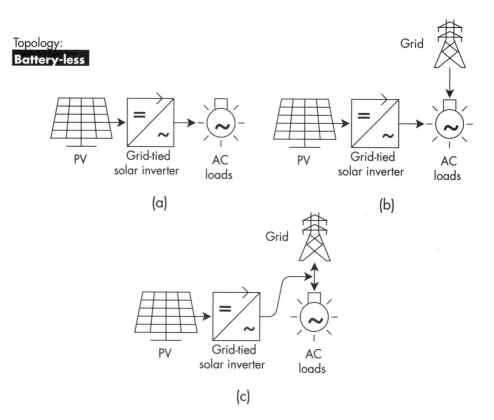

---

50. As there is no battery, this application is off-topic for this book; I only mention it for the sake of completeness.

2.5.4 Off-Grid Tiny House

Recently, people in the developed world have started building tiny houses[51] for practical, social, and environmental reasons. Some of these houses are off-grid, using DC sources and batteries. These houses use standard, AC-powered appliances.

This house (Figure 2.25) uses the DC + inverter topology (Figure 2.13(A)) and includes

- DC power sources: DC-coupled solar, DC genset with a current limited DC-DC converter;
- A battery;
- DC loads, such as low-voltage lighting;
- An inverter to power AC loads: outlets and small appliances.

If the house is pointed towards the noon sun and is covered in properly angled solar panels, solar power produces on the order of 10 kWh a day. By comparison, the average U.S. house uses 30 kWh a day. While you may be prepared to live in a tiny space, you must also be prepared to live with a small energy footprint. It's physically impossible for rooftop PV panels to gather enough energy to power, by themselves, all the energy-hungry products that let Americans live the lifestyle to which they have grown accustomed[52]. Instead, you must live like a German (10 kWh/day) or a Mexican (5 kWh/day). A 20-kWh battery (400-Ah Li-ion cells for a 48V system) is sufficient to power the tiny house over a few overcast days.

2.5.5 Off-Grid House

The classic off-grid house uses low-voltage batteries and an inverter to power all the loads. Various topologies are possible. Here I suggest three which differ on how solar power is applied:

- AC-coupled, before the charger (Figure 2.26(a)): Uses an off-grid solar inverter; the disadvantage is that it requires an automatic transfer switch to select the AC power source; uses the AC + inverter topology (Figure 2.13(B)).

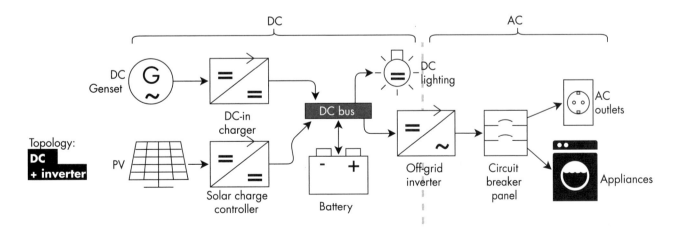

FIGURE 2.25 Off-grid tiny house.

---

51. Less than 100 m² of floor space.
52. Adding a generator would allow higher consumption but would be against the spirit of a tiny home.

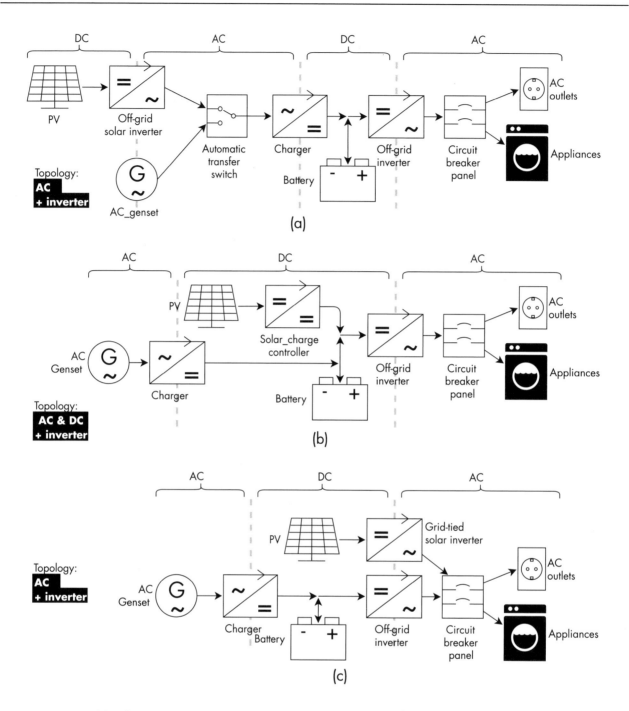

FIGURE 2.26   Possible off-grid house topologies: (a) AC-coupled solar before charger, (b) DC-coupled solar, and (c) AC-coupled solar after the inverter.

- DC-coupled (Figure 2.26(b)): Has the disadvantage that if a Li-ion battery shuts down, the solar charge controller won't be able to restart and recharge the battery (the generator must be turned on, at least initially); uses the AC and DC + inverter topology (Figure 2.13(C)).

- AC-coupled, after the inverter (Figure 2.26(c)): Uses a grid-tied solar inverter because it must not compete with the off-grid inverger; the disadvantage is that the generator must be used to charge the battery and, if the battery is empty, the solar power shuts down as well; uses the AC + inverter topology (Figure 2.13(B)).

In all these cases, the off-grid inverter can run autonomously.

A 60-kWh battery[53] is sufficient to power the average off-grid house over a few overcast days.

2.5.6 On-Grid House with Backup Power

In a house connected to the grid, batteries may be used simply for back-up in case the grid goes down.

This on-grid house (Figure 2.27) is similar to an off-grid one, except that there are two AC buses:

- Auxiliary loads: connected to the grid; loads go off when the grid goes down;
- Critical loads: loads remain powered when the grid goes down.

This house uses the bidirectional AC topology (Figure 2.13(d)) and includes

- Critical panel and loads (e.g., grandpa's medical ventilator);
- Auxiliary panel and loads (e.g., grow lamps in the basement);
- An off-grid inverger:
  - When the grid is OK, it charges the battery until full.
  - When the grid goes down, it disconnects from the grid, and powers the critical loads.

2.5.7 On-Grid, AC-Coupled Solar

The following two residential systems provide not only backup but also include solar power and may back-feed to the grid. This first house (Figure 2.28) is like the previous

FIGURE 2.27
On-grid house with
backup power.

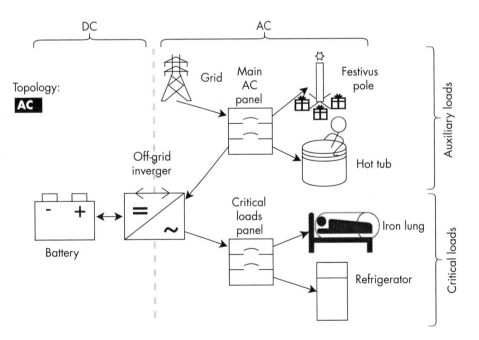

---

53. For a 48V system, this could be three 400-Ah LFP cells in parallel and then 16 in series.

FIGURE 2.28
On-grid house with
AC-coupled solar.

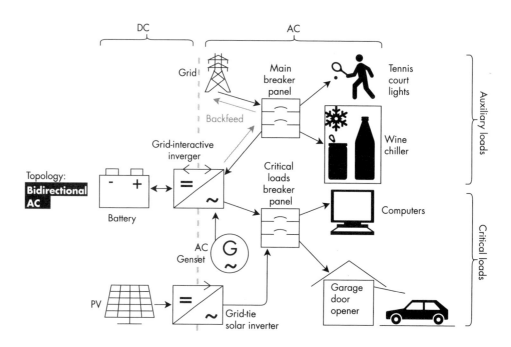

one, but it adds AC-coupled solar and an AC genset. It uses the bidirectional AC topology (Figure 2.13(d)).

The system uses a grid-interactive inverger for home use, with three AC ports:

- Grid (and main panel for auxiliary loads);
- Critical loads (critical loads panel, which is also connected to the solar inverter);
- AC genset.

If the inverger does not have an input for a genset, then an automatic transfer switch is used to select either the grid or the genset.

When the grid goes down, the inverger goes into the off-grid mode, powering just the critical loads. The solar inverter has no idea that the grid is down because it sees the AC reference from the inverger, so it continues to operate normally.

If the battery is full, the inverger stops drawing current from the grid or the solar inverter. The inverger may signal the solar inverter to shut down in one of two ways:

1. Through a communication bus;
2. By slightly lowering the AC frequency of its output; the solar inverter interprets this as a loss of grid power; its anti–islanding function kicks in, shutting down the output.

2.5.8   On-Grid, DC-Coupled Solar

This second house (Figure 2.29) uses DC-coupled solar instead: a solar charge controller is connected to the battery. It uses the bidirectional DC and AC topology (Figure 2.13(e)). If the battery is full, both the solar charge controller and the inverger notice this and stop charging the battery. If the battery is empty, charging is disabled, and the system cannot recover. Even if there is sunshine, the solar charge controller cannot start because it needs power from the battery, which is off. Even if the grid returns, the inverter may not start, because it too may be powered by the battery. Some measures may offer a partial solution (see Section 2.11).

FIGURE 2.29
On-grid house with
DC-coupled solar.

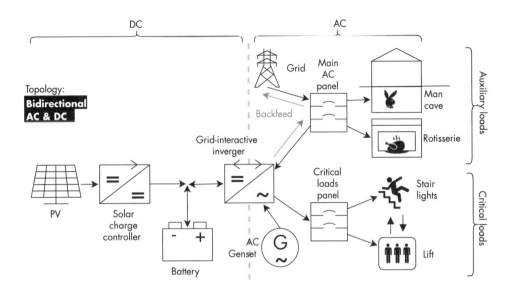

2.6 MARINE HOUSE POWER APPLICATIONS

Vessels may use Li-ion batteries for house power or traction (see Section 3.13.1). This section[54] discusses the former.

In sailboats, yachts, and other small to medium-sized vessels, house power is used for lighting, navigation and communication equipment, and other relatively low power loads[55]. It is not used for traction.

House power uses a battery that is charged by

- Docked: shore power, a cable plugged into an outlet in the dock;
- At sea: an alternator on the engine, a dedicated generator, solar panels, a wind generator, or a hydro generator powered by the motion of sailboat through the water.

Having a battery allows a generator to run only part-time, which increases its efficiency and reduces noise.

This chapter explores a variety of issues related to the design of a vessel's house power. The ABYC[56] is the de facto regulator of these systems, even though it is a private organization.

Li-ion has a bad reputation in the yachting world, due to some nasty fires (no one has been hurt to date)[57]. Due to the dire consequences of an on-board fire, sailors are unwilling to face such risk within the confined space of a sailboat; they may refuse to allow picturesque oil-burning lamps on board, or even to install an engine or an outboard motor in their sailboat. Those who are keenly aware of the dangers of a Li-ion in thermal runaway may refuse to have any Li-ion battery onboard other than laptop computer batteries.

If a Li-ion battery is installed in a vessel, it must be done with a high level of respect for the destructive power of the Li-ion technology. It must be implemented by experts in the Li-ion and marine fields and be subjected to peer evaluation. It

---

54. Many thanks to Bruce Schwab of OceanPlanet Energy and Rod Collins of Compass Marine for help with this chapter.
55. Low power is relative: house power can be used for high power loads such as 240-Vac compressor pump motors to fill diving tanks.
56. American Boat and Yacht Council; abycinc.org/.
57. See Section 5.2.2.2 for cautionary tales (beware of jack-of-all-trades captains who treat Li-ion as if it were lead acid).

requires rigorous testing, validation, and quality control at every stage of development, production, installation, and maintenance.

To be sure, batteries other than Li-ion may also cause a fire, as happened to Steve Fosset's Playstation catamaran in 1990, when an overcharged NiMH battery ignited [1].

2.6.1 Marine Devices

Marine applications use the following devices[58], in addition to the ones described at the start of this chapter (see Section 2.2). Do not use a battery combiner switch or a voltage sensing relay with Li-ion!

2.6.1.1 Terminology

In the marine industry, some terms have a different meaning from what we may be used to, such as in Table 2.5.

2.6.1.2 Shore Power

Docks may provide an AC power outlet to power any AC loads in the vessel, such as a battery charger (Figure 2.30). A variety of voltages, currents, and therefore outlet shapes are used (see Section 2.6.3.2).

TABLE 2.5
Marine Terminology

Term	Boat People	The Rest of Us
Isolator	Switch	A material that provides electrical isolation between two conductors
Solenoid	Contactor	Electromagnet
Galvanic isolation	Electrical continuity between two circuits, but with a small voltage drop	Electrical isolation between two circuits
ESD	Electric shock drowning	Electrostatic discharge

FIGURE 2.30
Marine shore power.

---

58. Major manufacturers of marine-grade batteries and electronic products include Balmar, Blue Sea, Eaton, Kisae Technology, Lithionics, Magnum Energy, Marinco, Mastervolt, Midnight Solar, NorthStar, ProMariner, Smartplug, Victron, Wagan, West Marine, and Xantrex.

2.6.1.3 Shore Power Corrosion Detector

A shore power corrosion detector[59] detects current in the ground wire, indicating that there is a ground fault in this vessel or nearby vessels. It also detects a miswired shore power outlet. It looks like a short AC extension cord that is inserted in line with shore power. Promptly correcting the issue prevents the rapid depletion of a sacrificial anode[60].

2.6.1.4 Isolation Transformer

Routing shore power through an isolation transformer solves two potential problems:

1. Ground fault in the vessel's AC power (Figure 2.31(a–c)): Without a transformer, this would result in a short circuit. The short circuit current requires a complete path. The ground fault forms one link in this path; since the output of the transformer is floating, it doesn't complete this path, so no current can flow as a consequence of a ground fault.

2. Wiring problems in the shore power outlet (Figure 2.31(d–f)): The vessel is powered by the floating AC output of the transformer, not by the incorrectly polarized voltage of the shore power outlet.

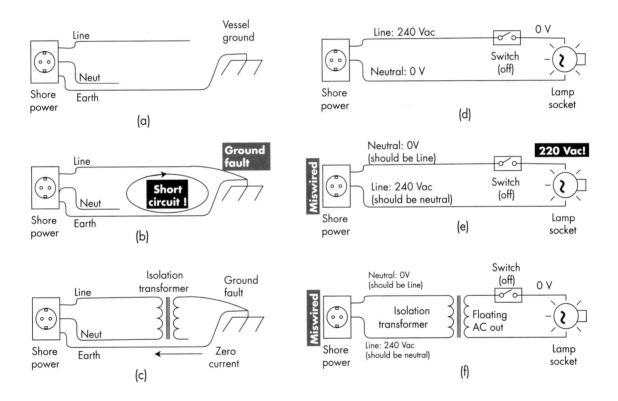

FIGURE 2.31 Advantages of an isolation transformer: (a) no ground fault, (b) ground fault without a transformer, (c) ground fault with a transformer, (d) correctly wired shore power; (e) miswired shore power without a transformer, and (f) miswired shore power with a transformer.

---

59. The GalvanAlert from Marinco.
60. A zinc slug mounted to a metal hull below the waterline. Any current into the water consumes the sacrificial anode, rather than corrode the hull or the propeller.

When using an isolation transformer, there are two ways to deal with the shore power ground (green wire) [1]:

1. *Polarization wiring* (Figure 2.32(a)): The ground wire is connected to the vessel's frame.

2. *Isolation wiring* (Figure 2.32(b)): The ground wire is left floating.

The advantage of the isolation wiring is that it completely disconnects the vessel from the dock and other vessels. This isolation prevents galvanic corrosion of the propeller and a metal hull. The disadvantages of the isolation wiring are that

- It's illegal in some jurisdictions; some regulations require that all AC loads be grounded to the ground of the source of AC power (shore power in this case); however, regulations are changing[61] to recognize the reality that an isolation transformer is safer than grounding to the AC source.

- There is a slight chance that a fault develops inside the transformer, energizing the metal case of the transformer; the solution would be to encase the transformer in insulating material, rather than metal, yet it appears that all the marine transformers use a metal case.

Isolation transformers are large, heavy, and expensive; yet the protection they offer is often considered to be worthwhile.

2.6.1.5   Galvanic Isolator

Without an isolation transformer, the earth ground from the shore power connector must be connected to the frame of the vessel. This connection may enable corrosion of the propeller and of a metal hull. A galvanic isolator (Figure 2.32(c)) reduces this effect.

A galvanic isolator is placed in series with the AC ground line (see Section 2.6.3.2) between the vessel and shore power.

FIGURE 2.32
Isolation transformer
grounding:
(a) polarization wiring,
(b) isolation wiring, and
(c) galvanic isolator.

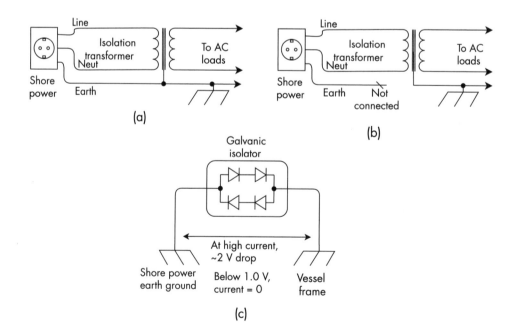

61.   The ABYC considers the isolation transformer to be a new source of power, allowing for isolation wiring.

The galvanic isolator conducts no current up to $+/-1.2V$ across it[62]. As long as the voltage between the shore power earth ground and the vessel's hull is less than 1.2V, the galvanic isolator does not conduct any current. This blocks the small current that would cause corrosion.

In case of a ground fault in the AC circuits inside the vessel, current flows to the vessel's ground, through the galvanic isolator, through the power cable, and to the earth ground in the shore power pedestal[63]. At high current, the voltage drop across the galvanic isolator is about 2V, which is not dangerous. What's important is that the fault current flows back to the shore power ground and not through the water. The galvanic isolator must be able to carry the maximum current that the shore power outlet can provide (typically ranging from 15 to 50 A). For high-power outlets (480V, 600 A), a standard galvanic isolator is impractical as it can't dissipate this much power (as much as 1 kW).

A galvanic isolator is not as effective as an isolation transformer because it only protects against the low currents created by dissimilar metals in an electrolyte. It does not block any high leakage current that flows through electric devices in this vessel or nearby vessels.

2.6.1.6 Solar, Hydro, and Wind Generators

A vessel may include generators that charge the battery from natural resources. A hydro generator is ideal when sailing. It can produce up to 600W at 10 knots. It does add a bit of drag. A wind generator is ideal when at anchor. Its power is rather low: even in gale force winds, it produces only 100W or so. A wind generator requires a dummy load into which to dump its energy if the electrical load is too small. Without it, it spins out of control. The size of the vessel limits the number of solar panels that can be installed. In practice, a small sailboat can generate about 500W from the sun.

2.6.1.7 Genset

A vessel may use either an AC or DC genset because

- An AC genset can power DC loads through a charger.
- A DC genset can power AC loads through an inverter.

An AC genset may be better because AC loads tend to require more power, especially at start-up (e.g., air conditioning). In any case, AC gensets are more commonly available. With a DC genset, the inverter must be able to provide the peak power of any AC load. However, with an AC genset, the charger must only provide the average power of any DC load because the battery provides the peak current. Therefore, AC gensets are preferable.

2.6.1.8 Battery Isolator

Some vessels use two batteries and two DC buses for redundancy. If the vessel has only a single charging source, when it is connected to both batteries, the two batteries are connected in parallel, defeating the purpose of having two separate DC buses.

---

62. The galvanic isolator places two diodes in series with the AC ground wire to shore power. Diodes have relatively constant voltage drop (of about 0.6V at low current, about 1V at high current).
63. With a voltage drop clamped to about 2V.

A battery isolator (Figure 2.33(a))[64] solves this problem by allowing a single charging source to charge two or more batteries, yet preventing current from flowing between batteries. Initially, the most discharged battery (the one with the lowest voltage) is charged. When this battery's voltage approaches the voltage of the other battery, the charging current flows into both batteries. A battery isolator won't let you select which battery gets charged first.

The typical battery isolator uses diodes that drop the voltage by about 1V and generate heat, making it somewhat inefficient. Battery isolators with MOSFETs do not have this problem. It is safe to use a battery isolator with Li-ion batteries.

2.6.1.9   Voltage Sensing Relay (VSR)

Although the voltage sensing relay[65] (Figure 2.33(b)) is marketed as being a "better battery isolator," it is quite a different device. The advantage of a VSR is that it doesn't use rectifier diodes, so it doesn't waste power in heat (unlike a battery isolator).

Normally, the VSR connects the charging source to just one battery. The VSR detects when this battery is being charged (it notices that its voltage is increasing). If so, it connects the two batteries in parallel, so both may charge. When it senses that the voltage is steady (the batteries are no longer being charged), it removes the parallel connection between the two batteries.

A VFR is very dangerous for Li-ion; the moment that it connects two Li-ion batteries in parallel at different voltages, a large current flows between them (see Volume 1, Section 3.3.6). A digital voltage sensing relay is just a more accurate version of a VSR, but is still dangerous with Li-ion.

2.6.1.10   Current Limiting Relay

A current limiting relay is an ugly contraption left over from Henry Ford's days. If the current is too high, it opens a relay. As soon as it opens, the current drops to zero. After a bit, the relay closes again. Some people use it as a sort of a resettable fuse; others use it to charge a battery. Just don't.

2.6.2   Battery Switches

Battery switches come in many forms.

FIGURE 2.33
Two DC buses with
battery combiner switch:
(a) battery isolator, and
(b) voltage sensing relay.

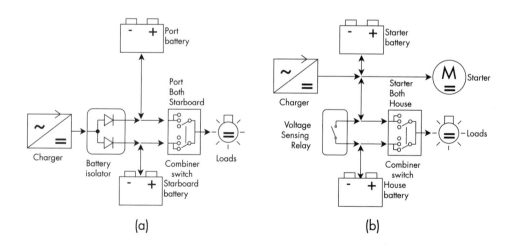

(a)   (b)

---

64.   Also known as a split diode.
65.   Also known as automatic charging relay or voltage sensitive relay.

2.6.2.1 *Battery Disconnect Switch*

This switch disconnects the battery in case of emergency (Figure 2.34(a)). If, at the time, the alternator was charging the battery at high current, it produces a high voltage the moment that the battery is suddenly disconnected, especially if there is no other load (Figure 2.34(b)). This voltage will damage the rectifiers in the alternator and electronic products still connected to the line. Therefore, using a battery switch is ill-advised.

If you must use a battery switch, and if the field winding of the alternator is externally accessible, use a battery switch with an alternator field disconnect[66] and wire it between the regulator and the alternator (Figure 2.34(c)). This auxiliary switch opens before the main switch opens, giving the magnetic field in the alternator time

FIGURE 2.34
Battery switches:
(a) battery disconnect,
(b) high-voltage spike
when first disconnected,
(c) battery disconnect
with alternator field
disconnect, (d) sequence;
(e) battery selector, and
(f) battery combiner.

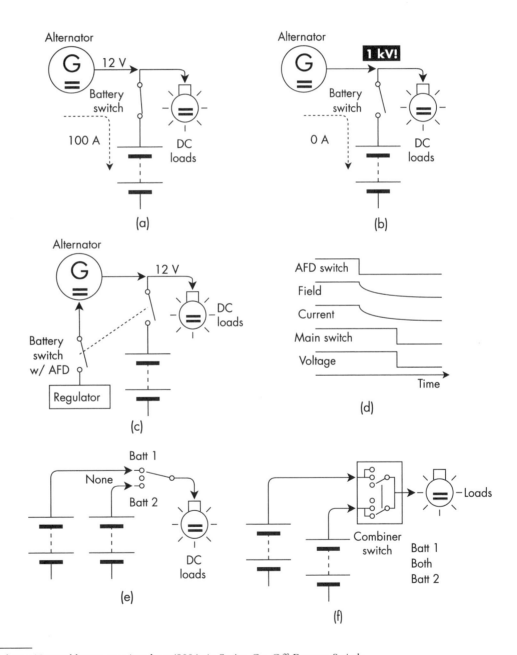

---

66. Blue Sea e-series 9004e, https://www.bluesea.com/products/9004e/e-Series_On-Off_Battery_Switch.

to decay, so that there is no current out of the alternator by the time the main switch opens; therefore, there is no high-voltage spike (Figure 2.34(d)).

2.6.2.2   Battery Selector Switch

A battery selector switch allows the selection of one and only one battery, or none (Figure 2.34(e)). Again, to avoid frying the alternator and electronics, use a selector switch with an alternator field disconnect[67]. It is safe to use a selector switch that selects one Li-ion battery, or the other, but not both.

2.6.2.3   Battery Combiner Switch

A battery combiner switch (Figure 2.34(f)) allows the captain to select one battery, the other battery, or both in parallel. Connecting to both batteries may be acceptable for lead-acid batteries because they are more forgiving and have a relatively high resistance to limit the current. It is not allowed with Li-ion batteries because closing a switch between two batteries at different voltages results in a large inrush of current.

2.6.2.4   Battery Equalizer

Some vessels have both a 12V bus and a 24V bus. Two 12V batteries are connected in series to power the 24V bus, while a tap between them powers the 12V bus (Figure 2.35(a)). Of course, this results in a significant unbalance between the two batteries because both batteries are charged equally from the 24V bus, while the lower battery also gets discharged by the loads on the 12V bus yet is never recharged by it.

A solution is to add a DC-DC converter (step-down or buck converter) to power the 12V bus from the 24V bus (Figure 2.35(b)). A better solution is to use a battery equalizer[68] (Figure 2.35(c)), a DC-DC converter powered by the two batteries that charges the lower battery from the 24V bus to maintain the same voltage on both batteries. A battery equalizer is more effective than a DC-DC converter because the equalizer only needs to provide the average load current of the 12V bus. This is because the lower battery functions as a buffer. The DC-DC converter must provide the peak load current and may not be able to handle peaks in load current.

2.6.3   Technical Considerations

Grounding in a vessel is tricky because there are different grounding requirements [2] for AC shore power, onboard AC circuits, DC, RF[69], lightning, and corrosion.

2.6.3.1   DC Grounding

For various electrical and regulatory reasons, the negative terminal of a house power battery is grounded to the vessel's metal frame. This has safety consequences:

- Any accidental connection between the positive DC bus and the vessel's grounded metal frame results in a short circuit; a fuse may prevent damage.
- Although the battery is considered to be low voltage, it can still cause electrical shock when touching the B+ bus.

67. Blue Sea e-series 11001, https://www.bluesea.com/products/11001/e-Series_Selector_3_Position_Battery_Switch_with_AFD.
68. Victron Energy Battery Balancer, Eaton Series 21000.
69. Radio frequency, for communication equipment.

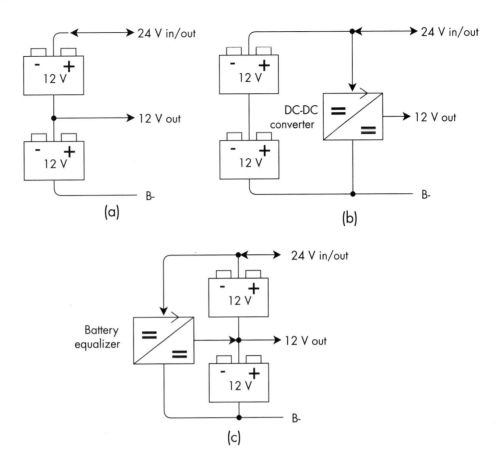

FIGURE 2.35
Generating 12V from 24V:
(a) center tap,
(b) DC-DC converter,
and (c) battery equalizer.

- For safety, the BMS should be isolated between the battery side and the communication side; grounding the battery defeats this isolation because the communication links are also grounded at the far end, away from the BMS; yet it is
  still advantageous if the BMS is isolated, so that the communication links are
  grounded only at one point, at the far end from the BMS; this prevents a ground
  loop and helps with noise immunity.

2.6.3.2   AC Grounding

The earth ground wire of shore power should be connected to the metal frame of
the vessel[70]. If the vessel has a metal hull, it is grounded through the engine block
(Figure 2.36(a)). A ground fault in the AC electrical circuits in the vessel results in high
current from the AC line wire, through the metal frame, through the ground wire,
back to shore (Figure 2.36(b)). Without this ground wire, a ground fault results in a
high current flowing through the metal frame, through the hull, through the water,
to earth (Figure 2.36(c)). This current through the water has killed many swimmers
in marinas. It's called electric shock drowning [3] (ESD)[71]. This is why, in the absence
of an isolation transformer, grounding the metal frame to the shore power ground is
essential for the safety of nearby swimmers.

---

70.   This may be optional when an isolation transformer is used.
71.   Not to be confused with electrostatic discharge.

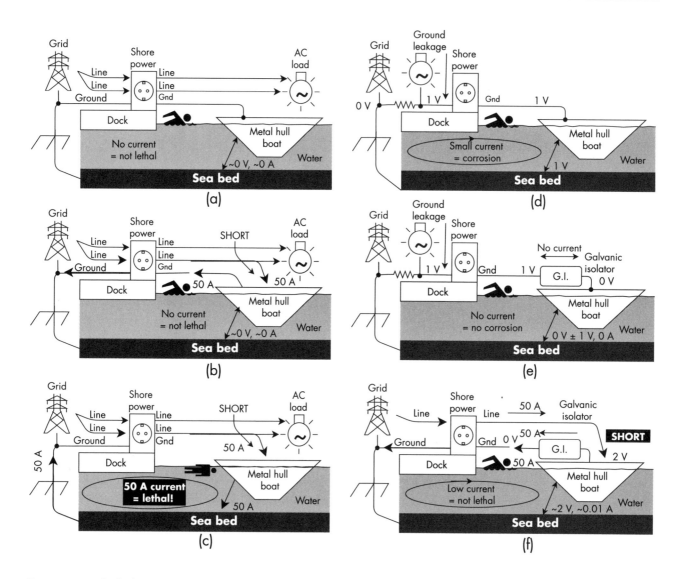

FIGURE 2.36   Left: shore power grounding: (a) grounded, (b) ground fault, grounded, and (c) ground fault, ungrounded. Right: galvanic isolator: (d) without isolator, (e) with isolator, no ground fault, and (f) with isolator and fault.

2.6.3.3   Corrosion of Propeller or Metal Hull

A disadvantage of grounding a metal hull to shore power ground is that it completes a path in which current may flow through the water. The voltage of the earth ground wire in the shore power outlet is not exactly 0V, due to leakage current from the vessels and other loads on this same circuit. This small voltage is too low to be a safety concern, but high enough to generate a small current between the water and a grounded propeller or a metal hull (Figure 2.36(d)). This corrodes the hull, especially in seawater.

Placing a galvanic isolator (see Section 2.6.1.5) in series with the ground wire to shore power (Figure 2.36(e)) solves this problem. If there is a ground fault in the vessel (Figure 2.36(f)), the fault current flows through the galvanic isolator.

2.6.3.4 Corrosion of Electrical Components

The marine environment is notoriously challenging to electronic products as salty water permeates everything, corrodes connections, and leaves conductive deposits on the surface of PCBs. Sooner or later, a standard BMS will fail in this environment, sometimes spectacularly. Electrical components must either be enclosed in sealed compartments or be otherwise protected to keep corrosion under control. Fuse-holders, switches, connectors, contactors must be rated for use in marine environments.

Ideally, the BMS should be completely sealed and use sealed connectors[72]. Today, no off-the-shelf BMS is explicitly promoted for marine use[73]. Instead, a standard BMS for traction batteries for passenger or industrial vehicles must be adapted for use in a marine environment:

- Place the BMS in a sealed enclosure.
- Install sealed glands in the enclosure, and run cables to the batteries through them.
- Place in a dry equipment area.
- Make gas-tight connections: properly crimped terminals are best, snug ring terminals are good.
- Coat all connections with anti-corrosion spray or paste.

Electronic assemblies must be directly sealed with potting compound or a thick silicone coating (standard conformal coating is insufficient). If that is not possible, they must be enclosed in a hermetically sealed enclosure; cables must pass through sealed glands; a protective vent mounted on the enclosure allows pressure equalization without breaking the seal[74].

2.6.3.5 Fuses

Marine Rated Battery Fuses (MRBF) are IP-rated[75]. They have an unusual, concentric shape. ANL fuses are not sealed and are light-duty; yet they are recognized by the ABYC. For high current, use Class T fuses because they have a high current interruption rating. For safety and reliability, the fuse must be mounted on the appropriate fuse-holder, not hung from wires (see Volume 1, Section 5.12.6).

2.6.3.6 Ignition Protection

Vessels may be exposed to fumes from fuel tanks or lead-acid batteries, at times in contained spaces; measures must be taken to avoid igniting any such fumes. Electrical components must be rated for ignition protection. In particular, any component that opens an electrical circuit (fuses [4], switches, contactors, relays, brushed motors, generators, appliances, and outlets) must be contained so that any arcing won't ignite fumes.

---

72. The most common automotive-grade connectors are not sealed.
73. Genasun makes one, but no longer sells it by itself.
74. Gore Tex Protective Vents: Lighting Enclosures, https://www.gore.com/products/gore-protective-vents-for-lighting-enclosures.
75. Ingress Protection, sealed.

2.6.3.7 Current-Limited Charging Sources

Charging current must be limited to avoid degrading a battery by charging at too high a current. Not all power sources on a vessel are current-limited; solar charge controllers and chargers are, generators and alternators usually aren't.

Attempting to charge a battery with a source that is not current-limited may degrade the battery. In practice, however, given that marine batteries tend to be large, the charging current is often no higher than 0.3 C. Therefore, the problem is not charging the battery too fast; rather the problem is that the source will overheat. In particular, the windings of an alternator that is designed for lead-acid batteries may be "cooked" by charging a Li-ion battery for a long time. I strongly recommend a voltage regulator that monitors the temperature of the alternator (see Section 2.2.1.1).

2.6.3.8 Voltage Limitation in Charging Sources

Charging sources on a vessel are designed for lead acid, not Li-ion. They may have a profile for lead acid (i.e., bulk, absorption, and float) or no voltage regulation at all. This is not ideal for Li-ion batteries. The life of Li-ion cells is reduced if they are kept at the rated maximum voltage (see Volume 1, Section 2.4.1.1). Yes, the BMS disables charging if the cells are full, but it's better to reduce the bus voltage and let the battery remain turned on.

The CV setting for charging should be set so that the cells in a balanced battery are never fully charged (e.g., 3.4V per LFP cell). The disadvantages are slower charging and reduced battery effective capacity. If fast charging is required, a charger can be configured for a higher voltage during the absorption stage, as long as it is also configured to limit the duration of this stage.

2.6.3.9 Reserve Capacity

The BMS may be configured for two different thresholds for low voltage cutoff:

- Reserve voltage cutoff: when the battery still has about 10% to 15% charge left in it;
- Low-voltage cutoff: when the battery is completely empty.

Using the reserve voltage cutoff improves the battery's cycle life. If the battery shuts down because it reached the reserve voltage, the captain may access the reserve capacity by forcing the BMS to reenable discharging and use the low-voltage cutoff instead.

2.6.3.10 Communications

The de facto communication standard for control and power products in vessels is NMEA 2000, based on the CAN bus. It is a closed standard that uses multiframe messages. An NMEA 2000 compatible BMS is integrated seamlessly into the vessel's control and power systems, allowing the captain to monitor the battery.

2.6.3.11 Multiple House Batteries

For weight distribution, a vessel may have two large house batteries, one on the port side and one on the starboard side. Two separate batteries should provide redundancy. However, if the two batteries are connected in parallel, when one battery is empty, both are empty. Therefore, the vessel has two separate DC buses; each battery is connected to just one bus. This way, even if one battery is depleted, the other one is still available to start the engine. If multiple Li-ion batteries are used, it is dangerous to connect one battery to a bus that is already connected to another battery (see Volume 1, Section 3.3.6). There are a few approaches to use two Li-ion batteries together in a safe manner using one or two DC buses.

2.6.3.12 Two Batteries, Single-Bus

There are a few ways to use two batteries on a single DC bus:

- The batteries are connected permanently in parallel (Figure 2.37(a)); electrically, they are really a single battery and, by definition, are not subject to the danger of inrush current.
- An array-compatible BMS manages each battery to keep it from connecting to a shared bus when not safe (Figure 2.37(b)) (see Volume 1, Section 6.3.1).
- An array master manages both batteries to keep them from connecting when not safe (Figure 2.37(c)) (see Volume 1, Section 6.3.2).
- The batteries are ganged and operate simultaneously (Figure 2.37(d)) (see Volume 1, Section 6.4).
- A bidirectional DC-DC converter between each battery and the bus limits the current and allows the battery to equalize (see Volume 1, Section 6.3.3).

2.6.3.13 Two Batteries, Dual Bus

Two separate batteries are more likely to be used with two separate DC buses. The following topologies are possible.

Two separate but equal DC buses (Figure 2.38(a)):

- Each battery is connected to its own DC bus.
- A battery isolator (with rectifier diodes) allows a single charging source to feed both buses and charge both batteries; a selector switch to let the captain select which battery is charged at a given time may be used instead of a battery isolator.
- Another set of rectifier diodes feeds a load from whichever battery has the highest voltage; the diodes prevent flow between batteries; the voltage drop in the diodes wastes heat, making this solution appropriate only for 48V systems and small loads.
- Alternatively, a selector switch (without a "both" position!) lets the captain select which DC bus feeds each load.

A main bus and a critical bus (Figure 2.38(b)):

- Each battery is connected to its own DC bus.
- All charging sources are connected just to the main bus and charge the main battery.
- A unidirectional DC-DC converter transfers power from the main bus to the critical bus (and not the other way around) to charge the critical battery.
- The main bus powers the main loads and the critical bus powers the critical loads; even if the main battery is discharged or otherwise disabled, the critical battery can still power the critical loads.

FIGURE 2.37
Safe use of two batteries
on a single DC bus:
(a) permanent parallel
connection, (b) batteries
with array compatible
BMS, (c) array master,
and (d) ganged batteries.

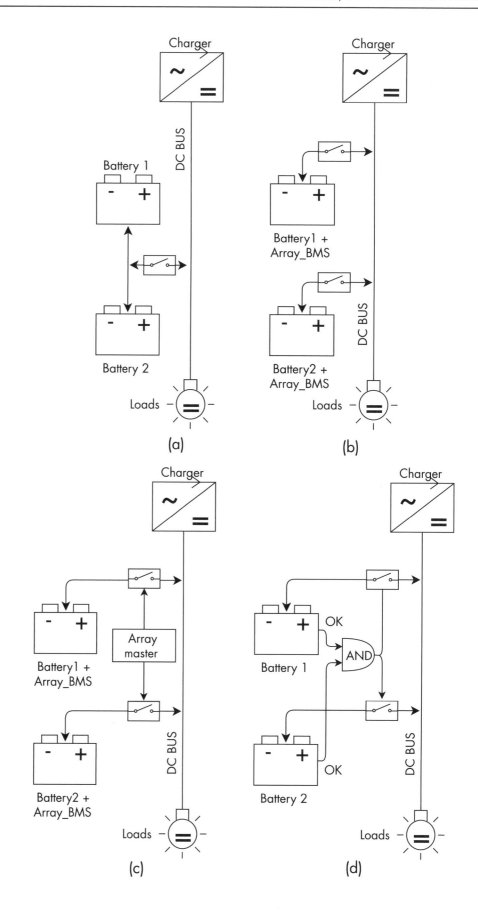

FIGURE 2.38
Safe use of two batteries
on two DC buses:
(a) two equal buses,
(b) main and critical
bus, and (c) charge
and load bus.

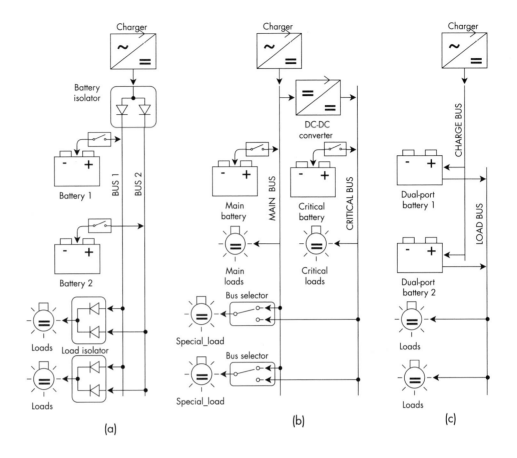

• Either bus may power some specific loads, according to the position of selector switches.

A charge bus and a load bus (Figure 2.38(c)):

• Dual-port batteries have their charge port connected to the charge bus and their discharge port connected to the load bus; each battery includes circuits to prevent current from flowing into the discharging port, or out of the charging port, therefore preventing current from flowing between the batteries.

• All charging sources are connected just to the charge bus and charge both batteries.

• All loads are connected to the load bus.

• If both batteries are in use, there is little redundancy, because both batteries discharge simultaneously; discharging could be disabled for one of the batteries to keep it full and in standby for emergencies; or batteries could be alternated from time to time.

In the first case, a single cutoff switch is between the common of the two batteries and the DC bus. In the last two cases, each battery has a cutoff switch.

2.6.3.14 Topology Comparison

Table 2.6 compares the advantages and disadvantages of each approach.

TABLE 2.6
DC Bus Topology
Comparison

Figure	Buses	Solution	Advantages	Disadvantages
2.37(a)	1	Permanent parallel	Simplicity; parallel connection not an issue	No cutoff between batteries; no redundancy
2.37(b)	1	Array BMS	Simplicity	Array BMSs are not readily available
2.37(c)	1	Array master	Sophisticated control	Array masters are not readily available
2.37(d)	1	Ganged	No special BMS required	If one battery is low, both are
—	1	DC-DC converters	No need for current limiting in the charging sources	Cost, size and availability of bidirectional DC-DC converters
2.38(a)	2	Dual DC bus	Redundancy; may keep one battery ready to start the engine	Requires battery isolators and selectors; captain must select batteries
2.38(b)	2	Main and critical bus	Redundancy; less likely that critical battery is empty	Requires DC-DC converter
2.38(c)	2	Charge and load bus	Simplicity	Few options for dual-port BMS; if one battery is low, both are

2.6.3.15   Critical Function Batteries

Mariners are afraid that they cannot start the engine or radio for help because all the batteries are discharged or otherwise not available. To address this fear, in addition to the house battery, the vessel may have one or more smaller batteries to power critical functions and provide redundancy:

- Engine starter battery: for engines or generators;
- Generator starter battery;
- Buffer battery: for lighting and navigation.

These batteries are usually 12V. They may be charged from a DC bus through a DC-DC converter, or from a charger powered by an AC source.

They are allowed to power only their respective critical functions. They may not power a DC bus so that they are not likely to be discharged when needed.

2.6.3.16   Control Freaks

The biggest challenge I found in the marine world is the captain who insists on being the energy management system. Such captains positively won't allow the BMS to shut down the battery current to protect the cells. They want to monitor all the parameters in the electrical system and turn on and off generators and loads as they see fit.

Some manufacturers of marine products acquiesce to this attitude and design Li-ion batteries without a protection switch. Many vessel designers do not incorporate Li-ion safety measures in their designs because they don't understand the importance of doing so. The battery manufacturers do not educate them; in turn, they do not educate their customers and captains.

A vessel is the worst place to have this attitude [5]: you could be left stranded in a dingy in the middle of the ocean, watching your yacht burn down because you let a battery discharge too deep, which then caught fire while recharging.

As a battery designer, you must overrule this mindset, and design a safe battery, one that the BMS can turn off if required to protect its cells.

Yes, do reduce the need for the battery to shut down (see Section 2.11). It's OK if the battery does shut down because there are people onboard who can act and rectify the problem. A little inconvenience is worth the price of safety on the high seas.

2.6.3.17 Yacht State

Roughly speaking, a yacht may be in one of several states, which affect how its Li–ion battery is used (Table 2.7).

In all of these yacht states, the battery can be in one of just these three operating modes:

1. Run: normal operation;

2. Storage: low duty operation around 50% SoC;

3. Off: brought to about 50% SoC and then disconnected.

The ideal battery should have a three-position switch allowing the captain to select the most appropriate mode given the state of the yacht, for the sake of maximizing the life of the battery:

· Run: the BMS uses the standard limits for charging and discharging;

· Storage: the BMS uses narrow limits, such as 60% and 40% SoC;

· Off: the BMS disconnects the battery and shuts itself down.

No marine BMS today offers this feature.

2.6.3.18 Lead Acid Versus Li-Ion

Typically, vessels use lead-acid batteries; Li–ion is now being introduced. One would think that heavy lead-acid batteries would be better than a light Li–ion battery because they can also be used as ballast. Yet Li–ion batteries are now preferred due to their long life because batteries are often placed in tight places, making them hard to service and maintain.

			Li-Ion Battery Use	Mode
Sailing			The battery is in full use.	Run
Docked	**No shore power**	**Short term**	The battery is in full use.	Run
		Long term	Solar power keeps the battery at around 50% SOC to power a few loads (e.g., bilge pump).	Storage
	With shore power	**Short term**	The battery is not needed, and should ideally be kept at about 50% SoC. The battery is available in case of power failure in shore power.	Storage
		Overwintering	A few loads remain in use (e.g., bilge pump). The battery is not needed, and should ideally be kept at about 50% SoC. The battery is available in case of power failure in shore power.	Storage
		Getting ready to sail	Shore power charges the battery fully.	Run
Dry storage	**No shore power**		The battery should be brought to 50% SoC and completely disconnected.	Off
	With shore power		The battery should be kept at about 50% SoC. The battery is available in case of power failure in shore power.	Storage

TABLE 2.7 Yacht State and Battery Mode

A significant advantage of Li–ion in marine applications is that a yacht may remain docked for long periods. Even if all loads are turned off, a lead-acid battery will self-discharge significantly during this time, but a Li–ion battery will retain most of its charge for years.

However, Li–ion has a bad rap in the yachting world due to some nasty fires (see Section 5.2.2); luckily, no one has been hurt to date.

2.6.3.19 Off-the-Shelf Li-Ion Marine Batteries

Ready-made Li–ion batteries for marine use are available. The batteries themselves can be of two types:

1. Monitor only: includes just a monitoring BMS; a communication link reports its state to the system and requests that the current be reduced or stopped.

2. Self-protected: includes a complete protectors BMS; communications are optional.

Monitor-only batteries are cheaper and lend themselves to parallel and series arrangement to achieve the desired voltage and capacity (see Volume 1, Chapter 6).

These batteries require a separate system for protection. The battery vendor may offer a ready-made BMS for use with these batteries.[76] Otherwise, the vessel designers must add a protector switch of some sort. The risk is that the vessel designer doesn't know how to do this, leaving the door open for severe damage to the batteries, at best.

Self-protected batteries are more expensive than unprotected batteries due to the cost of the protector switch. They may not be connected in parallel (to increase capacity) because their BMSs are unable to reconnect safely (see Volume 1, Section 6.3). Up to a certain number of them may be connected in series to achieve the desired voltage; check with the manufacturer.

2.6.3.20 Dual-Port Marine Batteries

Manufacturers of marine batteries offer both single-port and dual-port batteries. A dual-port battery routes the charge and discharge currents to separate terminals.[77] All the charging sources are connected to a charge bus and to the input of the dual-port battery. All the loads are connected to a loads bus and to the output of the dual-port battery (Figure 2.39(a)). A dual-port battery is cheaper but this arrangement has several disadvantages (see Volume 1, Section 5.12.2.3).

Two dual-port batteries may be used in parallel, as long as rectifier diodes are added to prevent current from flowing into the discharge port or out of the charge port (Figure 2.39(b)). This is particularly important if bus selector switches are used because they allow the two batteries to be used differently and reach different SoC levels. The diodes prevent the inrush of current when two batteries at different SoC levels are connected in parallel (see Volume 1, Section 3.3.6).

Boat designers do not use diodes (Figure 2.39(c)) because they are wasteful[78] and because they believe they are not needed. These designers assume that the SoC of

---

76. Such as the Lithionics Never-Die.
77. Lithionics Never-Die dual-channel BMS, available only for use with Lithionics batteries. Note: it does not prevent charging current into the discharging port, or discharging current through the charging port!
78. They drop the voltage and dissipate much heat during operation.

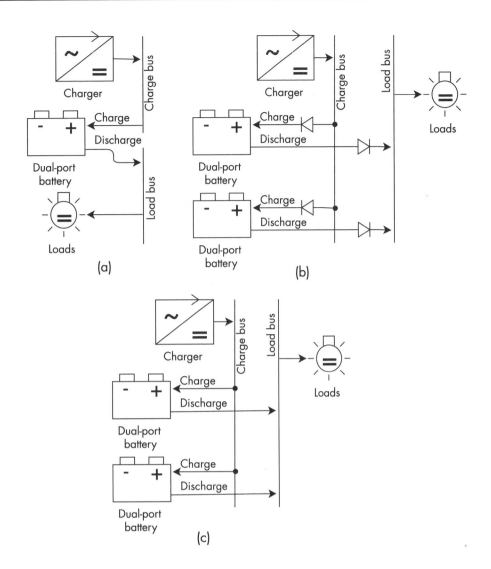

FIGURE 2.39
Dual-port batteries:
(a) single battery, (b) two
batteries with steering
diodes, and (c) two
batteries without diodes.

the two batteries will always match because, at any given time, only three cases are possible:

1. Both batteries are fully connected to both buses (Figure 2.40(a)).
2. One battery is disconnected from the charge bus (Figure 2.40(b)).
3. One battery is disconnected from the load bus (Figure 2.40(c)).

Indeed, in all these cases, the two battery voltages match each other because both batteries are connected to at least one bus; hence, their voltages must be equal to the voltage of this bus and, therefore, to each other.

However, there is a fourth case: one battery is connected to the charge bus and one to the load bus (Figure 2.40(d)). The SoC of the first battery increases as it's charged while the SoC of the second battery decreases as it's discharged. Therefore, no, it is not true that the voltages of both batteries always match.

Once the two batteries reach different voltages, there is an inrush current between them as soon as either one connects to the other bus. When a BMS tries to disable discharging, it can't do so because the battery can be discharged through its input port (Figure 2.40(e)). This is because the input port uses a circuit that can stop charging

FIGURE 2.40
Two dual-port batteries:
(a) connected to both
buses; (b) top battery
thinks it's disconnected
from the load bus,
(c) top battery thinks it's
disconnected from the
charge bus, (d) one battery
connected to each bus,
(e) inrush from top battery
upon reconnection, and
(f) inrush from bottom
battery upon reconnection.

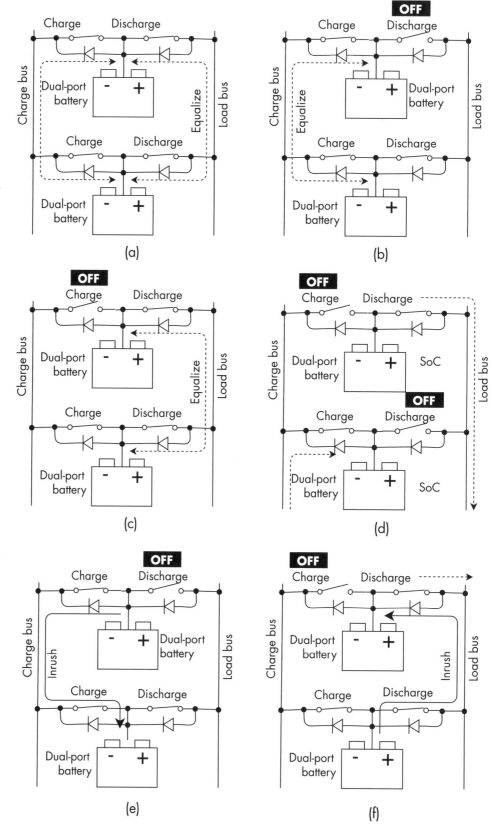

current but cannot stop discharging current.[79] Similarly, when a BMS tries to disable charging, it can't do so because the battery can be charged through the output port (Figure 2.40(f)).

Therefore, when two dual-port batteries are connected without additional diodes, they are no longer protected. This is disastrous.

When made aware of this issue, boat designers have used language such as "care must be taken" and "there must be a procedure"; in practice, this is ineffective because

- The captain won't remember these cautions when reconnecting batteries.
- The BMS reconnects its battery on its own, at a time when the captain is not looking.

In conclusion, I must strongly warn you against connecting two dual-port batteries in parallel.

2.6.3.21 Custom Li-Ion Batteries

A custom-designed Li-ion battery may be a better solution due to limitations with the available space or because the battery is particularly large.

Boat designers prefer LFP cells due to their perceived safety (see Volume 1, Section 2.2.1.2). Yet one may argue that an LFP cell from a shady cell manufacturer is less safe than an LCO cell from a reputable one.

Large prismatic Li-ion cells are most suitable, as they fit nicely in battery boxes and are easy to interconnect. Use 100-Ah or 200-Ah cells in parallel to achieve the desired capacity[80]. Connect parallel blocks in series to achieve the desired voltage (e.g., eight in series for 24V). The constant rocking of a vessel puts a different kind of mechanical stress on batteries than vibration does for land vehicles. Interconnect cells with braided cables and place resilient spacers between adjacent cells. While hard bus bars may be OK for a homemade EV car, they are inappropriate in marine batteries due to the stress they place of the terminals of large prismatic cells as they sway with the waves.

2.6.3.22 Battery Location

Place batteries in an accessible, cool (but not cold), dry place. For weight distribution, place an equal number of cells or batteries port and starboard. Do not place batteries in the engine compartment (too hot) or areas exposed to splashes (corrosion).

2.6.3.23 Battery Emergency Design and Procedures

The difference between surviving a burning Li-ion battery at sea and sinking is a matter of battery design, procedures, and training. Design the battery and its surrounding for an emergency (see Section 5.3).

For a battery that weighs up to 100 kg, consider ways to remove it and dunk it off board:

- Consider place the battery where it has direct access to the deck and therefore the sea, rather than inside the bowels of the vessel.

---

79. Ideally, each port in a dual-port battery should have back-to-back switches to block current in either direction; however, this is never done in practice. See Volume 1, Section 5.12.2.3.
80. "LiFePO4 Batteries on Boats," Rod Collins, marinehowto.com/lifepo4-batteries-on-boats/.

- Place an easily accessible kit close to the battery; include cable cutters, firefighter gloves, and a face mask.
- Let the battery be held down by straps or brackets and quick release fasteners that are removable without any tools so that the battery can be freed in 2.0 seconds[81].
- Fit the battery with fire-proof straps, so two people can lift it out and carry it outside, dunk it in the sea to cool down while securing the rope to the vessel to keep the battery from sinking; save the battery for postmortem analysis.

For a larger battery, consider ways to flood it:

- Install a fire hose near the battery, powered by some other source.
- Design a gravity-fed method to flood the battery with a large volume of water from a tank.

Prepare a set of emergency procedures and train the crew on the emergency procedures.

2.6.4   Yacht Wiring, Main and Critical Batteries

There are innumerable ways to wire the house power in a yacht; I would like to suggest three diagrams, one with a main and a critical battery, one with a dual-port Li-ion battery, and one with two ganged batteries. They all use the AC and DC and inverter topology (Figure 2.13(c)).

This first diagram addresses many of the issues with Li-ion batteries in vessels discussed in the previous sections (Figure 2.41): electric shock drowning, hull and propeller corrosion, depleted engine starter battery, and parallel connection of Li-ion batteries.

2.6.4.1   Buses

There are four DC buses:

- House: connected to the house battery and powered by high-current charging sources;
- Charge: powered by low-current charging sources;
- Load: powers the noncritical loads;
- Critical: connected to the critical battery and normally powers critical loads.

2.6.4.2   Batteries

This vessel uses three batteries, for redundancy:

- House:
  - Large battery with a nonprotector BMU;
  - Normally cycled fully (100% to 0% SoC);
  - Powers most house loads.
- Critical:
  - Small battery, uses a protector BMS;

---

81. Sinopoly has stated that cells over 200 Ah should not be used on a vessel due to the increased chance of vibration failure.

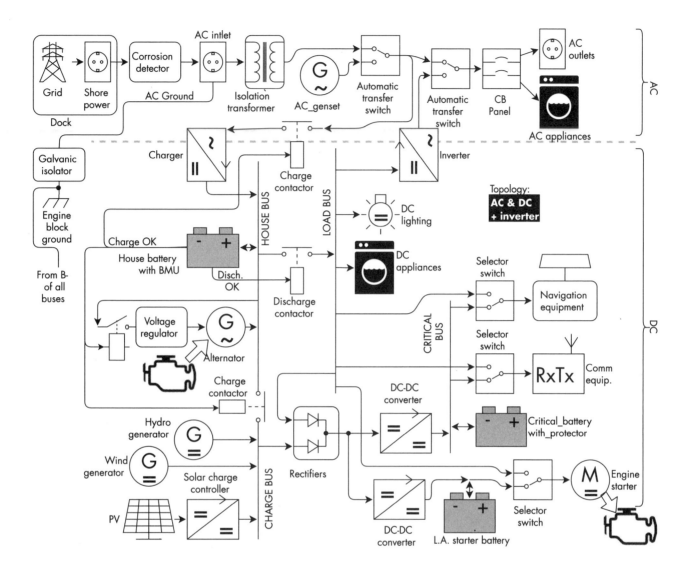

FIGURE 2.41   House power for yacht using a single-port BMS.

  · Normally kept full;

  · Powers critical loads.

· Starter:

  · Can be a lead–acid battery;

  · Dedicated battery to reduce the chance it's discharged;

  · Using a separate battery isolates the electrical disturbances generated by the starter motor from affecting the sensitive electronics such as radio equipment.

The BMS in the house battery controls charging and discharging through external switches (see Volume 1, Section 5.12.2.4):

  · Disables charging in multiple ways:

    · Opens a contactor between the charge bus and the house bus.

    · Disables the alternator by opening a relay that powers the voltage regulator.

- Opens a contactor on the AC input of the charger.
- Disables discharging by opening a contactor between the house bus and the load bus.

It is best if the system is designed to minimize the chance that the BMS disables charging or discharging, relying instead upon a careful selection of charging voltages and automatic activation of a generator (see Section 2.11.1).

The BMS in the critical battery controls charging and discharging through internal switches. If the starter battery is lead acid, it has no BMS.

2.6.4.3   Grounding

The vessel has a single grounding point (at the engine block) that grounds the propeller and the metal hull (if any). All the negative buses (B-) connect to this point.

The ground from shore power goes through a galvanic isolator and then to this common grounding point. If an isolation transformer is used, it is possible to use isolation wiring (see Section 2.6.1.4), by not connecting the ground from shore power. If so, the galvanic isolator is not needed.

2.6.4.4   AC Circuits

Shore power goes through a shore power corrosion detector before entering the vessel. Optionally, AC power goes through an isolation transformer to remedy any polarization issues with shore power and to avoid short-circuit current in case of a ground fault. An AC genset provides AC power when there is no shore power. An automatic transfer switch selects shore power or AC genset power, whichever is available. Its output feeds the charger through a contactor that the house battery can open to stop charging.

The load bus powers an inverter to power the AC loads when AC power is not available. A second automatic transfer switch selects how the AC loads are powered. By default, it selects the AC sources (shore power or AC genset, from the first automatic transfer switch). If those are absent, it selects the inverter. AC loads include outlets and AC appliances.

2.6.4.5   Charging

Multiple sources may charge the house battery:

- The charger;
- The alternator, driven by the engine;
- The charge bus, fed by low-power generators (solar, wind, or hydro).

The charger is powered after the first transfer switch and before the second transfer switch, so that the inverter may not power it. Letting the inverter power the charger would be pointless because the charger ultimately powers the inverter.

The BMU can stop charging by opening a contactor on the input of the charger, disabling the alternator (by removing power to the regulator), and disconnecting the charge bus[82].

---

82.   .I am seriously concerned about the risk of injury while removing a heavy battery and dunking in the water off board. Yet I believe that is preferable to a burning vessel sinking.

Either the house bus or the charge bus, whichever has the highest voltage, powers a DC-DC converter; this converter charges the critical battery and performs three functions:

- Current limiting:
  - It avoids the current inrush that would occur with a direct parallel connection between the house battery and the critical battery.
  - It set the CC level to charge the critical battery.
- Voltage conversion:
  - It may raise the voltage if doing so is required to fully charge the critical battery from a mostly discharged house battery.
  - It allows the house and critical buses to work at different nominal voltages, such as 24V and 12V, respectively.
- Discharge prevention:
  - Power may flow only from the house bus to the critical bus; this prevents the discharge of the critical battery by powering noncritical loads.

The load bus powers a DC-DC converter that, in turn, charges the starter battery. This converter is optional since a small lead-acid battery can handle being paralleled directly to a Li-ion battery for charging.

2.6.4.6   DC Loads

The load bus powers regular DC loads, which may include lighting, appliances, electric thrusters[83], and windlasses[84]. Normally, the critical bus powers critical DC loads (e.g., communication and navigation electronics). However, each device has a switch to select power from the load bus instead. The engine starter motor has a selector switch to select either the starter battery or the load bus.

2.6.5   Yacht Wiring, Charge and Load Bus

The following diagrams use only two DC buses: charge and load.

2.6.5.1   Yacht Wiring, Dual-Port Battery

This second diagram uses two buses and a dual-port battery. It is noticeably more straightforward than the previous one (Figure 2.42). A dual-port battery routes the charge and discharge currents to separate terminals[85].

Even if the battery disables discharging, it still can power the solar charge controller through its input because a dual-port battery cannot stop discharge current from going out of its input. Therefore, the system can recover from the sun alone. However, in the long run, the solar charge controller would overdischarge the battery. Therefore, the battery must be disconnected during long-term storage.

---

83. Maneuvering thrusters that push the vessel laterally to improve maneuverability and allow the sailboat to remain more vertical. As they do not provide forward propulsion, their batteries are not considered traction batteries.
84. A winch used to hoist anchors and sails.
85. Lithionics Never-Die dual-channel BMS.

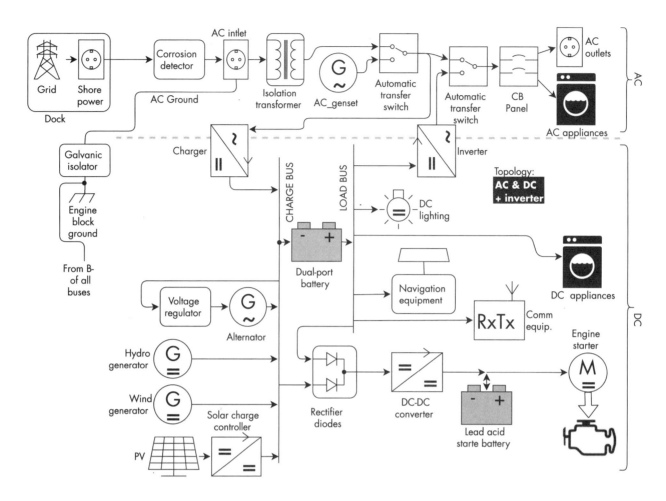

FIGURE 2.42   House power for a yacht using a dual-port battery.

2.6.5.2   Yacht Wiring, Two Ganged Batteries

This third diagram (Figure 2.43) uses two ganged batteries (see Volume 1, Section 6.4) that work in unison so that they are guaranteed to be always at the same SoC. Hence, there is no inrush current when they are connected directly in parallel.

This is hardly any different from having a single battery; yet psychologically, it feels better to have two batteries on a vessel, on the theory that if one battery "dies," it can be disconnected, and the second battery can still be used. There is a significant risk that, at some later point someone reconnects the first battery with no regard to its voltage. While large warning signs and posted procedures may help, nothing prevents someone from placing a battery back into service by reconnecting it to a second battery at a different voltage.

2.7   RECREATIONAL VEHICLE (RV) APPLICATIONS

In U.S. parlance, a recreational vehicle (RV) is a vehicle with living space and amenities usually found in a home: a motorhome, a camper, a converted bus, or a trailer (Figure 2.44(a)). In the United Kingdom, they are called caravans. In New Zealand[86], they are called campervans.

---

86. I singled out NZ because of its high rate of travel by campervan: in 2018, motorcaravans and trailer caravans represented 15% of registered vehicles (NZ Transportation Agency).

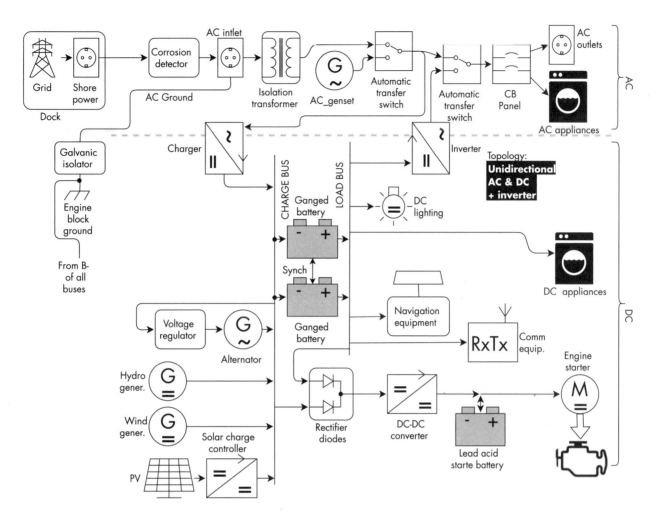

FIGURE 2.43 House power for yacht using two ganged batteries.

Typically, RVs park for a while at a dedicated camp with full hookup, where they have access to water, sewage, WiFi, and electricity. RVs may also boondock: stop at locations without facilities. Some use an RV as a permanent residence, parked on private property or public roads or parks. In either case, a house battery is essential to avoid the cost and noise of continually running a genset.

2.7.1 RV Devices

RV applications use the following devices, in addition to the ones described at the start of this chapter (see Section 2.2).

2.7.1.1 Terminology

Note that RV people use some terms differently from the rest of us. For example:

- Solenoid, separator = contactor (relay);
- Pedestal = shore power;
- Converter, RV converter, power converter = charger.

FIGURE 2.44
Recreational vehicle:
(a) on shore power, and
(b) on generator power.

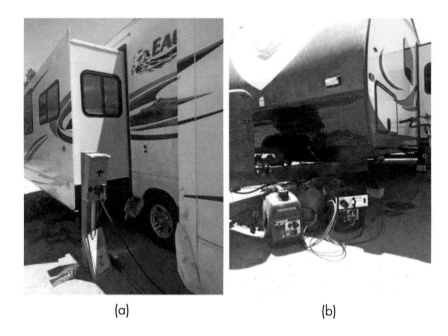

(a)   (b)

2.7.1.2 Shore Power

At an RV campground, each site offers shore power to the RV through a pedestal (Figure 2.45(a)). The pedestal includes a circuit breaker and one or more outlets (Figure 2.45(b)). The configuration depends on the country:

- In the United States, there may be a set of outlets: 120 Vac single-phase 20 A, 120 Vac single-phase 30A and 240V split phase, 50A.
- In the United Kingdom, there is a 220-Vac, 16-A, single-phase outlet.
- In continental Europe, there is a 220-Vac, 6-A, single-phase outlet.
- In New Zealand, there is a 230-Vac, 15-A, single-phase outlet.

A miswired outlet is a real possibility, which can result is the risk of tingling or electrical shock. RV people carry a combination surge protector and polarity tester to confirm that the outlet is wired correctly (Figure 2.45(c)). The user plugs a power cord to the RV's AC power inlet and the shore power pedestal. These devices may introduce a short delay (typically about 20 seconds) after they are powered, to give

FIGURE 2.45
RV shore power: (a) RV
plugged into a pedestal;
(b) U.S. outlets and
circuit breakers; and (c)
surge protector/tester.

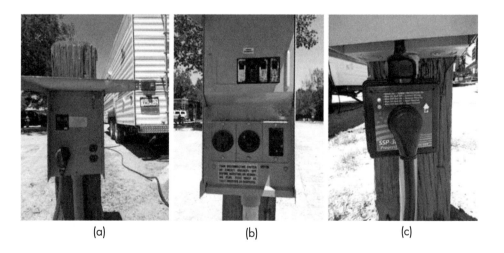

(a)   (b)   (c)

time to the user to notice the problem and unplug before power is passed to the RV. This delay may be unexpected.

In the United States, RV people also carry adapter cords, for example, to power a 30-A RV from a 50-A outlet.

2.7.1.3 Generator

Typically, RVs use an AC generator (a genset), either portable (Figure 2.44(b)) or built-in. A portable generator is placed on the ground outside the RV. Some parks do not allow the use of such generators, especially at night, due to noise. The user plugs a power cord to the RVs AC power inlet and the generator. The fact that the same inlet is used to power the RV, whether from shore power or a genset, ensures that only a single AC source powers the RV.

A built-in generator is more convenient and quieter. The RV has an AC power selector switch to select the built-in generator or shore power. The house DC bus powers the generator's starter motor, meaning that the generator can't be started if the house battery is empty.

The continuous power of these generators ranges from 1 to 4 kW, although higher power is available in the short term to start appliances. This power is less than what is available from a 240-Vac outlet in a shore power pedestal. The output voltage of a generator sags under heavy load, which some AC loads in the RV do not appreciate; if the generator has a circuit breaker, it may trip.

2.7.1.4 Inverter

An inverter[87] is used to generate AC power from the house DC voltage if neither shore power nor a genset is available. The inverter is off-grid, so any other AC source must be disconnected when the inverter is in use. Typical power is 1 to 3 kW.

2.7.1.5 Automatic Transfer Switch

An automatic transfer switch is simply a relay that selects the source for AC power: inverter or shore power/genset. The AC power inlet controls its selection:

- Shore power or genset present: relay on, selects shore power or genset.
- Shore power or genset absent: relay off, selects inverter output.

A short delay after switching away from the inverter helps by giving the genset time to stabilize[88].

2.7.1.6 Batteries

RVs have two batteries:

1. Starter battery (see Section 2.9): also known as cranking battery, vehicle battery, truck battery, chassis battery, SLI.
2. House battery: also known as leisure battery, coach battery.

House batteries may consist of 6V or 12V lead-acid batteries. Li-ion batteries are rare (see Section 2.7.2.4).

---

87. AIMS Power, Ampeak, Go Power, Magnum, Nature Power, Potek, Power Teck On, True Blue, Vertamax, and Xantrex.
88. Progressive Dynamics sell them as a "patented Automatic transfer switch" with a "Generator detect and delay."

RV experts maintain that two 6V batteries in series are much better than a 12V battery based on various dubious arguments. The actual reason is that 6V deep-discharge batteries for golf-cart use are readily available and better suited to RV use than a single 12V starter battery. This is not because of the voltage, but because of the difference between deep-discharge and starter technologies. Other than that, for a given total volume and technology, two 6V batteries in series are the same as a single 12V battery. There is one more advantage: lifting two 6V batteries one at a time is easier on one's back than lifting a single 12V battery.

2.7.1.7   Alternator

The vehicle's alternator charges the starter battery. If the vehicle DC system is connected directly to the house DC system, the alternator tries to power the loads in the house. However, a standard alternator is not designed for continuous operation at high current; it may overheat and be damaged.

2.7.1.8   Charger

An AC-powered battery charger is used to charge the house battery. It is designed to charge lead-acid batteries through two or more stages. The charger must not be operated without a battery because it is not designed to supply the current peak that occurs when an appliance is turned on. The charger supplies only the average current drawn by all the loads, while the battery provides the peak current, acting as a buffer. Some chargers for RVs have two outputs, one for the starter battery, one for the house battery.

2.7.1.9   Inverger

In some systems, a single inverger[89] may be used instead of a separate charger and inverter[90]. While an inverger for residential use is grid-interactive, one for RV use is off-grid; just like an inverger for residential use, this inverger includes an automatic transfer switch.

2.7.1.10   Battery Selector Switch

Some RVs use a battery selector switch (see Section 2.6.2.3) to select which battery feeds the house DC bus: house battery, starter battery, or both. This allows starting the engine if the starter battery is empty or starting the genset if the house battery is empty. It also allows the charging of both batteries. However, it makes it easier to discharge both batteries unintentionally. Therefore, a switch that has a "both" position is not recommended in an RV.

2.7.1.11   Solar and Wind

Solar panels may charge the house battery through a solar charge controller. This device is similar to a solar charge controller for residential use, except that it is low power and uses a low-voltage solar panel. A dual-charge controller charges both the starter battery and the house battery[91]. A wind generator may be used as well, although it doesn't provide much power. The power from these sources is not sufficient

---

89.   Bidirectional charger/inverter. See Volume 1, Section 1.2.
90.   Xantrex Freedom SW.
91.   Morningstar Sunsaver Duo.

to power the RV continuously, although the energy stored in the house battery can power the RV intermittently.

2.7.1.12 *Loads*

In general, critical functions use DC power: water pumps, air blower for a propane heater, and reading lights. DC loads are powered through a fuse box, similar to the one in a car; although DC breakers could be used, fuses are more common in RVs.

A circuit breaker panel distributes the AC power to AC loads: air conditioner, microwave oven, satellite TV receiver and TV, clothes washer, dishwasher, and other household appliances. Optionally, a load manager may automatically manage high-power AC loads to prevent overload of the AC source[92].

Refrigerators for RV use do not use a compressor. Instead, they use heat to drive a cooling mechanism with no moving parts. In a two-way refrigerator, the heat can come from AC power or cooking gas. In a three-way refrigerator, it can also come from DC power.

For better integration and ease of wiring, some companies offer multifunction combo boxes, for example:

- AC breakers, DC fuses, charger[93];
- Dual battery charger, with MPPT solar charge controller[94].

2.7.2 Technical Considerations

RV technology is at the confluence of automotive and residential designs.

2.7.2.1 *Electrical Systems*

RVs have two electrical systems:

1. High-voltage AC (120 or 240 Vac) for the living space;
2. Low-voltage DC (12V or 24V), which is somewhat compartmentalized as:
   - Vehicle: related to the basic truck or bus, powered by the starter battery;
   - House: related to the living space, powered by the house power battery.

If a pickup truck is topped with a camper, the truck and the camper have separate DC system (vehicle and house) that are joined by an umbilical cord cable. Otherwise, the DC system is fully integrated, even though it's compartmentalized.

DC appliances may be 12V or 24V. Most RVs use 12V because 12V appliances are more available and cheaper. The advantage of 24V is that the currents are lower, allowing for smaller gauge wiring.

A mixed system using both 12V and 24V is possible, although not common. It uses a DC-DC converter between the two voltages. It may also use a 12V tap in a 24V battery, and a battery balancer to maintain the voltage of the 12V bus midway to the 24V bus (see Section 2.6.2.4). The most likely mixed system would use a 12V system for the vehicle and 24V for the house.

Only one AC source may power the high-voltage system at a given time: the grid, an AC genset, or an inverter.

---

92. Xantrex Freedom Sequence.
93. Progressive Dynamics Energy Management System.
94. Redarc BCDC.

2.7.2.2 Battery Isolation

The point of having two batteries is so that house loads do not deplete the starter battery. This allows the engine to be started even if the house battery is empty (see Section 2.6.3.9).

A few approaches are used to isolate the two batteries in an RV:

- *Battery isolator*[95]:Two diodes route the current from the charging source (usually the alternator) to each battery; the charging current can reach both batteries, yet no current can flow between batteries (Figure 2.46(a)).

- *Electronic battery isolator or smart battery isolator*[96]:This is the same, but has a lower-voltage drop and may include an input for the ignition voltage to enable connecting both batteries for charging (Figure 2.46(b)).

- *Separator solenoid*: Contactor between the batteries, closed when the ignition is on, allowing the alternator to charge both batteries; in a camper, it's placed in the truck, before the 2-pin connector to the camper (Figure 2.46(c)).

- *Voltage sensing relay*[97] (VSR): Contactor between the batteries closed when the voltage on the vehicle battery is increasing, indicating that it's being charged, allowing the alternator to charge both batteries (Figure 2.46(d)).

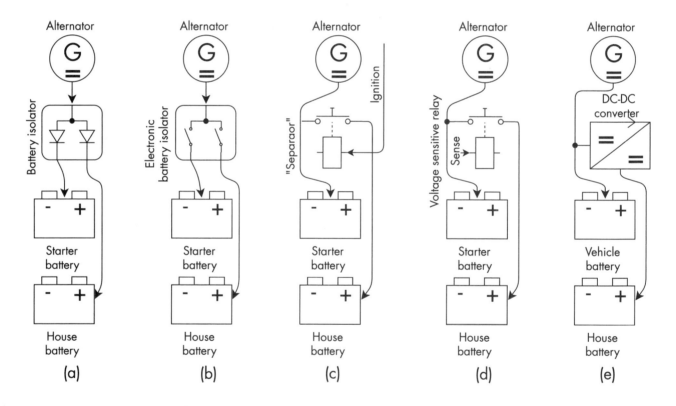

FIGURE 2.46   Isolation of two batteries: (a) battery isolator, (b) electronic battery isolator, (c) separator solenoid, (d) voltage sensing relay, and (e) DC-DC converter.

---

95. Also known as split diode. See Section 2.6.1.8.
96. Redarc Smart Start.
97. Also known as automatic charging relay or voltage sensitive relay.

- *DC-DC converter/charger between the batteries*: The alternator charges the starter battery, which, in turn, charges the house battery through the converter[98] (Figure 2.46(e)).

2.7.2.3 Grounding

For safety, the AC ground and the negative ends of both DC systems are grounded to the metal frame of the RV.

2.7.2.4 Li-Ion

Generally, in an RV, a Li-ion battery does not offer significant advantages over a lead-acid battery. As RVs start using more appliances and more powerful ones, the additional energy stored in a Li-ion battery of a given size is advantageous.

LFP batteries are preferred over other Li-ion chemistries for their perceived safety because people sleep in an RV: they may not be awake if there is a fire. Conveniently, the voltage of four LFP cells in series nicely matches the voltage of a 12V system, and eight cells in series match a 24V system (see Volume 1, Section 6.6.2.1).

If an RV has two Li-ion batteries, many of the standard techniques in RVs may cause excessive current between the two batteries. Therefore, the following measures are necessary:

- To isolate the batteries, a diode-based battery isolator (Figure 2.46(a)) and a DC-DC converter (Figure 2.46(e)) are appropriate; all other solutions are not acceptable.
- Pay particular attention to the umbilical cord between a truck and its camper: mating those connectors must not connect the two batteries directly in parallel.
- Do not use battery selector switches with a "both" position (see Section 2.6.2.3).
- Use an alternator whose regulator limits the alternator's current and temperature (see Section 2.2.1.1).

Chargers and alternators set a float charging voltage that would reduce the life of an LFP battery. For example, a 14V bus voltage divides to 3.5V per cell, which degrades the cells if maintained; reducing the DC voltage to 13.5V (for a 12V system) or 27V (for a 24V system) improves the life of the LFP battery considerably.

As always, Li-ion batteries must include a protector BMS. The starter battery must be able to supply the cranking current to start the engine, which is too high for a protector BMS with a solid-state protector switch. The house battery doesn't need to work as hard; even starting the generator or an appliance does not require as much current as cranking the engine. Therefore, a standard protector BMS with solid-state switches should be fine for the house battery.

A few companies[99] are offering Li-ion batteries for RVs that are a drop-in replacement for lead-acid batteries (12V) as well as 24V batteries, but no 6V batteries. Other companies offer batteries with BMS plus external protector switch. This has the advantage that several battery modules can be connected in series and parallel to achieve the desired voltage and capacity. A protector switch is added to complete the battery and protect all the modules[100].

---

98. LSL Products Trik-l-start and Amp-l-start.
99. GreenLife, NexGen, Smart Battery, Victron
100. Victron Lynx Ion BMS 1000A.

The eye-watering price of these ready-made batteries may encourage the do-it-yourself person to make a Li-ion battery out of individual cells and a protector BMS. Use these guidelines:

- Use large prismatic LFP cells for ease of interconnection and safety.
- Parallel two or more cells into a block to achieve the desired capacity.
- Place four such blocks (for 12V) or eight such blocks (for 24V) in series.
- Use a large, DC-rated fuse in series in an appropriate location.
- Add a single-port protector BMS for LFP batteries, rated for the maximum current; a smart BMS with state of charge evaluation that supports a display would be preferable.

Li-ion is suitable for high-voltage batteries, opening a new opportunity, an RV that uses only high voltage AC devices. A high-voltage battery could be coupled to a compatible inverger to power the RV that uses standard residential lighting and appliances (see Section 4.6.2).

2.7.3   Typical RV Wiring

Here is a suggested diagram for an RV (Figure 2.47). This diagram assumes a pickup truck with a camper and an umbilical cord between the two:

- It uses a battery isolator, which is but one of the possible solutions to isolate the two batteries.
- The generator is permanently installed in the vehicle.
- Solar power is included, but not wind.
- It uses a separate charger and inverter, rather than an inverger; there is no loop between the charger and the inverter because the charger is powered before the automatic transfer switch, that is, the inverter's output cannot feed back to the charger.
- A three-way refrigerator may be AC or DC powered.

This system uses the AC and DC + inverter topology (Figure 2.13(C)).

2.8   ELECTRICAL AUXILIARY POWER UNITS (eAPUs)

Auxiliary power units (APUs) are fuel-powered generators used in vehicles to power its electrical system when the engine is off. Electric APUs (eAPUs) perform the same function, but use batteries instead of a generator[101]. Although these batteries reside in vehicles, they are not traction batteries because they are not used to propel the vehicle (see Section 3.1).

The vehicle's starter battery (see Section 2.9) could be used for the same function. However, a starter battery is designed to deliver short bursts of high power and stores little energy, while an eAPU stores much more energy and is designed to deliver continuous power. Also, an eAPU incorporates additional functions such an inverter and climate control.

---

101. Many thanks to Brian Banghart for help in this section.

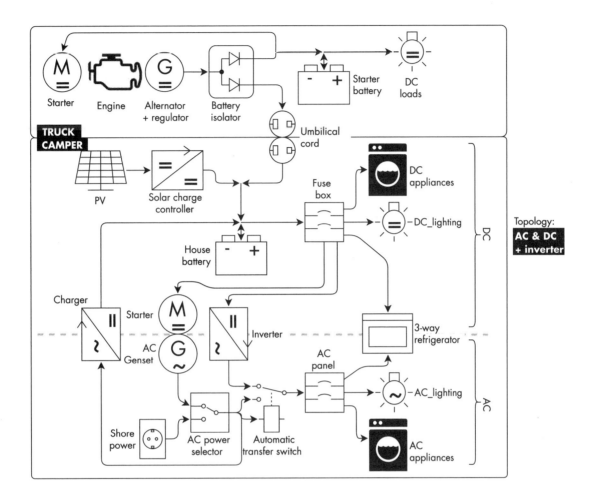

FIGURE 2.47   RV electrical diagram.

Among vehicle eAPU applications, I picked three representative ones[102]: long haul trucks, utility trucks, and airliners.

2.8.1 Long-Haul Trucks

While long-haul truck drivers park to rest and sleep, they keep the engine running to power accessories and climate control. The U.S. Department of Energy (DOE) estimates that a truck idles 1,800 to 2,500 hours per year. For every hour idling, the truck uses as much fuel as 6 to 8 km of driving. Idling trucks pollute and are noisy.

A truck APU allows the engine to be turned off during a stop by providing power to those accessories. Doing so reduces costs, noise, and pollution. The APU provides power for

- AC loads: TVs, microwave ovens and computers, about 1 to 3 kW;
- Charging ports for consumer electronics (e.g., USB[103]);
- Climate control: on the order of 4 kW [6].

---

102. Food trucks use high power AC appliances powered by an AC genset. As ice cream trucks do not stop long enough to set-up a genset, they can use a house battery, just like a utility truck.
103. Universal Serial Bus, a standard computer interface.

Normally, the engine drives the air conditioner or provides heat. When the engine is stopped, there is no climate control. Either the truck must use electric heat and an electrically powered air conditioning (so that it can be powered by either the alternator or by the eAPU), or the eAPU on its own must provide independent climate control. An eAPU[104] (Figure 2.48) does so silently and at little operating cost, though at a significant upfront cost.

While traveling, the eAPU is charged by the truck's alternator, with a minor effect on the truck's fuel economy. At a designated truck stop, the eAPU may also be charged from shore power such as an EV charging station.

Typically, eAPUs use lead-acid batteries, although more Li-ion units are becoming available to address the need for more energy storage and to support faster charging as well as prolonged periods of air conditioning. The energy ranges from 1 kWh for a lead-acid unit to 100 kWh for a Li-ion unit. Internally, the battery voltage is typically 24V or 48V. Various features are available, such as a state of charge indicator and Bluetooth for a smartphone app.

This eAPU uses the AC and DC + inverter topology (Figure 2.13(C)). In an eAPU that uses Li-ion, the BMS is likely to be a nonprotector type, and to control the charging sources and loads through separate switches for each device, using the external switch topology (see Volume 1, Section 5.12.2.4):

- The Charge OK line enables the charger when charging is desired.
- The Discharge OK drives a relay that powers the DC-DC converter for the USB outlet and contactors that power the inverter and climate control.

The contactors draw noticeable power, which would discharge the battery slowly when the APU is not in use; a power switch turns off the APU when not in use for long periods.

2.8.2   Utility Trucks

Construction and repair workers use utility trucks in the field. The truck may need to provide electrical power for power tools, a cherry-picker lift, and other electrical devices. Typically, the truck's engine is kept running because the vehicle's starter battery cannot power these tools, at least not for long. An eAPU[105] allows the engine to be turned off (Figure 2.49).

FIGURE 2.48
eAPU for long-haul trucks.

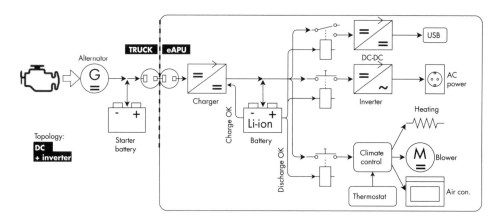

---

104. Also known in this industry as an idle eliminator.
105. Also known in this industry as an auxiliary power system (APS).

FIGURE 2.49
Auxiliary power system
for a utility truck.

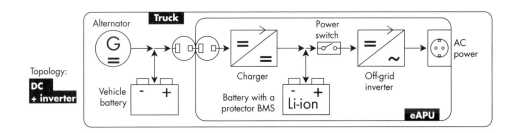

A charger converts the voltage from the truck's DC system to the higher voltage of the large Li-ion battery[106] (typically 48V). An off-grid inverter generates AC power. This device uses the DC + inverter topology (Figure 2.13(A)).

A standard protector BMS is sufficient, as the current involved is relatively low: charging current is on the order of 1 C, and discharging current is on the order of 0.2 C.

Standby power is low for the BMS, but not for the inverter. A power switch lets the user turn off power to the inverter when not in use.

2.8.3   Aviation

Private and commercial aviation may use eAPUs whose function depends on the size of the aircraft. As the weight of the battery is a significant concern, Li-ion can be of great help.

2.8.3.1   Airliner

When an airliner is parked at the gate, it is connected to shore power (Figure 2.50(a)) that provides 28-Vdc and 400-Hz AC[107] (Figure 2.50(b)). This source could be either the airport building or a ground power unit, a truck with a generator.

Before the airplane is disconnected from shore power, the pilot starts the airplane's fuel-powered APU. The APU generates 28 Vdc to start the main engines and 115 Vac, 400 Hz for house power. The pilot then switches from shore power to APU power[108].

FIGURE 2.50
Airliner: (a) shore power,
and (b) electric APU
and engine APU.

(a)

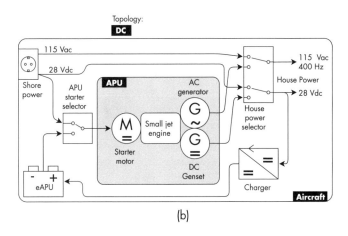

(b)

---

106. Since this is a new technology, there is not a long history of lead-acid use.

107. A voltage of 28V happens to be the voltage of a full 24V lead-acid battery, even though airliners do not use lead-acid batteries.

108. You may have noticed that the cabin lights blink off at that moment.

If shore power is not available to start the fuel-powered APU, the airplane uses an eAPU, which is simply a starter battery[109]. The airplane uses the eAPUs also to power refueling operations and navigation lights [7] and for emergency backup power.

An eAPU typically uses NiCD cells for lightness, power density, and safety. Li-ion cells are rare in civilian aircraft, due to lingering safety concerns. Boeing was among the first manufacturers to use Li-ion in commercial aircraft: the Boeing 787 Dreamliner uses a Li-ion battery with eight LiCo cells in series to generate 28V nominally, 32V maximum. This APU came into the spotlight when it caught fire (see Section 5.2.3.1) [7].

Electric APUs for aviation must provide a high level of reliability, performance, traceability, and airworthiness. They are developed in direct cooperation with the airplane manufacturer to meet the needs of the specific aircraft.

As the eAPU is used as a starter battery, it must be able to provide a high discharge current. Conversely, the charging current is relatively low. Therefore, a dual-port topology may be appropriate; yet the current is too high for a standard protector BMS with solid-state switches. The BMS would use a contactor to control discharging into the starter motor and may use solid-state switches for all the other loads and for charging. Ultimately, using a separate contactor (or some other type of switch) for each charging source and load may be more practical, using the external switch topology (see Volume 1, Section 5.12.2.4).

2.8.3.2 Small Aircraft

A small aircraft does not have a fuel-powered APU to provide power while the engine is off. It does have an engine starter battery, but it's not suited to provide house power for long periods.

Batteries for small aircraft are lead acid, but not the type used in cars: they are smaller (lower capacity) lightweight, and compact, with plates that are closely spaced, thin, and fragile. They are either 12V (for a nominal bus voltage of 14V) or 24V (for a nominal bus voltage of 28V).

The voltage regulator for the alternator may include a temperature sensor to be mounted on the battery to decrease the regulated voltage slightly as the temperature increases. A voltmeter in the instrument panel monitors the DC bus voltage. The pilot monitors this meter to spot any issues with the DC system.

Jump-starting a small aircraft from another battery is particularly dangerous because the pilot is tempted to take off before the battery has had a chance to charge fully. If the engine shuts down in flight, the pilot may not be able to restart it with the little charge left in the battery. The only safe procedure is to charge a starter battery with a charger over a few hours, fully. In any case, jump-starting damages the delicate aircraft battery (see Volume 1, Section 3.3.6).

NiCd starter batteries are uncommon in small aircraft. NiMH and Li-ion batteries may power individual devices, but the starter battery does not use Li-ion. Li-ion starter batteries would seem to be ideal: low internal resistance, fast charging, low weight, better performance at cold temperatures; yet safety concerns have kept them from being widely adopted[110].

Companies like True Blue offer Li-ion LFP batteries specially designed for aircraft. They use a 500 or 750-A protector BMS, and a flame-proof, rugged, metal

---

109. Note that this is a different functionality than for the eAPU used in long-haul trucks.
110. LFP cells, and especially LTO cells, are considered inherently safer than most other Li-ion chemistries.

container allows venting. Jump-starting these batteries is not advisable either, due to their low internal resistance. At best, the BMS protects the cell and cuts off the battery current. Do not even consider a Li–ion battery without a protector BMS: an aircraft is not a fitting place to play the "I don't want a BMS" game.

These starter batteries are small: 0.2 to 1 kWh. They cannot provide house power for an extended period: either the engine must be on, or the aircraft must be connected to shore power.

An eAPU would be useful to provide house power when on the ground with the engine off and away from shore power. However, my research shows that such products are not yet available[111]. An eAPU for a small aircraft might look like that shown in Figure 2.51(a).

2.8.3.3 Ground Power Unit

Ground power units are mobile devices that provide 28V power to an aircraft while on the ground. Typically, they are gensets. Battery-based ground power units normally use lead–acid batteries; some may use Li–ion batteries for faster charging and more efficient discharging (Figure 2.51(b)).

As they remain on the ground, these units can use less safe Li–ion chemistries. Eight NMC cells in series would generate a nominal voltage of 29V, 33V fully charged, and about 25V under load while the engine is cranking. The protector switch in the BMS must withstand the high current required to start an engine.

FIGURE 2.51
Small aircraft: (a) electric APU, and (b) Li-ion ground power unit.

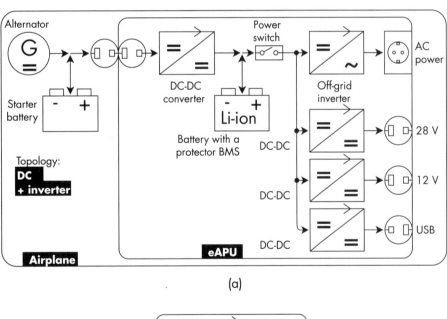

(a)

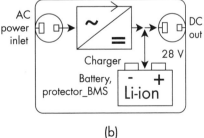

(b)

111. There's an idea for a product!

2.9 ENGINE STARTER BATTERIES (SLI)

Engine start batteries (also known as starter, lighting, ignition (SLI)) power the 12V electrical system of a car. Most SLI batteries are lead acid, although Li-ion is starting to be used. Some hobbyists are attempting to use super-capacitors instead of batteries.

2.9.1 Lead-Acid Starter Batteries

Lead-acid starter batteries are optimized for low internal resistance to produce the high current required to start the engine. Despite this, they survive a connection to the full voltage of another battery during a jump start. Lead-acid starter batteries degrade relatively quickly, in 2 to 5 years.

2.9.1.1 Lead-Acid Battery Ratings

Lead-acid starter batteries are rated by

- Use: starter type (rather than deep cycle).
- Type: sealed VRLA (rather than flooded, wet cell); VRLA includes AGM and gel types.
- Cranking amps (CA): the current that the battery can produce at 0°C to start the engine; also known as marine cranking amps (MCA); the voltage of the battery sags down to 7.2V.
- Cold cranking amps (CCA): same as above, but at −17°C.
- Capacity: nominal capacity, measured with a slow discharge at C/20; this determines how long the battery can power the parasite electrical loads with the car turned off, such as a car alarm; this is not the same as the effective capacity at high current because lead-acid batteries become less efficient at higher currents[112].
- Reserve capacity (RC): same as above, but at a specified current, higher than C/20; for example, "84 Ah at 25 A"; therefore, this value is less than the value of nominal capacity; this determines how long the battery can power the electrical loads with the car turned on but while the engine is off, such as headlights and a radio.
- Energy: some batteries are also rated in kWh, which is the energy delivered by the battery when discharged at C/20; this rating is as unimportant as the capacity rating.

Most users care more about cold cranking ratings than about the reserve capacity.

2.9.1.2 Charging an Empty Lead-Acid SLI Battery

A fully discharged lead-acid starter battery should only be charged at low current, with either an AC-powered battery charger or a portable car battery charger. It should not be jump-started, and it should not be charged by the car, because

- Jump-starting a car may damage its lead-acid battery due to the high current into it.
- The alternator in a car is designed to maintain the charge in an SLI battery; it is not designed to charge a fully discharged lead-acid battery because it is not

---

112. As described by the Peukert law. Search online for Peukert coefficient.

current-limited; the high charging current may damage the battery and over-heat the alternator.

2.9.2 Li-Ion Starter Batteries

Li-ion starter batteries are not yet widely used, due to their cost and short lifetime when connected to a car designed for lead-acid batteries.

2.9.2.1 Technical Considerations for Li-Ion Starter Batteries

Li-ion starter batteries promise better performance than lead-acid batteries: longer-lasting, high power, and high energy. The low weight of a Li-ion battery is only an advantage in a few applications, such as aviation or race vehicles. However, Li-ion starter batteries are not drop-in replacements for lead-acid starter batteries:

- There is a poor match between the regulated voltage in a car and voltages achievable with a discrete number of Li-ion cells in series.
- The low internal resistance of Li-ion cells results in too high a current when charged by an alternator or jump-started, which can cause damage to the cells, the protector switch, or the alternator.

A car regulator controls the voltage to about 14V, although it could range between 13V and 15V (or worse). This range is fine for a lead-acid battery, whose voltage varies significantly with the state of charge. Conversely, the voltage of Li-ion cells remains more constant over a wide range of SoC.

The voltage of four Li-ion cells in series does not match the voltage of the 12V system in a car designed for a lead-acid battery (Table 2.8).

The voltage in a standard car is too high for LFP cells; hence, a significant portion of the capacity of an NMC or LCO battery remains unused. A modern car regulates the voltage more tightly, in the range of to 13.8V to 14.2V. Even so, the maximum cell voltage, 3.55V, is too high for an LFP cell. Yes, an LFP cell can be charged to 3.55V safely, but, if it is kept at this voltage, its life is rapidly reduced.

2.9.2.2 Drop-In Replacement Li-Ion Batteries

While Li-ion starter batteries are commercially available, none of them appear to be compatible with the electrical system in standard cars. These 12V Li-ion starter batteries use four LFP cells in series.

Some have specifications such as[113]

- Maximum charge current: 1C;
- Fully charged voltage: 14.6V (3.65V per cell);
- Charge cutoff voltage: 15.6V (3.9V per cell);
- Capacity: 30 to 50Ah;

TABLE 2.8
Mismatch Between Four
Li-Ion Cells in Series and
a 12V Automotive Bus

DC bus voltage	13V	14V	15V
Cell voltage	3.25V	3.5V	3.75V
LFP	~50% SoC	OK for a short time	Too high
NMC, LCO	~10% SoC	~50% SoC	~80% SoC

---

113. KOK POWER, KOK–CSB–02.

- Discharge current for 3 to 5 seconds: 10 C.

Things to note:

- The maximum charging current is quite a bit lower than what the alternator can produce (2 to 4 C).
- The maximum cell voltage is quite higher than the safe level for LFP cells.
- It can produce enough current to start the engine.

Other batteries are inadequate to start an engine due to their specifications[114]:

- Charge voltage: 14.6V;
- Charge cutoff voltage: 15.6V;
- Maximum discharge pulse current: 60A.

Things to note:

- The discharge current is insufficient to start an engine.
- Again, the voltage is too high for LFP cells.
- The fine print: "Must use LFP charging circuit. Do not use a lead–acid charger to charge this battery."

Some of these batteries have a BMS, and others don't. Starter batteries without a BMS are overdischarged when the car is parked for a month or cranked for too long. Their cells are "cooked" by the high voltage in the DC system in a car.

Do not connect multiple Li–ion starter batteries (with a BMS) together:

- In parallel: because the BMS is incapable of safe array operation (see Volume 1, Section 6.3.1);
- In series: because the maximum voltage of the protector MOSFETs may be too low (see Volume 1, Section 6.3.4.2).

2.9.2.3   Design of a Drop-In Replacement Li-Ion Battery

It is possible to design a drop-in replacement Li–ion starter battery that is compatible with the electrical system of a standard car. This battery should have the following specifications:

- Self-contained, only two power posts, no other connections;
- Compatible with the voltage in a 12V car system with no reduction in battery life;
- 500-A discharging current for cranking, with minimal voltage drop;
- Minimal operating current when not cranking to allow long standby time with the vehicle off;
- 50-Ah minimum reserve capacity

The battery that meets these specifications (Figure 2.52) would use

- LPF, 26650 size power cells from K2, A123, or the like;
- 20 cells in parallel to form a 52-Ah block;

---

114. batteryspace.com LF-LA12V32.

FIGURE 2.52
Starter battery circuit.

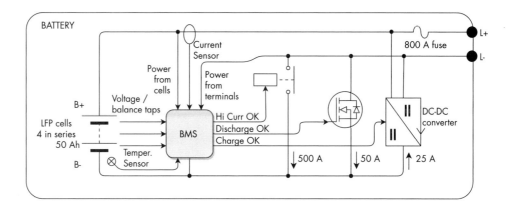

- Four blocks in series to get to 12.8V nominal;
- A BMS with:
  - Sensing of cell voltage, terminal voltage, temperature, and current;
  - Bypass balancing;
  - Three outputs:
    - Charge OK:
      - Turned on if all cells are below 3.35V and the temperature is in the range 0°C to 50°C;
      - Turned off when any cell reaches 3.6V or if outside the temperature range.
    - Discharge OK:
      - Turned on if all cells are above 3.0V and the temperature is in the range −20°C to 60°C;
      - Turned off when any cell voltage under load drops down to 2.0V, if the temperature is not in this range, or if the current exceeds 500A, for short circuit protection.
    - High current discharge OK: on if Discharge OK is on and the battery current exceeds 50A.
  - Powered by either the cells or the terminals, whichever has the highest voltage.
- An N-channel MOSFET[115] between B- of the cells and the L- output to handle low current discharge while wasting no power in standby; controlled by the Discharge OK line; capable of 50A of continuous current and with a voltage rating of 60V to handle a load dump[116].
- A contactor in parallel with the MOSFET, controlled by the high current Discharge OK line; turned on, with nearly zero voltage drop, when the BMS detects high current (due to cranking).
- A step-down DC-DC converter with CCCV output[117] limited to 25A to charge the cells at 0.5 C, controlled by the Charge OK line, placed in parallel with the contactor. The input can handle 60V. If the BMS is off, it defaults to trickle-charging the battery until the battery voltage is high enough to power the BMS.

---

115. N-channel MOSFETS have a lower on-resistance than P-channel but require a positive drive voltage on the gate. Therefore, the best placement is in series with the negative lead of the battery.
116. A voltage spike from the alternator when its load current drops suddenly.
117. Unlike the typical buck converter, the common is positive, and the input and outputs are negative.

The maximum voltage on the car's 12V bus won't harm the cells because the DC-DC converter isolates the cells from the bus. The cells are not kept at the maximum voltage because the BMS shuts off the DC-DC converter when the battery is full. A jump-start does not harm the cells because the DC-DC converter limits the current into the battery.

In case of a short circuit, the BMS shuts off the Discharge OK line. This turns off the MOSFET and the contactor before the fuse has a chance to blow. In case of reversed polarity jump-starting, the current flowing through the reverse diode in the MOSFET blows the fuse.

A standard automotive starter relay[118] works as a contactor because it is closed for a short period. There is no need for an expensive EV quality contactor. This relay is cheaper and performs better than a solid-state switch (see Volume 1, Section 5.12.3).

2.9.2.4 *Design of a Vehicle with a Li-Ion Starter Battery*

A vehicle designed to be compatible with a Li-ion starter battery (Figure 2.53) would perform better, and the battery would be simpler and cheaper. The Li-ion battery is not degraded by a direct connection to the 12V bus because it is tightly regulated to a lower voltage and the alternator current is limited.

Letting the starter relay act also as the high discharge current protection switch reduces the cost. This battery has two power ports:

1. Starter: high current discharge, protected outside the battery by the starter relay;

2. House: low current charge and discharge, protected inside the battery by MOSFETs.

It also has

- A line to disable the starter relay if required to protect the cells;

- A charge current limit (CCL) line to the voltage regulator to regulate the current from the alternator;

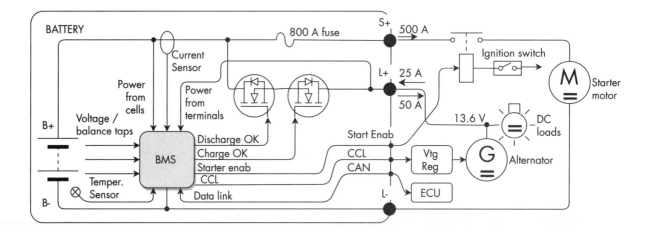

FIGURE 2.53   Li-ion battery integrated in vehicle.

118. Rated for ~75A continuous and hundreds of A peak.

• A CAN communication link to the vehicle to report the state of the battery.

The voltage regulator sets the DC bus voltage from the alternator to 13.6V, to maximize the life of the four LFP cells in series. The BMS monitors the charging current, and, if required, limits the current from the alternator to 0.5 C, through the CCL line and the voltage regulator.

If discharging is enabled, the battery grounds the starter enable line to power one end of the starter relay coil. The ignition switch powers the other end of the coil.

2.9.2.5   Protector BMS Selection

Very few protector BMSs with a MOSFET protector switch are able to handle the high current of a starter battery (Table 2.9).

Otherwise, one may use a BMU that drives a contactor.

2.9.3   Super-Capacitor Starter Batteries

Some hobbyists have replaced starter batteries with super-capacitors; this doesn't work so well.

Car batteries have two functions:

1. Starting the engine;

2. Powering the car's electrical system when the engine is off (or if the alternator is not working).

Super-capacitors perform the first job very well, especially at low temperatures. They perform the second job poorly because, after the car is turned off, the car's parasitic loads discharge the super-capacitors in just a few hours, after which the car cannot be started.

One would think that combing super-capacitors with a battery would provide the best of both worlds: high cranking current as well as high energy storage. However, it doesn't work this way (see Section B.2).

2.10   OTHER APPLICATIONS

Among other applications with large, low-voltage batteries, here I mention uninterruptible power supplies, pole-mounted batteries, microgrids, and battery modules.

2.10.1   Uninterruptible Power Supplies

A UPS can range from a small device to power a home computer to a large facility to power a computer server farm or a hospital. This chapter discusses UPSs that use a low-voltage battery; a later chapter will discuss those that use a high-voltage battery (see Section 4.2.4). An electronic generator is just a UPS with a larger battery.

TABLE 2.9
Protectors for SLI
Batteries

Brand	Model	Current	Communications	Notes
Shenzhen Volt-cell technology/Shenzhen Dongruilong Industrial Development Co., Ltd.	VL4S350A-1000A	350A continuous; 450–1,350A for 5 seconds	None	

A UPS with a low-voltage battery may use many topologies, of which these are the most common:

- Online, also known as double conversion or voltage and frequency independent (VFI);
- Standby, also known as off-line or voltage and frequency-dependent (VFD).

These are described superficially below. A later chapter will go into depth into these and other topologies (see Section 4.2.4.2). Both use the AC + inverter topology (Figure 2.13(B)). It is different from the topology in an alternative energy home in which a single device, an inverger, handles both charging and discharging.

2.10.1.1 Online Topology

In an online UPS (see Section 4.2.4.4) (Figure 2.54(a)), the grid powers a charger that charges the battery and powers an inverter, which in turn power the AC loads.

The UPS operates as follows:

- *AC power present:* the battery is full; the inverter powers the AC loads.
- *AC power absent:* the battery continues to power the inverter; the AC loads are entirely unaffected by the power loss.
- *AC power returns:* the charger recharges the battery while also powering the inverter; therefore, the power of the charger must be high enough to power both the loads and charging.

The inverter type is off-grid because the AC loads are isolated from the grid. The inverter does not need to synchronize its output waveform to the AC grid waveform. It does not back-feed to the grid.

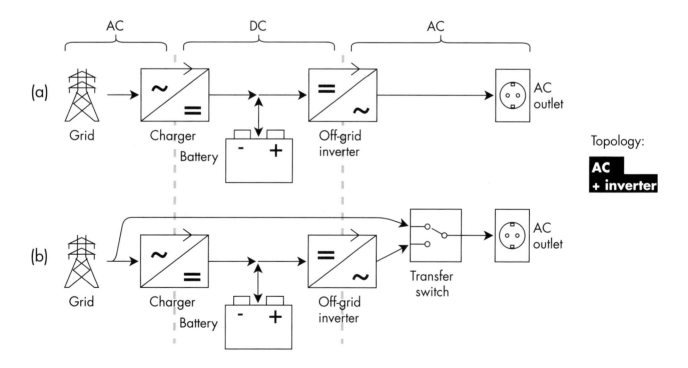

FIGURE 2.54   Uninterruptible power supply: (a) online, and (b) standby.

The AC voltage at the output of a UPS is unaffected by disturbances on the grid, including a loss of AC power; the inverter continues to operate as if nothing happened. The AC voltage from the inverter usually is cleaner than the AC power from the grid. However, cheap UPSs generate a squarish approximation to a sine wave. While the conversion is more efficient, many loads misbehave when powered by such a nasty waveform.

You may ask: Won't continuously charging and discharging damage the battery? The answer is that it won't, because, once the battery is charged, it sits idle. The battery current is zero because the inverter pulls from the charger only as much current as it needs, and the charger does not "push" current, so there is no excess current going to the battery (see Volume 1, Section 1.3.2).

You may ask: Can a battery charge and discharge at the same time? The answer is that the battery charges when it needs to be charged, and it discharges when there is no AC power; those occur at different times (see Volume 1, Section 1.4.1).

The UPS continues operating while a technician replaces its battery because the charger still powers the inverter. Of course, if there were an AC power failure during the battery replacement, the loads would not be powered.

2.10.1.2 Standby Topology

The online UPS topology is inefficient because the load is powered through two converters: the charger and the inverter. The standby topology (see Section 4.2.4.3) solves this by including a transfer switch[119] to select the source for the AC output (Figure 2.54(b)):

- AC power present: the switch selects the grid to powers the AC loads.
- AC power absent: the switch selects the inverter to power the AC loads.

If the load power is well characterized and it varies a lot, a smaller charger can be used, as it only needs to provide the average power; the battery provides the peak power; the inverter must be sized to power the peak power.

2.10.1.3 Battery Type

A UPS uses an energy battery. Normally, the battery is kept fully charged. During a power failure, the battery is discharged slowly. Given that the typical UPS has a discharge time of on the order of 1 hour (under the maximum load that its inverter can support), the battery uses either Li-ion energy cells or deep discharge lead-acid batteries.

2.10.2 Pole-Mounted Batteries

Pole-mounted batteries supply continuous power to low-power devices such as telemetry sensors, transmitters, radio repeaters, rural broadband WiFi hot spots, LED lighting, signage, remote cameras, cathodic protection, radar road speed signs, and rural traffic lights. Off-grid pole-mounted batteries are charged by a small solar panel. In a smart grid, they are charged by the grid itself, to power grid monitoring equipment even when the grid fails.

A weather-protected, vandal-resistant enclosure houses this battery. Typically, the enclosure and the solar panel are mounted on a pole, hence the polemount name.

---

119. Not to be confused with a static switch or a bypass switch. See Section 4.2.4.8.

The battery is 12V, 24V, or 48V, from 30 to 1,000 Ah. The solar panel is sized to match the battery to keep it charged. These batteries must endure a wide range of temperatures. Normally, this application uses a lead–acid battery; discharge should be limited to 50% to prolong battery life. Therefore, a 10-W load requires a 12-V, 100 Ah battery and a 40-W solar panel to account for overcast days.

Li-ion batteries are now becoming available, offering a lower maintenance cost. Four LFP cells in series are used in 12V batteries to match the voltage of a lead–acid battery; 24V and 48V batteries may also use other chemistries. Although LTO cells would be ideal thanks to their low-temperature performance, they are not selected due to their high cost.

A solar charge controller designed for a lead–acid battery is not a good match for a Li-ion battery because it continues to charge the battery when full, which reduces battery life. Instead, the solar charge controller's voltage must be well matched to the Li-ion battery. Charging must be disabled once the battery is full.

2.10.2.1 Grid-Tied, Pole-Mounted Batteries

A grid-tied, pole-mounted battery is different. The power utility mounts them on a power pole to buffer the grid and supply extra power at times of high demand and to provide short-term, local power in case of a grid failure. These batteries are much larger (on the order of 40 kWh) and high power [8].

2.10.3 Microgrid

A microgrid includes generators, distribution, control, and loads, just as a full-sized grid does. The only difference is that a microgrid does so on a much smaller scale. A microgrid may use high-voltage batteries (see Section 4.4) or low-voltage ones. In this chapter, we discuss the latter.

A microgrid that is completely low-voltage DC (Figure 2.55(a)) is like an off-grid telecom site (see Section 2.4.3). A microgrid with AC sources and loads (Figure 2.55(b)) is like an off-grid house (see Section 2.5.5).

A microgrid may be subdivided into multiple cells, connected through a high-voltage backbone, which can use DC (Figure 2.56(a)) or AC (Figure 2.56(b)). High voltage allows the use of smaller wires for a given amount of power transferred.

Each microgrid cell is likely to include a battery to supply or receive the power difference between supply and demand. Depending on the desired time scale, the

FIGURE 2.55
Simple microgrid
examples: (a) DC, and
(b) AC and DC.

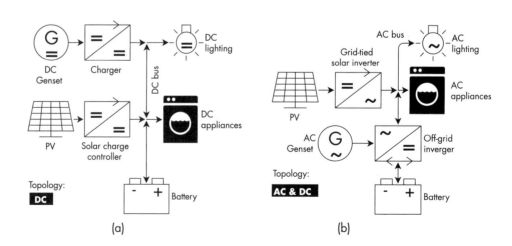

FIGURE 2.56
Multicell microgrid
interconnection:
(a) high-voltage DC, and
(b) high-voltage AC.

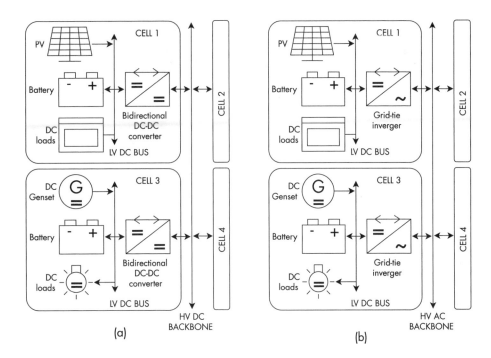

battery is used as a buffer battery, an energy battery, or both (see Volume 1, Section 5.1.4):

- Instantaneous compensation: a small buffer battery, kept at mid-SoC levels;
- Daily compensation (e.g., solar production in the day, high usage in the evening): a large energy battery, charged fully whenever possible, and significantly discharged when required;
- Backup: a battery whose size and performance matches the expected duration of the loss of charging power.

2.10.4 Battery Modules

These modules (Figure 2.57(a)) are not complete batteries. They are a component in a modular battery: two or more of them are combined in series and parallel to achieve a battery of a given size.

These modules do not include a complete BMS (they only implement a slave function) and have no protector switch. A master with a protector switch is added to the array of modules to complete the battery (see Volume 1, Section 6.2). Each module communicates the state of its cells to the master via a cable. Each module includes (Figure 2.57(b))

- Cells;
- A slave, powered by the cells, that monitors the cell voltages, temperature, and possibly the current; it also balances the cells;
- An isolated data link to the master;
- A fuse.

Modules do not use a circuit breaker because it would be dangerous; the user could turn it back on without checking if it's safe to do so, resulting in an inrush of current between modules.

FIGURE 2.57
Battery module:
(a) diagram, and (b)
valence U-charge module.

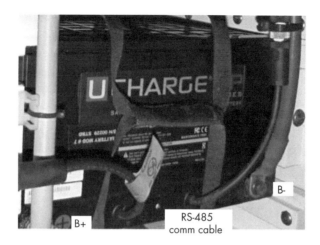

(a)

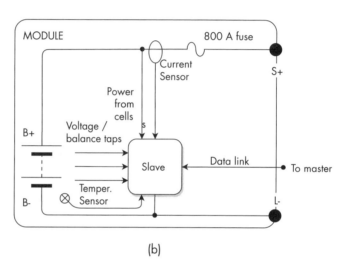

(b)

If properly used, these modules are quite effective and safe; yet they are often misused, making them frighteningly dangerous: people connect modules carelessly or skip the protector switch (see Section 5.2.2.2).

2.11 STRATEGIES FOR BATTERY SHUTDOWN

The ability of a Li-ion battery to shut down is good for its cells but not for the application:

- For telecom sites because they require 100% up-time;
- For solar charge controllers and invergers because they are powered by the battery and cannot restart without one (see Sections 2.2.3.4 and 2.2.5.2); complex workarounds are only partially effective.

In the next sections, we will explore strategies to

- Avoid shutdown in the first place.
- Find ways to handle the disruption should shutdown occur.
- Find ways to recover from a charge disable.
- Find ways to recover from a complete shutdown.

2.11.1 Avoid Shutdown

To keep a Li-ion battery from shutting down

- Use well-proven, reliable technology in a well-engineered engineered battery.
- Use the battery in a way that keeps its cells well within their safe operating area, to avoid:
  - Overtemperature or undertemperature: use climate control to keep the battery temperature within range.
  - Overcurrent: use individual circuit breakers for each source and load on the DC bus; if any one piece of equipment experiences a fault, it opens its breaker, avoiding an overcurrent shutdown of the battery.
  - Cell undervoltage or overvoltage.

In case of overvoltage, a dual-switch, single-port battery avoids a complete shutdown by only disabling charging while keeping discharging enabled. The following two sections assume a single switch battery, one that has to shut down that single switch in case of overvoltage, disabling both charging and discharging.

2.11.1.1 Overvoltage Shutdown, Balanced Battery

The most straightforward way to prevent the battery from turning off at the end of charge is to

- Top-balance the battery at the time of manufacture and then let the BMS keep it in balance.
- Configure the CV setting of all the charging devices to a slightly lower value so that, at the end of the charge, the cell voltages are below the cell voltage limit configured in the BMS[120].

When the battery voltage reaches the charger's CV setting, all the cells reach the same voltage simultaneously. The BMS does not shut down the battery because this voltage is lower than the maximum cell voltage limit. Lowering the maximum cell voltage has the added advantage of increasing the life of the battery.

2.11.1.2 Overvoltage Shutdown, Unbalanced Battery

While charging an unbalanced battery, the most charged cell reaches the maximum cell voltage, and the BMS shuts down the battery to avoid overcharging it (see Volume 1, Section 3.2.1). After the BMS removes some charge from this cell, it reenables the battery. This results in the battery turning off and on continuously until the battery is fully balanced.

There are two ways to avoid shutdown of an unbalanced battery:

1. Increase the balancing current.
2. Decrease the charging current.

The first approach is to increase the balancing current so that it exceeds the charging current; balancing is turned on and off so that the average of the balancing current matches the charging current (see Volume 1, Section 4.7.5.1). This requires a BMS able to bypass the entire charging current.

---

120. For example, with 16 LFP cells in series, configure the BMS for 3.6V and the charger for 3.4V × 16 = 54.4V.

High current balancing requires handling a lot of power, which is a problem, especially for bypass balancing. For example, for a 48V battery used in conjunction with a 100-A charger, this is a total of 5 kW of balancing power. While this is doable, it may be prohibitively expensive, and the required cooling may be excessive.

If the battery is nearly balanced, the charge current can be quite low: near the end of the CV phase, the current approaches zero asymptotically. Therefore, a BMS with a balancing current that is not as high as the maximum current that the charger can provide may be sufficient to avoid shutting down the battery during balancing.

The second approach is to decrease the charging current to match the balance current (see Volume 1, Section 4.7.5.2). In the unlikely case that the charging source allows external control of the output current, the BMS can instruct that source to reduce the charging current down to the same level as its balancing current. Both the charging current and balancing remain turned on continuously because the two currents are equal. A solar charge controller or an inverger may allow controlling the output current through messages on a communication link (see Section 2.13.3).

2.11.1.3   Undervoltage Shutdown

The possibility of an undervoltage shutdown may be minimized by

- Providing multiple, reliable charging sources;
- Using a large battery so that there is enough time for corrective action (in a telecom site, it may take days for a technician to reach the site);
- Turning on a genset when the SoC starts getting low;
- If possible, telling the loads to reduce the discharge current when the SoC starts getting low;
- Turning off noncritical loads when the SoC starts getting low;
- Turning off all loads before the battery shuts down;
- Using loads with a low-voltage cutoff with a higher cutoff voltage than the voltage where the BMS disables discharging.

For example, in a telecom site, it may be possible to disable noncritical loads first, while continuing to power critical loads as long as possible (Figure 2.58(a)). The Discharge OK line enables noncritical loads until 20% SoC, while the Critical discharge OK enables the critical loads until 0% SoC.

Turning off the loads, or using loads with low-voltage cutoff, keeps the battery from disabling discharge (Figure 2.58(b)). Devices that require a battery (i.e., solar charge controllers and invergers) remain powered, ready to restart charging when possible (i.e., when the sun returns, or when the grid comes back on) (see Section 2.11.4).

2.11.2   Shutdown Disruption Mitigation

There are a few ways to minimize the impact of a battery shutdown:

- Start some form of backup power before the battery shuts down.
- Power the loads with auxiliary storage during the short transition while the battery is off.
- Provide independent control of charging and discharging.

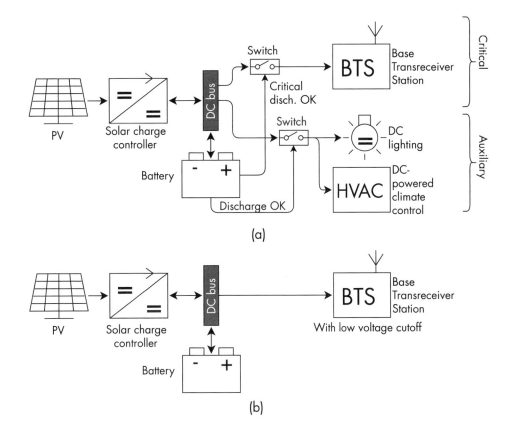

FIGURE 2.58
Load reduction:
(a) separate switches for
critical and noncritical
loads, and (b) low-
voltage cutoff.

2.11.2.1   Warning Starts Backup Power

If the BMS notifies the system that it is about to shut down the battery, the system may be able to prepare for it by turning on another power source, such as a genset. For example:

- The BMS issues a high-temperature warning.

- The system controller reacts by turning on a genset.

- The BMS issues a temperature fault and turns off the battery.

- The genset continues to power the loads.

2.11.2.2   Auxiliary Storage During Switching Transition

Under certain conditions, a single-switch battery may leave the load unpowered for a split second:

- The battery is full; its contactor is open to prevent overcharge.

- The power source suddenly goes away.

- The BMS takes some time (100 ms or so) to react and turn on the contactor to allow discharge.

During this transition, the load is not powered. Short-term storage may be added to the DC bus to power the load during the transition. One may assume that a large capacitor on the DC bus could provide this short-term storage (Figure 2.59(a)).

FIGURE 2.59
DC bus storage: (a) large
capacitor, and (b) small
lead-acid battery.

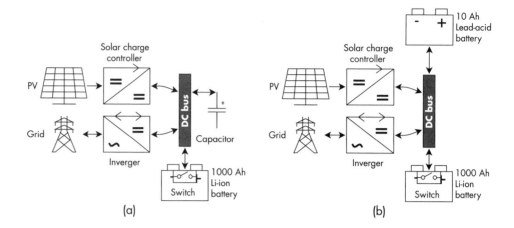

(a)   (b)

Holding the bus within 2V while powering a 100-A load for 100 ms requires a
5 F[121] capacitor. This is a large and expensive capacitor. It requires a precharge circuit.

A better solution is to connect permanently a small lead-acid battery to the DC
bus (Figure 2.59(b))[122]. This battery could consist of Cyclon small cylindrical spiral
cells with a 2.5-Ah capacity. The battery loses only 0.1% of its charge during the 100
ms during which it powers a 100-A load.

2.11.2.3   *Independent Control of Charging and Discharging*

A problem with a battery with a single switch is that when the switch is opened to
prevent overcharging, it also prevents discharge (see Volume 1, Section 5.12.2.1). If the
power source disappears suddenly, the BMS takes some time to react and turn on the
switch. During this time, the loads are not powered.

A solution is to use a dual-switch battery (see Volume 1, Section 5.12.2.2) or
a dual-port battery (see Volume 1, Section 5.12.2.3). When the battery is full, even
though charging may be disabled, discharging remains enabled; upon loss of the
power source, the battery is immediately able to power the loads.

2.11.3   Recovery from Discharge Disable

Should a system shut down because there is no source of power and the battery is
empty, when any source of power returns, the system must be able to recover on
its own. If this source is the sun, the system cannot recover because a solar charge
controller is powered by the battery. Similarly, if this source is the grid, the system
cannot recover because an inverger is also powered by the battery.

The system must include a different power source so that the solar charge
controller and inverger may recover automatically. Here are some possible solutions:

- Loads with low-voltage cutoff;
- A lead-acid battery on DC bus;
- A dual-port battery;
- An auxiliary, low-voltage power supply on the DC bus: powered by the battery
  or powered by the sun or the grid.

---

121. Farad.
122. This suggestion was not received well by people intent on moving away from lead acid: why go back to lead-acid?

While the following discussion only mentions a solar charge controller, it applies to an inverger as well.

2.11.3.1 Loads with Low-Voltage Cutoff

This solution assumes that all the loads include a low-voltage cutoff. Is this a reasonable assumption? It depends on the application:

- Telecom: whether BTS devices for telecom applications include a low-voltage cutoff is hard to say, as their specifications only list a minimum voltage and are silent about what the current draw is below this voltage.
- Residential: whether invergers and solar charge controllers include a low-voltage cutoff is irrelevant since these are the devices that we do want to power when the battery has disabled charging.
- Marine: loads in a vessel are unlikely to include a low-voltage cutoff.
- RV: certainly, some loads in an RV do not include a cutoff (e.g., lamps).

Still, let us assume that all the loads do include a low-voltage cutoff and that their threshold voltages are higher than the minimum voltage required to power a solar charge controller and an inverter, which, in turn, is higher than the minimum battery voltage.

When the battery voltage drops below a certain level, the loads shut down. As the battery is still turned on, it powers the solar charge controller. At this low current, the remaining battery charge lasts a long time. If the sun returns before the BMS shuts down the battery, the system recovers (Figure 2.60).

2.11.3.2 Lead-Acid Battery on DC Bus

This solution is the same, but there is also a 50-Ah lead-acid battery connected to the DC bus (Figure 2.59(b)). Even after the Li-ion battery shuts down, the lead-acid battery continues to power the solar charge controller a bit longer (Figure 2.61).

This solution helps if the minimum voltage of the Li-ion battery is higher than the minimum voltage of the solar charge controller. It doesn't help much, though: after the Li-ion battery shuts down, the bus voltage drops more rapidly because the small lead-acid battery is discharged faster than the larger Li-ion battery. Note that the solar charge controller is powered only for a while longer than the BMS in the Li-ion battery.

FIGURE 2.60
System recovery when the sun returns, using loads with low-voltage cutoff.

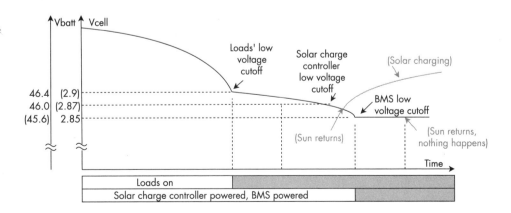

FIGURE 2.61
System recovery when
the sun returns, using
a lead–acid battery.

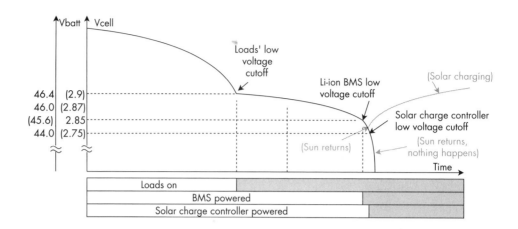

2.11.3.3   Dual-Port Battery

A dual-port battery powers a solar charge controller as long as there is even a little bit of charge left in the battery (Figure 2.62). The sequence is:

- Normally, both contactors are on; the solar charge controller charges the battery and powers the loads.
- If the battery is full, the BMS opens the charge contactor, preventing over-charge; the discharge contactor is still on, powering the load.
- At night, the battery continues to power the load; when the battery is no longer 100% full, the BMS turns on the charge contactor, which powers the solar charge controller.
- If the battery is nearly empty, the BMS opens the discharge contactor; the loads are no longer powered; yet the charge contactor remains on, continuing to power the solar charge controller.
- When the sun returns, the solar charge controller restarts because it is still powered.
- Should the sun not return, the solar charge controller slowly depletes the battery; eventually, the voltage is so low that a fault occurs and the charge contactor opens; the battery is bricked; only manual intervention with a separate power source can recover it.

This only works because the solar charge controller uses little power. It doesn't work with the typical inverger that includes a breaker panel for critical loads because they require the full battery current.

FIGURE 2.62
Recovery from
discharge disable using
a dual-port battery.

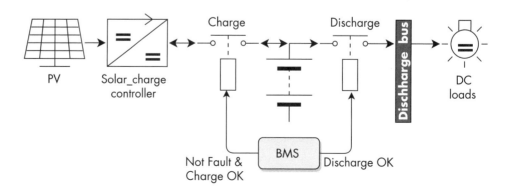

A dual-port battery with MOSFETs works a bit differently because it is physically impossible for the BMS to stop the current from exiting the battery through the charge port. Even after a low-voltage fault, the battery continues to power the solar charge controller. Eventually, the cells are overdischarged, and the battery is damaged. Therefore, do not use a dual-port battery with a MOSFET protection switch in a system that includes a solar charge controller.

2.11.3.4   Auxiliary Power Supply from Battery

After the BMS disables discharging, there is still a little bit of charge left in the battery, powering the BMS. It may be possible to use this charge to power the DC bus with a voltage that is too low to power the loads[123], yet high enough to power the solar charge controller[124].

This problem is hard to solve. Here I propose three workarounds. Admittedly, these are not elegant solutions; yet this is the best I have been able to devise.

A small DC-DC converter inside the battery can generate this voltage (Figure 2.63). For example, if the battery minimum voltage is 45V, the load minimum voltage is 24V, and the solar charge controller can operate down to 10V, the DC-DC converter can feed 12V to the bus. The typical solar charge controller consumes between 10W and 40W in standby, which, for a 48V battery, is up to 16 Ah overnight. A 1,000-Ah battery should be able to handle this.

The advantage of this circuit is that the solar charge controller is always powered. The disadvantage is that some load may try to power itself from the bus, overpowering the small DC-DC converter.

After some time, when a cell voltage drops too low, the BMS issues a low-voltage fault and removes power to the DC-DC converter and prevents further depletion of the battery charge. Recovery requires manual intervention.

The second solution assumes that, even if the battery is empty to the point that the BMS had to shut itself off, there may be enough charge left in the cells to power the solar charge controller. A DC-DC converter powers the solar charge controller for a short while, allowing it to start up. The time is limited to avoid overdischarging the cells. Unfortunately, once powered, solar charge controllers can take their sweet time to start producing power from the sun, as much as 5 minutes!

FIGURE 2.63
Powering a solar charge controller from a disabled battery with a low-voltage supply.

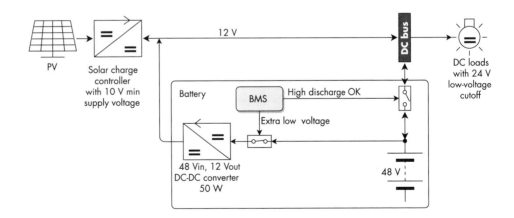

---

123. Assuming that they include low-voltage cutoffs that are therefore turned off.
124. This discussion mentions only a solar charge controller, though the same concepts apply for an inverger.

The converter is powered either just once when the sun returns, or a few times an hour. In the first approach, the DC-DC converter is turned on just once, when the sun returns, while the solar charge controller starts up. When the PV panels produce a voltage, a signal triggers a one-shot timer that powers the DC-DC converter for 1 minute or so (Figure 2.64(a)). Using a one-shot timer limits the discharge from the battery. Discharging the battery at such low power and for such a short time won't harm the cells.

This approach has several disadvantages:

- It is possible that there is enough sun to trigger the one-shot timer, but not enough to start the solar charge controller; if so, the next trigger may not come until the next day.

- It is also possible that the sun goes away soon after it triggers the one-shot timer.

- The BMS is off, so it can't disable the timer if the battery is completely depleted; therefore, over time, this circuit will overdischarge the battery.

- The battery is not self-contained (stand-alone) because it has an input to detect the presence of the sun.

- A load on the DC bus may overload the DC-DC converter; the resulting low voltage may not be able to power the solar charge controller.

- The duration of the pulse may be too short for the solar charge controller to start.

The third solution overcomes the disadvantages of the previous one by powering the solar charge controller with regular pulses, one of which may occur while there is sunshine (Figure 2.64(b)). Of course, this approach discharges the battery faster. Therefore, it is more likely to overdischarge it. The rate is low to minimize this danger, every 30 minutes or so.

2.11.3.5 Auxiliary Power Supply from the Sun or the Grid

The previous solutions use the little charge that may be still left in the battery. The following solutions use power from the charging sources instead. Ideally, the sun should power a solar charge controller, not the battery. Similarly, the grid or the battery, whichever is available, should power an inverger. Unfortunately, the designers of these products chose to power them from the battery.

These solutions work around this limitation by implementing the power supplies that are missing in those devices (Figure 2.65(a)):

- The PV panels power a small DC-DC converter, which, in turn, powers the solar charge controller through the DC bus; this allows the solar charge controller to recover on its own when the sun returns.

- The grid powers a small power supply, which, in turn, powers the inverger through the DC bus; this allows the inverger to recover on its own when the grid returns.

These solutions don't tax an empty battery because power comes from the charging source, not from the battery. The output voltage of these power supplies is too low to power the loads[125], but high enough to power the solar charge controller and the inverger. For example, if the power supplies generate 11 V, this voltage would

---

125. Assuming that they have low-voltage cutoffs that turned them off.

FIGURE 2.64
Powering a solar charge
controller from a disabled
battery: (a) power pulse
when the sun returns, and
(b) repeated power pulses.

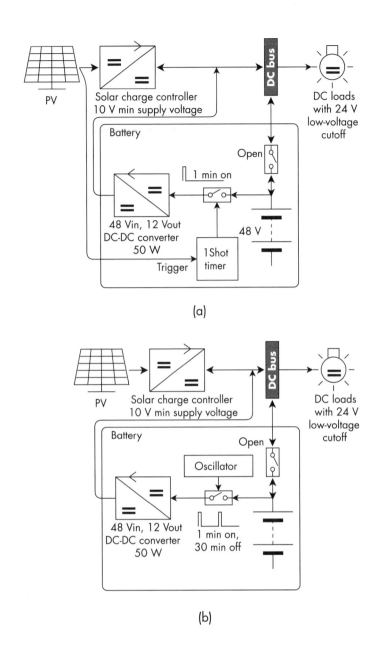

be too low to power the 48V loads, yet it would be high enough to power the inverter and the solar charge controller. Diodes in series with the output of the power supplies disconnect them from the DC bus when it is powered by the battery, the solar charge controller, or the inverger[126].

Alternatively, a small, low-voltage solar panel could power the bus directly when the sun returns, to power the solar charge controller (Figure 2.65(b)).

This solution can also be used with combo products that combine a solar charge controller and load control in a single unit, which is powered by the battery. When the Li-ion battery disables discharging, some combo products are no longer powered, even if there's sun (Figure 2.66(a)).

---

126. The diodes are reverse-biased because the DC bus is higher than the output voltage of the power supplies.

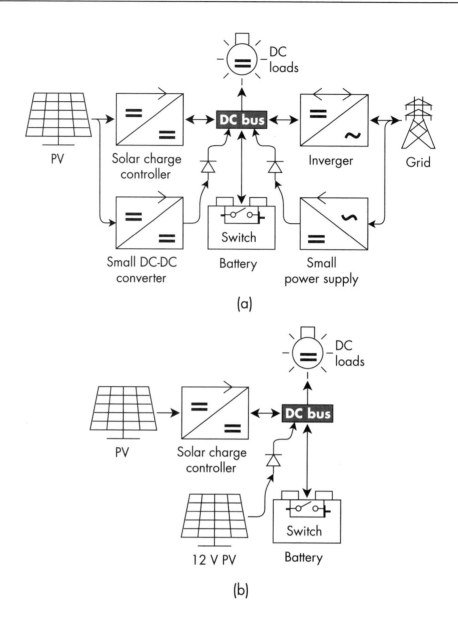

Adding a power supply (Figure 2.66(b)) overcomes the design choice made by the designers of these combo boxes. This small DC–DC converter powered by the PV panels adds the function that should have been included in those products. It powers the combo box so that it can recover when the sun returns.

2.11.4   Recovery from Complete Battery Shutdown

If the battery shuts down completely, the BMS is not powered, and charging is disabled. An external source must power the BMS so that the battery may recover.

This circuit allows the BMS to be powered by two sources, whichever one is available (Figure 2.67(a)):

1. The cells, but only until there is an undervoltage fault (Figure 2.67(b));

2. The battery terminals, when a charger is connected to the battery (Figure 2.67(c)).

FIGURE 2.66
Combo product:
(a) when the battery
disables discharging, the
system is off, even if
the sun is up. and
(b) solution using external
DC-DC converter.

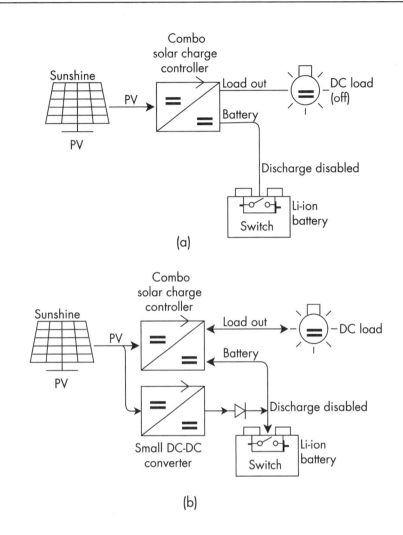

Two diodes allow whichever source has the highest voltage to power the BMS. At the same time, the diodes prevent a direct connection between the cells and the battery terminals. This method allows the battery to recover when a voltage is applied to its terminals. This may be any source that does not require a battery to start charging, such as a DC genset, a charger, or a lead-acid battery. A charger will not work if it requires CAN messages from the BMS before it start generating an output voltage; that's because the BMS is off.

2.12 BATTERY TECHNOLOGY

This section discusses technology inside the battery. The next section will discuss technology in the system (outside the battery).

2.12.1 Safety

The voltage of these batteries is low enough that they are not considered to be a significant shock hazard[127]. However, these batteries can produce an intense current, which can cause severe damage in case of a short circuit (e.g., a wrench dropped into a battery). This section lists some ways to minimize this danger.

---

127. Don't lick a terminal!

FIGURE 2.67
Recovery from complete
battery shutdown, using
two power sources:
(a) circuit, (b) during
discharge, the BMS is
powered by the battery,
and (c) during charge, it is
powered by the charger.

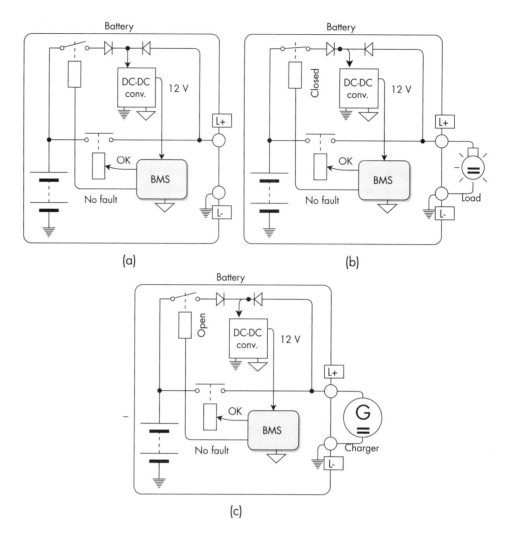

2.12.1.1 Fuses

A fuse may help in case of a short circuit does occur. If the fuse is in the path of the short circuit current, it protects the battery from further damage[128]. The most effective placement of a fuse is mid–pack because this is where it is statistically most likely to be inside the path of the short circuit current.

If a fuse blows, the full battery voltage appears immediately across the open fuse. This voltage may damage the BMS, depending on how it's wired. A BMS with two isolated banks, one on each side of the fuse, is not damaged. A bulletproof BMS that can handle any combination of miswiring won't be damaged either, regardless of the position of the fuse. Otherwise, the BMS will be damaged.

If a large battery consists of multiple modules in series, each module must have a fuse. The fuse must be rated for the lowest current of all of the following:

- The maximum continuous and peak current that the battery can deliver;
- The average current rating of the current–carrying components in the battery;
- The current handling of the wires (cables) connected to the battery;
- The maximum current the load can draw.

---

128. A short across a single cell is an example of a short circuit whose path does not include the fuse.

If feeding multiple loads, power each with its own circuit, each fused for the current of the particular wiring and load. For example, to feed a small DC-DC converter connected with 18 AWG wires, use something like a 1-A fuse, placed on the cells' end of the wire (not on the load end).

The fuse must also be rated for the voltage (do not use a 32V fuse on a 48V battery) and must be DC-rated. For parallel arrays, the fuse voltage must exceed twice the maximum bus voltage (e.g., 120V for a 48V system); if a battery module is connected backwards, its fuse will blow and the voltage of two batteries in series (twice the 48V) will appear across it; 100V may arc across a fuse rated for 60V!

The fuse must have a breaking capacity (also known as interrupting rating) that matches or exceeds that short circuit current. Otherwise, the fuse may not open fully in case of a short circuit. To get the short circuit current, divide the battery voltage by the total series resistance of the string of cells. Roughly speaking, this current is some 30 times the battery's rating (so, for a 300-A battery, the short circuit current is on the order of 10 kA).

Consider, for example:

- Auto-link type fuses:
  - LittelFuse BF2, 20A to 300A, 58 Vdc, 1-kA breaking;
  - LittelFuse Mega and Eaton AMG, 30A to 200A, 32 Vdc, 5-kA breaking.
- Concentric fuses, mounted on a single bolt:
  - Eaton CBBF, 30A to 300A, 50 Vdc, 2-kA breaking.
- Bolt-down, axial, blade fuses, 100A to 1,000A, at least 120 Vdc, up to 100-kA breaking:
  - LittelFuse TLS Series: Compact Current Limiting Telecommunications Fuse 170 Vdc, 1A to 125A.

2.12.1.2 Circuit Breakers

Circuit breakers are slower than fuses, but are resettable, and may be used as a safety disconnect. For example:

- Eaton CB184, CB185, 25A to 150A, 48 Vdc, single circuit, undefined breaking capacity;
- Other brands include Airpax, Carlin, E-T-A.

2.12.1.3 Isolation

A short circuit requires a complete path as does a continuous electrical shock. Floating a battery removes one segment of this path, preventing short circuits and electrical shocks (see Volume 1, Section 5.14). Therefore, large, low-voltage batteries should be isolated from earth ground for safety. Even though some applications connect a battery terminal to earth ground, the battery itself should be floated.

2.12.1.4 Tap Wires and Thermistors

Make sure that tap wires cannot contact other cell terminals in a way that cuts through their insulation; use small gauge wires so that they fuse easily, yet are large enough to carry the balancing current. Wire insulation rated for 150V is sufficient for these low-voltage batteries. The same goes for thermistor wires.

2.12.2   Grounding and Electrical Noise

Grounding of a battery affects safety, galvanic corrosion, and EMI immunity[129]; ground loops affect communication links (see Volume 1, Section 5.10.3). Large stationary batteries are connected to high-power electronics that generate strong electrical noise.

2.12.2.1   Grounding

The negative terminal of a battery is not ground, unless it is truly connected to earth ground. The battery should be floated, giving the system designer the freedom to decide whether or not the battery should be grounded and, if so, where. The string of cells in a battery may be floated (Figure 2.68(a)), or it may be grounded on the negative side (Figure 2.68(b)), or on the positive side (Figure 2.68(c)).

Electrically, it makes little difference whether or where a battery is grounded. Whether grounding is required, possible, or to be avoided depends on safety, maintenance, and radio interference issues.

Note that we are talking about grounding the string of cells, not the battery enclosure. A metal battery enclosure should be grounded for safety.

2.12.2.2   Electrical Noise Suppression

As a battery designer, you are not in control of the strong electrical noise generated by high-power products (such as chargers and invergers). All you can do is to try to attenuate that noise and minimize its effects. Besides the general remedies discussed earlier (see Volume 1, Section 5.16), the following solutions are applicable to large, low-voltage batteries.

For signal lines:

- Routing: do not run signal lines and digital links along high current bars and cables.
- Shielding: try running the signals through shielded cables; ground the shield only at one end and do not let it contact any other metal.
- Isolation: add a bus isolator in line with a CAN bus or other digital link; add relays, opto–isolators, or small solid state relays in line with digital level control lines.

For power lines:

- Filters: try running the high power lines through a low-frequency, high-power filter[130].

FIGURE 2.68
Battery grounding:
(a) none (float), (b)
negative, and (c) positive
(industry standard).

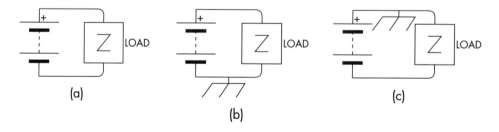

(a)   (b)   (c)

---

129. Electromagnetic interference. See Volume 1, Sections 4.12.3 and 5.16.

130. Such as FMER SOL from Schurter (https://www.schurter.com/datasheet/FMER_SOL), CNW filters from REO (https://www.reo-usa.com/us_en/products/emc-filter/high-current-filter.html), FD series from Curtis instruments (https://www.curtisind.com/products/rfi-power-line-filters/dc-filters/).

- Common mode transformer: run the high-power lines through magnetic cores[131]; this is not as effective as a filter, but it is much cheaper.
- RF chokes: ferrite beads on power cables are not effective because they add resistance in the 100-MHz range, while the switching frequency is in the 20-kHz range (with high-frequency harmonics and ringing in the 1-MHz range).
- Shielding: run long power connections through shielded power cables, like the ones used in electric vehicles.
- Grounding: while respecting safety concerns, try playing with how the various devices are grounded and to what; some have had success replacing a bearing in a motor with an isolated bearing.

2.12.3 Cell Selection

In most cases, large, low-voltage batteries use large prismatic cells (see Volume 1, Section 2.2.2.3). It is more practical to connect multiple 100-Ah or 200-Ah cells in parallel rather than using 400-Ah and 1,000-Ah cells. Small prismatic cells that meet European standards for electric vehicles are now entering the market (see Volume 1, Section 2.2.2.4). Large cylindrical cells may be considered, even though they do not offer sufficient advantages in these applications. Small cylindrical cells are less than ideal because they require too much labor to reach the desired capacity. Pouch cells should only be considered for high-volume applications to justify the upfront design time. However, blocks of pouch cells in ready-made frames are a good option.

2.12.4 BMS Selection

Many off-the-shelf BMSs designed expressly for low-voltage batteries are offered today. They range from inexpensive Chinese centralized protectors to sophisticated solutions for large arrays. Chinese analog protector BMSs offer great value. For a demanding application, consider an advanced BMS that is configurable, uses sophisticated algorithms, and offers extensive communication options.

Unless otherwise noted, Tables 2.10 and 2.11 list BMSs for 16 cells in series because this is what is recommended for a 48V LFP battery. All of the listed manufacturers also offer BMSs for fewer cells, in case your application uses less than 16 cells in series.

2.12.4.1 Analog Protectors (PCMs)

Table 2.10 lists some of the available analog protector BMS (PCMs) (see Volume 1, Section 4.3) for low-voltage batteries. Features in parenthesis are optional. When a slash separates two values for current, the first value is for charging.

2.12.4.2 Digital Protectors

Table 2.11 lists some of the available digital protector BMS (see Volume 1, Section 4.4.1) for low-voltage batteries. Unless otherwise noted, all handle one or more thermistors. Features in parenthesis are optional. When a slash separates two values for current, they are for charging and discharging, respectively. Consider also a semi-custom BMS from Clayton Power.

---

131. Such as Cool-blue inductive absorbers, Nanoperm rings from Magnetec: http://www.coolblue-mhw.com/

TABLE 2.10
Analog Protectors for
Low-Voltage Batteries

Brand	Model	Current	Notes
Ayaa	BMS-LB16S150-1501	100A	Thermistors
Ayaa	PCM-L08S60-537	20/70	13 cells
BestTechPower	HCX-D230, HCX-D140, HCX-D174, HCX-D272	15	
BestTechPower	HCX-D245	20	
BestTechPower	HCX-D132	25	
BestTechPower	HCX-D100, HCX-D107, HCX-D163, HCX-D167, HCX-D268	35	
BestTechPower	HCX-D166	50	
BestTechPower	HCX-D131	50/80	
BestTechPower	HCX-D276	55	
BestTechPower	HCX-D266	60	
BestTechPower	HCX-D083, HCX-D223V1	80	
BestTechPower	HCX-D138V1, HCX-D170	100	
BestTechPower	D170	200	
EVPST	EVPST-BMS-3	40	Capacitor balancing
Leadyo	PCM-L13S20-A22	5/20	13 cells
Leadyo	PCM-L14S25-424 (A-1)	25	14 cells
Qwawin	PCM-L16Sxxx-xxx	12 to 350	25 models
SuPower Battery	60V 16S xxA Lion	30 to 150	Many models
XJR Technology	BMS-WH16S1515LI10065	15	
XJR Technology	BMS-XTM16S0520LI145	20	Direct on pouch cells
XJR Technology	BMS-LH16S3030LI6957, BMS-WH16S3030LI14260	30	

2.12.4.3 Digital BMUs

These digital BMUs (Figure 2.69) are specifically designed for large, low-voltage batteries.

Table 2.12 lists some of the available digital BMUs for large, low-voltage batteries. Unless otherwise noted, they all use a centralized topology, include SoC evaluation, balancing, and temperature sensing. Features in parentheses are optional.

In addition to these BMSs, also consider BMUs for higher voltage listed in the following chapters; however, those may be more expensive and not feature the communication links desirable in a large, low-voltage battery.

For higher balancing current, you may add a stand-alone equalizer, which is a capacitor-based charge transfer balancer, available from Chinese companies[132]. Capacitor-based charge transfer balancing doesn't work well, as it is at best 50% efficient and take an infinite amount of time to balance (see Section A.5.2.2).

2.12.4.4 Battery Array BMUs

Table 2.13 lists off-the-shelf BMUs are designed specifically for an array of low-voltage batteries connected in parallel to a DC bus. Each battery requires its own BMS. The array also requires a master to manage the batteries as a single unit.

---

132. Unknown parentage, sold by IC GOGOGO on AliBaba.

Brand	Model	Cells	Current	Communications	Notes
BestTechPower	HCX-D270	8 to 16	25	I2C, RS485, RS232	
	HCX-D338	6 to 20	80	SMBUS, UART, RS485, CAN	
	HCX-D328	6 to 16	100	RS485, RS232, CANBUS, SMBUS, I2C	
	HCX-D170V1	6 to 96	2001350	(SMBUS, I2C, LCD)	(Contactor)
Energus	S516	4 to 16	301600	(Bluetooth, USB, CAN)	GUI
Leadyo	PCM-L13S25-B37(A-2)	13	30	USB	
Ligoo Anhui Wicom New Energy Tech.	YD-UP-21	12 to 16	40	RS232, Ethernet	Enclosed. For telecom 48V only
RoboteQ	BMS1060	11 to 15	100	CAN, RS485, USB	
Shenzhen Dongruilong Industrial Development Co., Ltd./Shenzhen Volt-cell technology	7S15A-80AH090065010S	7	15/80	None	
	6S100AF126062023SCD-Canbus	16	100	CAN	
	10S30AH065035015S	10 to 13	5/30	None	
Shenzhen Smartec	PCM-L15S60-D15	15	60	Bluetooth	
	PCM-L15S30-D45	15	30	RS232, Bluetooth	
	PCM-L15S20-B41 (A-3)	15	29	Bluetooth	
XJR Technology	BMS-LH16S5050LI9587		20 to 60	UART, Bluetooth	
	BMS-LH16S150150LI1814S		80 to 150	USB, Bluetooth	

TABLE 2.11   Digital Protectors for Low–Voltage Batteries

FIGURE 2.69
Digital BMUs for low-voltage batteries:
(a) Elithion Vinci LV,
(b) REC Q16S (courtesy of REC BMS), and (c) Ewert Orion Jr (courtesy of Ewert Energy).

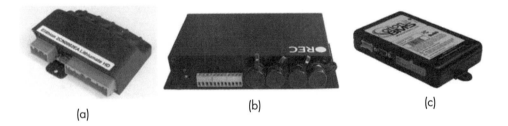

(a)   (b)   (c)

2.12.5   Battery Circuits

The design of large, low-voltage Li–ion batteries requires some special considerations.

2.12.5.1   BMS Power Supply

In a low-voltage battery, the method of powering the BMS depends on its type:

- PCM (Figure 2.70(a));
- Directly powered BMU (Figure 2.70(b));
- BMU with isolated DC-DC converter (Figure 2.70(c)).

Eventually, a PCM overdischarges the cells because it is powered directly from them (see Volume 1, Section 5.8.1); recovering from a BMS shutdown is nearly impossible because there is no other way to power it. In any case, recovery would be dangerous as, by then, the cells are overdischarged.

A BMU may be powered through two diodes, one from the cells and one from the terminals, which allows recovery after the BMS shuts down. In case of undervoltage, a relay disconnects power from the cells and prevents overdischarge.

Brand	Model	Additional Features
Electrodocus	**SBMSO+DSSR20**	**Shunt Sensor, Doesn't Support Contactor**
Elithion	Vinci LV	GUI, CAN, isolated, dedicated lines (RS485 ModBus, Ethernet ModBus, WiFi, XanBus), contactors, MOSFETs, latching relay driver, precharge, ground fault detection, Hall effect, shunt current sensor (charge transfer balance, lead acid)
Ewert Energy Systems	Orion JR	Hall effect (CAN)
Ligoo	EK-FT-12	Master/slave topology, isolated, buzzer, 2-A balancing, Hall effect current sensor, LCD, CAN, RS485
Lithium Balance	LiBAL s-BMS	CAN, RS232, dedicated lines; contactor driver; current shunt, certified
Nuvation	Low voltage	Ground fault detection; RS485 ModBus, Ethernet ModBus, contactor's driver
REC	Active BMS	4 cells (12V) only; charge transfer balancing
REC	Q16S	CAN, RS-485, contactor's driver; no temperature sensing (precharge)
Volrad	V-IQ 12; V-IQ 24	12: 6 to 12 cells, four temperature sensors; 24: 12 to 24 cells, eight temperature sensors; CAN, ModBus; Hall current sensor; 1.2-A balance

Brand	Model	Notes
Elithion	Vinci LV	One Vinci LV per battery, plus an array master
Nuvation	Low voltage	One low voltage BMS per battery, plus a grid battery controller

Powering an isolated BMU with an isolated DC-DC converter isolates the cells from any low-voltage control circuit. If there are no signal or communication lines, then there is no need to isolate the BMS power supply. Table 2.14 compares the three solutions.

2.12.5.2   Protector Switch, Precharge

The battery must include a protector switch (see Volume 1, Section 5.12). It should include precharge to prevent an inrush when first connected to a capacitive load (see Volume 1, Section 5.13). Yet very few batteries and practically no loads include precharge. In most cases, there is no precharge, or if there is, it's not effective.

2.12.5.3   Circuit with a Basic BMU

This circuit for a high-power, single-port, high-current, low-voltage battery uses an isolated BMU that implements just the basic BMS functions (Figure 2.71).

The BMU is assumed to have the following outputs capable of driving contactor coils:

- Charge OK: normally open, grounded when charging is allowed;
- Discharge OK: normally open, grounded when discharging is allowed;
- Not fault: normally grounded, open when there's any fault, but especially in case of cell undervoltage;
- Precharge: optional; normally open, grounded during the precharge phase.

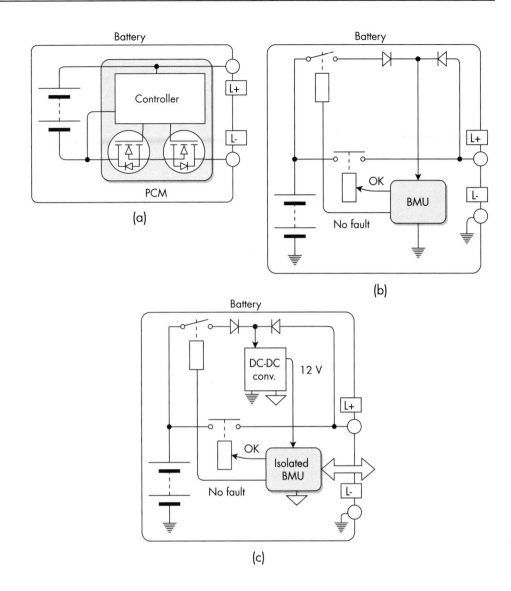

FIGURE 2.70
Power supply:
(a) PCM, (b) directly
powered BMU, and
(c) BMU with isolated
DC-DC converter.

This circuit has the following characteristics:

· It controls charging and discharging independently.

· It has an isolated data link.

· It can be turned off entirely by pressing the stop button (e.g., before transportation).

· Then it can be woken up by pressing the start button; yet it prevents a user from forcing the battery on by holding down the start button.

· It optionally includes precharge.

· It optionally includes a way to power an inverter or solar charge controller.

Principles of operation:

· An isolated 48V to 12V DC-DC converter (MOD1) powers the BMU and the relays; it is powered either by the string of cells or by a charger connected to the output terminals:

  · When the cell voltages are OK, the converter is powered through relay K3 (Not fault), normally closed switch SW2 (Stop), and diode D3.

TABLE 2.14
BMS Power Supply
Comparison

	Power	Isolation	Overdischarge	Shutdown Recovery
PCM	Directly from cells	None	Possible	Very hard
BMU, direct power	From cells or terminals	Unlikely	Protected	Power to terminals
BMU + DC-DC	From cells or terminals	Possible	Protected	Power to terminals

FIGURE 2.71
Low-voltage battery
using a generic BMS.

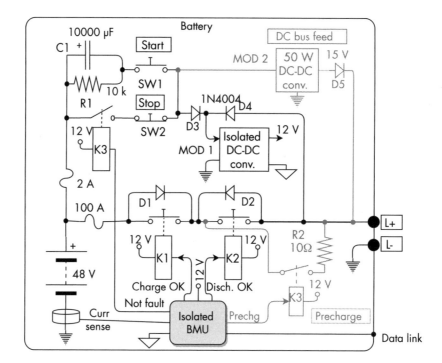

· When a charger is connected to the battery, the converter is powered through diode D4.

· The two diodes, D3 and D4, power the converter with whichever voltage is higher.

· The converter generates a 12V power supply that is isolated from the battery.

· Optionally, when discharging is first enabled, the battery precharges the load:

  · The BMU turns on relay K3.

  · The battery charges the load capacitance through D1, K3, and R2.

  · The BMU monitors the current with its current sensor.

  · When the current drops to a safe level, the BMU turns on the contactors K1 and K2 and then turns off K3.

· The BMU can control charging or discharging separately:

  · Normally, both contactors are on, and the string of cells is connected to the battery terminals

  · If the BMU decides that the battery cannot charge, it opens the Charge OK output; contactor K1 opens, and the battery can discharge through D1 and K2.

  · If the BMU decides that the battery cannot discharge, it opens the Discharge OK output; contactor K2 opens, and the battery can charge through D2 and K1;

the BMU remains powered because the fault relay (K3) is still closed; optionally, DC-DC converter MOD2 powers the DC bus with a low voltage, too low for the loads, but high enough to power a solar charge controller or inverger; diode D5 keeps the bus voltage from back-feeding into the output of the DC-DC converter when the battery is fully on.

- If the BMU decides that the battery can neither charge nor discharge, it opens both outputs; contactors K1 and K2 open, stopping the battery current in both directions.

- If the battery is discharged completely, the BMU issues a fault, which turns off relay K3; this, in turn, removes power to the BMU, and the battery goes to sleep.

- By pressing the Stop button (SW2), the user puts the battery directly to sleep (e.g., before transportation to a new site):

  - Doing so removes power to the DC-DC converter, shutting down the BMU and therefore the battery.

  - As there is no power, relay K3 opens; therefore, power is not restored when the user releases the stop button.

- To wake up the battery, the user may connect it to a charger:

  - The charger's voltage powers the DC-DC converter (MOD1) through D4; the BMU wakes up, turning on contactor K1; the charger can then charge the cells through rectifier D2 and contactor K1.

  - Alternatively, to wake up the battery, the user may press the start button (SW1):

  - Capacitor C1 charges as it powers the DC-DC converter for an instant; the voltage pulse applied to the DC-DC converter is enough to power it for just long enough for the BMU to wake up and turn on relay K3.

  - Afterward, the converter is powered from the cells through relay K3 and diode D3.

  - However, if there is a fault, the BMU does not turn on K3, and the battery shuts down again.

  - Continuing to press SW1 has no effect because capacitor C1 produces a voltage pulse is of limited duration; this prevents a user from forcing the battery on by holding down the start button.

  - Once the user releases the start button, capacitor C1 is slowly discharged through resistor R1; afterward, the functionality of the start button is restored.

  - After the battery is charged for a bit, the cell voltages are sufficiently high for the BMU to enable discharging; the BMU turns on K2, bypassing diode D2 and therefore stopping it from wasting power.

Component selection:

- The 12V DC-DC converter must be isolated and must generate enough current for the contactor coils, the BMU and other loads.

- The rectifier diodes D1 and D2 must be able to handle the continuous discharging and charging currents, respectively; they usually are not expected to have to dissipate high power because, typically, they only are used for a short time: D2 is only used until the battery is charged enough to allow discharging, and then K2 is turned on, bypassing D2; same for D1 and K1; use either two chassis mounted rectifier diodes or just a single common cathode diode pair (see Volume 1, Section 5.12.5.2).

- SW1 and SW2 could be integrated into a single component, a center off momentary switch, normally closed on one side and normally open on the other side.

- The value of capacitor C1 must be significantly higher than the capacitance on the input of the 12V DC-DC converter; otherwise, the voltage pulse won't be high enough to power the converter.

- The value of precharge resistor R2 is not critical; 10Ω is typical (see Volume 1, Section 5.13.5).

- The precharge relay, K3, only needs to be able to handle a low current; a 10-A relay is usually sufficient.

2.12.5.4   Circuit with Specialized BMU

In the previous sections, I described how to add functions to a generic BMU to make it suitable to a low-voltage application. A BMU explicitly designed for such applications includes most of those functions, simplifying the wiring significantly (Figure 2.72).

The BMU includes

- A small DC-DC converter to power itself, fed by either the external terminals of the battery or the internal string of cells (disabled if any cell is completely empty);

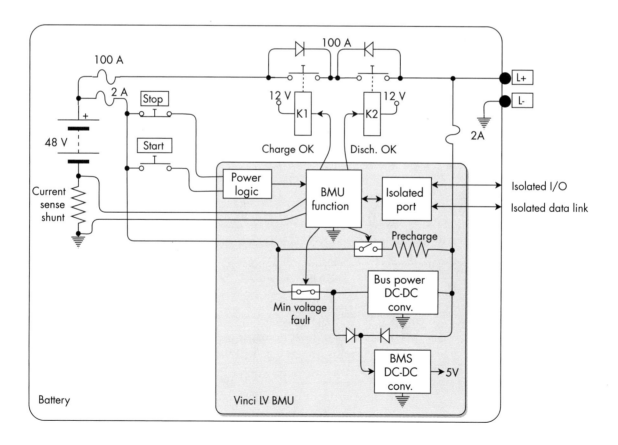

FIGURE 2.72   Low-voltage battery using a specialized BMS.

- Support for start and stop buttons;
- A timer to prevent the user from forcing the BMU on by pressing the start button continuously;
- A precharge circuit;
- A small DC-DC converter to power a solar charge controller or inverger;
- A data link, isolated from the battery.

Besides the BMU, all that is required to make a complete battery are the cells, power components, fuses, and push-button switches.

2.12.6 Traction Battery Conversion to 48V

A hobbyist may try to convert a used traction battery to 48V for residential solar applications. This project is impractical because the traction battery is 360V so it must be disassembled and the cells rearranged to achieve 48V. However, this is either too messy or too expensive, depending on the cell arrangement:

- Series-first is too expensive (Figure 2.73(a)); the BMS needs to monitor each and every cell, and be compatible with strings in parallel; a BMS that can handle a 13S8P battery costs about $3,000 to $4,000, more than the recycled traction battery (see Volume 1, Section 3.5.1).
- Parallel-first is too messy (Figure 2.73(b)); many wires must connect adjacent cells directly in parallel; the wires must be large enough to handle the total battery current, in case there is a weak cell (see Volume 1, Section 3.4.1).

FIGURE 2.73
Traction battery reuse options: (a) series-first (expensive), and (b) parallel-first (messy).

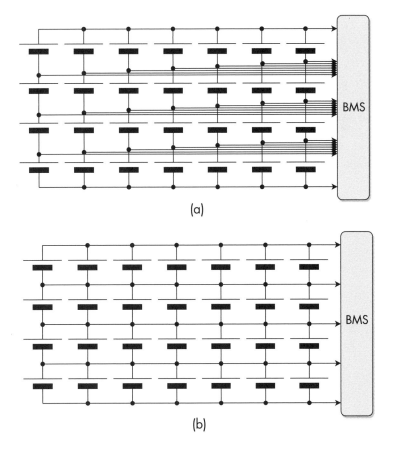

In any case, who wants to lug down to the basement a 400-kg traction battery that looks like an assembly from an armored combat vehicle?

Nissan and Sumimoto formed a joint venture, 4R Energy Corporation, to find second uses for Nissan Leaf EV traction batteries. A significant application is 48V batteries. We received a used Leaf battery and rearranged the modules into a 6S8P arrangement. The Nissan BMS could be reused as it is, as a single series string. Instead, we used our BMS, which supports multiple strings in parallel (Figure 2.74).

The project cost far more than the value of the used battery. This solution is not economically viable: the labor is too intensive, the BMS adds too much to the cost, and, in the end, the quality of the previously used cells is an issue.

The project proved that reconfiguring a traction battery into a 48V battery is uneconomical. Instead, these traction batteries should be used as is, for applications that use a 360V bus. This requires knowledge of the proprietary communication interface for the battery, which Nissan did not share with my company.

Next, since we had access to harvested Nissan modules, we decided to create a brand new 48V battery array (Figure 2.75). It consists of eight battery modules, each one light enough to lug down to a basement.

The eight batteries are connected directly in parallel to form an array (see Volume 1, Section 6.3). Each battery is stand-alone and can be used on its own. If you want to try reusing Nissan Leaf modules, Batrium makes cell boards that fit them perfectly.

2.13 SYSTEM INTEGRATION

While the previous section discussed technology inside the battery, this section discusses technology in the system outside the battery, including the battery interface.

FIGURE 2.74
Nissan leaf traction battery reconfigured for 48V:
(a) complete battery,
(b) harvested module,
(c) modules rearranged as 6S8P, and (d) new BMS compatible with strings in parallel.

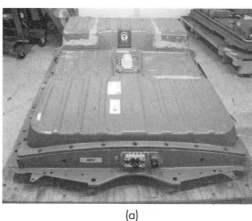
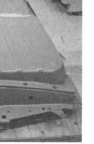
(a)

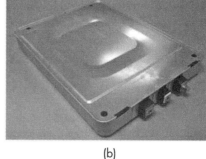
(b)

(d)

(c)

FIGURE 2.75
Battery array using
harvested Nissan Leaf
battery modules: (a) stack
of modules with voltage
sensing harness, and
(b) complete battery.

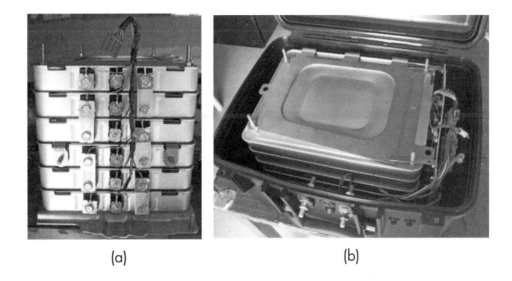

(a)   (b)

2.13.1 Protection Through External Switches

The protection switch could be implemented by the system, outside the battery, using the external switch topology (see Volume 1, Section 5.12.2.4).

Switching devices are distributed throughout the system, some interrupting current from the charging sources, some interrupting current to the loads. For example, in a residential application (Figure 2.76), contactors are placed on each charging source (e.g., a solar inverter, a genset, the main panel from the grid) and controlled by the Charge OK signal from the BMS in the battery. Similarly, contactors are placed on loads and controlled by the Discharge OK signal from the BMS. For safety, the battery must include a contactor that the BMS can open in case of fault or if the system disobeys the BMS.

Bidirectional devices (devices that can be either a charging source or a load) cannot be controlled this way. With an inverger, there is no way for an external switch to control charging from the grid independently from discharging to the grid.

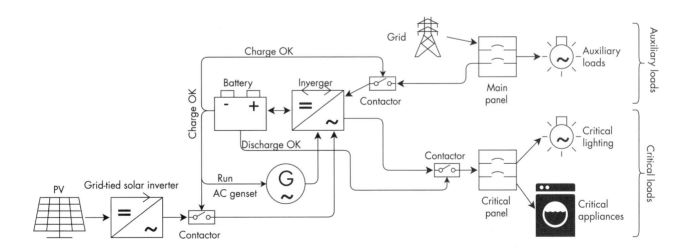

FIGURE 2.76   Residential application with external switches.

Contactor coils rated for low-voltage DC may be powered by the same supply that powers the BMS (Figure 2.77(a)), or by the battery voltage. If the latter, to maintain battery isolation, a helper relay is required (Figure 2.77(b)); the BMS drives the coil of the helper relay, and the contacts of the helper relay drive the coil of the contactor.

2.13.2 Protection Through External Control

The BMS may be able to control each device connected to the battery through dedicated lines or communication links using the external control topology (see Volume 1, Section 5.12.2.5). Bidirectional devices may be controlled this way as well.

For example, in a residential application (Figure 2.78), the battery uses a CAN bus to talk to the inverger and control its current in either direction. Note how much simpler this system is than the one that uses external switches (Figure 2.76).

Once more, for safety, the battery must include a contactor to be opened in case of fault or if the system disobeys the BMS.

A large battery may communicate with the system controller in a few different ways, including a digital data link, wireless communications, or even plain dedicated digital or analog lines.

Typically, the system designer chooses a suite of high-power products (such as an inverger and a solar charge controller) based on performance, rather than based on the communication protocol. Once this selection is made, the power converters determine the communication protocol for the entire system.

Ideally, a BMS should be directly compatible with these power devices (Figure 2.79(a)). Otherwise, a translator is required between the BMS and the power devices, which can be cumbersome. This is an example of such as convoluted communication (Figure 2.79(b)): BMS-CAN bus to ModBus gateway, to the ModBus-Conext ModBus Converter, to the XanBus-Schneider Electric Conext XW inverger[133].

2.13.3 Communication Links

A BMS may communicate over a serial link such as RS-485, RS-232, Ethernet, CAN, or super-sets of these (see Volume 1, Section 5.10).

FIGURE 2.77
Isolated contactor drive:
(a) direct, and
(b) through helper relay.

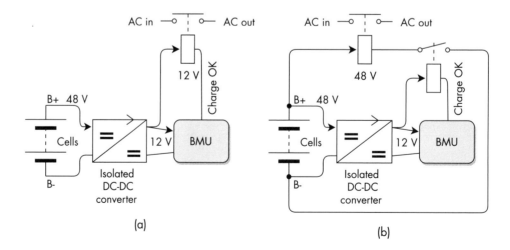

(a)   (b)

133. XanBus and Conext are trademarks of Schneider Electric.

FIGURE 2.78
Residential application
with external control.

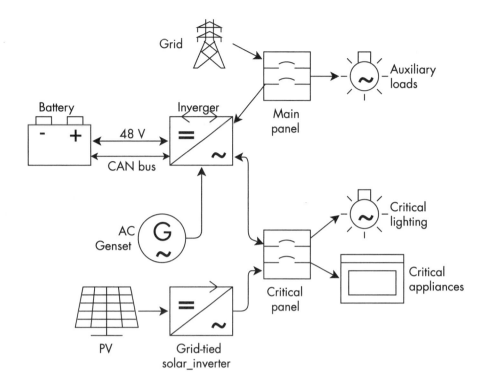

FIGURE 2.79
Example of
communication between
a BMU and a Schneider
Electric Conext XW
inverter: (a) convoluted
translation, and
(b) direct: the BMU
speaks the same language.

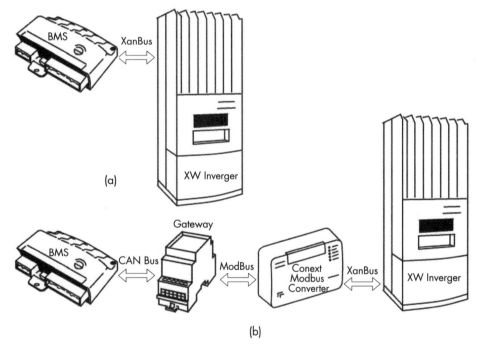

2.13.3.1 Physical and Data Layer

Some products expand standard buses. Schneider Electric's XanBus adds a 15V supply
to CAN, specifies a data rate (125 or 250 kbps), specifies an RJ45 connector:

- Pin 5 — CAN_H; 4 — CAN_L;
- Pins 1, 2, 7 — NET_S = +15 V supply; Pins 3, 6, 8 — ground.
- SMA places two CAN buses on the same cable, specifies an RJ45 connector:

- Main CAN bus — Pin 4: CAN_H; Pin 5: CAN_L;
- Sync CAN bus — Pin 3: CAN_H (SYNC_H); Pin 6: CAN_L (SYNC_L);
- 2 — Ground.

Outback power defines a connector for RS485.

2.13.3.2 Application Layer

The application layer specifies how the data are structured. Many devices use a standard protocol, while other devices use proprietary, CAN-based protocols. For example:

- Schneider Electric uses a proprietary protocol, XanBus, based on the NMEA 2000 standard [9]; each communication is quite verbose and requires multiple CAN messages to communicate even the simplest data.
- MasterVolt uses a proprietary protocol, MasterBus, also based on the NMEA 2000 standard.
- SMA uses a more reasonable protocol: the BMS needs to send only three messages and only in one direction.

Few BMS support these protocols because some of these companies make it rather difficult for BMS designers to develop such support. Therefore, these companies offer devices that translate from their proprietary protocol to a standard one, such as ModBus. However, their functionality may be limited. For example, there may not be a way for the BMS to control the power device to reduce or stop battery current.

References

[1] D'Antonio, S., "The Ins and Outs of Shore power Transformers," *PassageMaker*, www.passagemaker.com/technical/the-ins-and-outs-of-shorepower-transformers.

[2] American Boat & Yacht Council, abycinc.org.

[3] Electric Shock Drowning Prevention Association, www.electricshockdrowning.org/.

[4] BlueSea Systems, "DC Circuit Protection," www.bluesea.com/support/articles/Circuit_Protection/98/DC_Circuit_Protection.

[5] "The Final Voyage of Sandpiper — A Cruising Couple Must React Fast to Save Themselves When an Onboard Fire Quickly Envelops Their Sailboat About 40 Miles Offshore," https://seminar.stormtrysailfoundation.org/the-final-voyage-of-sandpiper.

[6] "Diesel APU EcoPower 3500," https://www.mcc-hvac.com/wp-content/uploads/2014/11/APUs.pdf.

[7] Boeing, "Batteries and Advanced Airplanes," 787updates.newairplane.com/787-Electrical-Systems/Batteries-and-Advanced-Airplanes.

[8] "Toronto Hydro Installs Pole-Mounted Battery," T & D World, https://www.tdworld.com/asset-managementservice/toronto-hydro-installs-pole-mounted-battery.

[9] National Marine Electronics Association, "Standard for Communication Between Electronic Equipment in Marine Vessels," www.nmea.org/.

CHAPTER 3

TRACTION BATTERIES

3.1 INTRODUCTION

A traction battery propels an electric vehicle (EV), which could be anything from a radio-controlled model car to an industrial tractor, from an airplane to a submarine, from a self-driving lawnmower to a Hyperloop train. Electric traction is poised to overtake internal combustion engine traction in land vehicles. Experimental off-land vehicles are starting to use electric traction as well.

The traction batteries discussed in this book range from 300 Wh to 300 kWh and have a voltage ranging from 12V to 700V (see Section 3.2.1.1). Today, Li-ion is the best battery technology to power electric traction.

3.1.1 Tidbits

Some interesting items in this chapter include:

- Regardless of the vehicle, the maximum load on a battery is ~2Ω (Section 3.2.1.1).
- Thou shall not murder batteries, except that UAVs[1] have a license to kill (Section 3.5.2).
- Solar-powered passenger cars will always remain a dream (Section 3.2.5.1).
- All motors are AC motors, even DC motors (Section 3.3.3.1).
- Drinking tea inspired a Japanese charging technology (Section 3.9.4.5).
- Plugging in an EV for charging can reduce pollution from the power company (Section 3.9.1.8).
- The rules for student races would strain the patience of a saint (Section 3.10.4).
- The Hyperloop contest is pushing technology away from an eventual vacuum train (Section 3.11.4.1).

3.1.2 Orientation

This chapter starts with technical considerations for traction batteries and a discussion of some of the devices and topologies typically used in electric vehicles. It then covers a range of applications, discussing practical issues that arise in each when using Li-ion and gives possible solutions. Applications are divided into sections in order of size:

- Unmanned: robots, UAVs;

---

1. Unmanned aerial vehicles, colloquially knows as drones.

- Light EVs: e-bikes and others;
- Small passenger EVs: motorcycles, golf carts, small commuter cars;
- Small industrial: forklifts, micro-pickup trucks;
- Passenger cars: EVs, HEVs, PHEVs;
- Race cars: dragstrip, land speed, formula, solar;
- Public transportation: from tuk-tuks to trains;
- Heavy-duty: industrial and agricultural;
- Off-land: marine, submarine, and airplanes.

It then discusses battery technology and how the traction battery is integrated into a vehicle. This chapter focuses on the traction batteries and specifically its BMS. It is not a complete guide on vehicle design.

3.1.3 Types of EVs

EVs may or may not have a battery and are generally classified as: electric vehicle, hybrid electric vehicle, and plug-in hybrid electric vehicle. These variations result in six permutations, five of which are possible (Table 3.1).

The initials HEV are only used for battery vehicles, even though a batteryless diesel train is just as much a hybrid EV as a hybrid car. Table 3.2 compares these vehicles.

The focus of this book is on battery EVs. It only glances at batteryless EVs (see Section 3.11.2). Later, we will examine the traction system topologies of different types of battery EVs (see Section 3.4) (Figure 3.1).

3.1.3.1 BEV

A BEV is powered entirely by electricity (Figure 3.1(a)). A BEV is advantageous for many reasons:

- Environmental: locally it produces no emissions.
- Simplicity: there's not much to one.
- Low noise: so low that a noisemaker may be required.
- Safety: for example, in mining operations.

3.1.3.2 HEV

An HEV is powered entirely by fuel, typically gas but sometimes hydrogen. It includes a battery and a motor/generator to

- Improve its fuel efficiency.
- Allow the use of a generator that is not able to generate the peak power required by the vehicle (e.g., a fuel cell).

TABLE 3.1
Types of EVs

	Batteryless	With Battery
Electric vehicle	Batteryless EV	Battery electric vehicle (BEV)
Hybrid electric Vehicle	Batteryless HEV	HEV, hybrid car
Plug-in hybrid EV	Not applicable	PHEV, plug-in hybrid

Type		Power Source	Charger	Battery	Motor	Engine/Generator	ZEV	Example
Batteryless EV		Electric rails or lines			✓		✓	Electric train, trolley bus
Batteryless hybrid		Gas			✓			Diesel electric train
BEV		AC power	✓	✓	✓	✓	✓	Nissan Leaf
HEV	**Full HEV**	Gas		✓	✓	✓		Toyota Prius
	MHEV	Gas		✓	✓	✓		Honda Insight
	Micro-hybrid	Gas		✓		✓		Daimler Smart
	FCEV, FCV, HFCV, gas	Hydrogen		✓	✓	✓	✓	Toyota Mirai
PHEV		Gas or AC power	✓	✓	✓	✓	~	Chevy Volt
HEV: hybrid EV; HFCV: hydrogen fuel cell EV; MHEV: mid-hybrid EV; PHEV: plug-in hybrid EV; ZEV: zero-emission vehicle.								

TABLE 3.2 Comparison of EV Types

FIGURE 3.1
Some EVs: (a) BEV: GEM utility vehicle, (b) micro-hybrid: Smart Fortwo, (c) full HEV: as I was writing this book, an 85-year-old person smashed a Prius into my office and nearly pinned me between my desk and the next desk, and (d) PHEV: Nancy plugging in a Ford Escape that our company converted to PHEV.

(a)

(b)

(c)

(d)

- Use power sources that cannot be used directly for traction (e.g., nuclear reactor).
- Use fuels that do not lend themselves to being used in an internal combustion engine (e.g., hydrogen).

An HEV, other than a micro-hybrid, combines these three technologies to use each at its fullest:

- The motor: for its high power;
- The fuel-powered engine/generator: for its energy;

· The battery: to even out the load to let the engine/generator operate at a steady low power, regardless of the demand from the vehicle at a given moment.

It is more efficient than a standard ICE car because the engine/generator either is turned off (the battery propels the vehicle on its own) or runs in an optimal range (e.g., for an engine, at a speed and torque where the efficiency is maximum).

There isn't a well-established line between full hybrids (Figure 3.1(b)) and mild-hybrids (Figure 3.1(c)); roughly, the motor in a full hybrid can provide 25% or more of the traction power, a mild hybrid less than this. Today, some passenger cars, race vehicles, buses, trucks, and large industrial vehicles use hybrid technology. A micro-hybrid is different in that the motor is only a starter motor, and the battery is tiny (Figure 3.1(d)). It is more efficient simply because the engine is turned off when the car is stopped. Gas-powered golf carts are micro-hybrids; as soon as you step on the throttle, the motor starts the engine, and then the engine moves the golf cart. When you release the throttle, the engine stops. As the starter motor is not a traction motor, the battery is not a traction battery. Still, I discuss micro-hybrids in this chapter out of convenience.

3.1.3.3   PHEV

PHEVs use power from either a fuel such as gas or electricity from a power inlet (Figure 3.1(e)). Plug-in hybrids combine the advantages of BEVs, HEVs, and ICE cars:

· Low emissions (like a BEV);
· Fuel efficiency (like an HEV);
· No range anxiety (like an ICE car).

Generally, all PHEVs are passenger cars; Formula Hybrid race cars could also be considered PHEVs because the team is allowed to charge the battery with an external charger before a race.

3.2 TECHNICAL CONSIDERATIONS

This section discusses technical issues common to all EVs. Issues specific to a particular EV application will be discussed in the sections for that application.

3.2.1   Voltage Range

The voltage of traction batteries may range from 12V for a small e-bike to 700V for an industrial vehicle. A higher voltage for a larger vehicle is advantageous because it allows the use of smaller wires and connector contacts. However, its BMS is more expensive.

3.2.1.1   Ideal Resistance of Battery Load

Curiously, a scattergraph of power and voltage of many traction batteries reveals a trend: most sit near a load line of 2Ω (gray band in Figure 3.2). The implication is that the nominal voltage of a traction pack should be selected so that, at full power, it sees a load equivalent to about 2Ω.

In other words, given the power of a vehicle, the ideal voltage of its traction battery is approximately:

$$Voltage[V] = \sqrt{(2\Omega * power[W])}$$

(3.1)

FIGURE 3.2
Traction battery voltage versus application power.

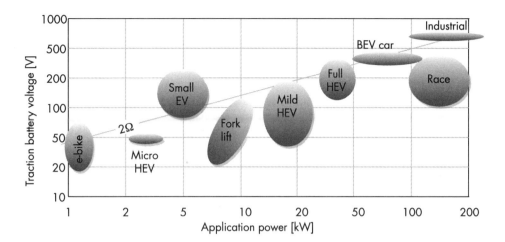

For example:

- 1-kW e-bike: $\sqrt{(2\Omega * 1kW)} = 44.7V \approx 48V battery$

- 250-kW truck: $\sqrt{(2\Omega * 250kW)} = 707V \approx 700V battery$

3.2.2 Regenerative Braking

A standard vehicle brakes[2] by converting its momentum to heat within its brakes. An EV may convert this momentum to electrical energy instead, through regenerative braking (also known as regen).

This energy can be stored in a battery (if the EV has one), it can power other devices in the vehicle (e.g., the lift fans in a hovercraft), or it can be wasted in heat through high-power resistors (e.g., in a train).

Popular culture tells us that "This car's battery is charged by braking" as if regen could miraculously power an EV by itself. Not quite; generally, regen accounts for a minor portion of the battery charging. The majority comes from a charging source (e.g., a generator, an AC power outlet). Most of the energy is spent overcoming friction, moving the vehicle through the air or water, and is not recoverable.

In any case, regen itself is not very efficient: on the order of 50% to 75%. That means that, if you start with 1 kWh of energy stored in the battery, use it to start and accelerate a car, and then brake the car until it stops, only 0.5 to 0.75 kWh of energy is stored back in the battery. This round-trip energy inefficiency is due to several steps along the way:

- While accelerating: the battery's inefficiency during discharge, the motor driver's inefficiency while operating at high current, the inefficiency of the gears and the traction mechanics, and the energy lost moving the car through the air and rotating the wheels against the road.
- While braking: the same steps, in reverse order.

While the efficiency of regen may be surprisingly low, it is still admirable when compared to the efficiency of a standard ICE car: 10% to 20% when accelerating, and 0% when braking.

---

2. Note the spelling difference between "a car breaks" (it no longer works) and "a car brakes" (it slows down).

3.2.3   Operating Modes

Electric vehicles may operate in one of two modes:

- *Charge Depleting mode (also known as the EV mode)*: The traction battery provides energy; it usually is fully charged before use and is depleted during use, with the possible exception of the occasional recharge from regenerative braking.

- *Charge Sustaining mode*: The traction battery provides power and is kept at around 50% SoC during use.

Different types of vehicles operate in various modes:

- *BEV*: Charge Depleting mode;

- *HEV*: Charge Sustaining mode;

- *PHEV*: Starts in Charge Depleting mode until the battery is nearly empty, at which point it switches to Charge Sustaining mode.

These graphs plot the SoC (Figure 3.3(a)) and the energy (Figure 3.3(b)) for the three types of EVs. The two charts look different because a buffer battery (see Volume 1, Section 5.1.4) in an HEV has a lower capacity than an energy battery in a BEV or a PHEV.

Note how

- The PHEV switches from the Charge Depleting mode to the Charge Sustaining mode when the traction battery is nearly empty.

- The BEV stops when the traction battery is empty, while the other two continue running on gas.

- All three batteries occupy approximately the same SoC range (Figure 3.3(a)); however, the energy in the HEV battery is much less than for the BEV and PHEV batteries (Figure 3.3(b)).

- In the Charge Sustaining mode, the swings in SoC are much wider for an HEV than for a PHEV because the HEV battery has a lower capacity (Figure 3.3(a)).

- After the PHEV switches to the Charging Depleting mode, it lets the battery SoC increase a bit, to operate it in a better range rather than the low SoC at which the battery degrades faster.

- The swings in energy are the same for HEV and PHEV, although the PHEV slowly moves into a region of higher energy (Figure 3.3(b)).

3.2.4   Traction Battery Shutdown

When a traction battery in a BEV shuts down, the vehicle coasts to a stop. This is no different from what happens when an ICE vehicle runs out of fuel, or its engine breaks down; yet most people see the former as catastrophic and the latter as just part of life. In both cases, the vehicle warns the driver, who should take action: refuel, recharge the car, or correct the problem. The difference is that a portable gas tank can refuel an ICE car and gas stations are everywhere. On the contrary, an EV with an empty battery usually must be towed to the nearest charging station. Today, there aren't many ways to recharge a stalled EV in place.

FIGURE 3.3
Operating modes for
BEV, HEV, and PHEV:
(a) SoC, and (b) energy.

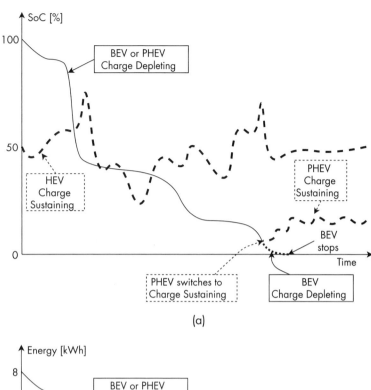

(a)

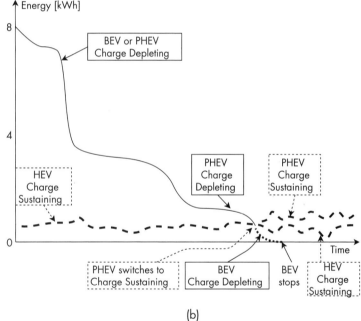

(b)

3.2.4.1 Stopped on the Railway Tracks

All too often, inexperienced EV designers argue that they won't allow the BMS to shut off the traction battery in a passenger EV because this could happen when the car is in a railway crossing, with a baby in the back seat. This emotional argument doesn't hold because:

1. The BMS issues a warning before shutting down the battery; the vehicle must respond by notifying the driver and by reducing the available power to let the driver move the vehicle to a safe place.

2. Research contradicts the conventional wisdom that cars tend to stop on tracks and be hit by a train[3].

3. The likelihood of a fire from an unprotected Li-ion battery is considerably higher than the likelihood of a protected battery stopping an EV on a railway crossing.

Recently, I heard a similar argument from the designer of a hovercraft: "I will not install a contactor between the battery and the motor because I can't let the BMS shut off power to the suspension air when I am going 40 mph over a field of rocks"[4]. This is when my duties switch from an engineer to a counselor, trying to make the customer see that avoiding a battery fire, as a result of operating without a BMS, is the most crucial concern. In a properly engineered hovercraft, the BMS warnings can trigger an orderly shutdown by forcing the vehicle to land.

Do not allow emotional arguments to derail you from the nondebatable requirement that the BMS must be allowed to shut off the battery current.

3.2.4.2 Strategies to Delay Battery Shutdown

It is possible to delay shutdown by operating the vehicle at low power because a battery is more likely to shut down when operated at high power, due to voltage sag. When the BMS issues a low-voltage warning, the vehicle should limit the battery current, which makes the vehicle sluggish. This warning may be sufficient to let a robot mosey back to a charging station, or a passenger EV reach the next freeway exit.

Reduced power and a check engine light give the EV driver feedback, to indicate that there is an issue with the traction system. The vehicle may reduce power either suddenly, when the BMS issues a warning, or gradually, based on the value of the DCL[5] or state of power from the BMS.

3.2.5 Vehicles Powered by Onboard Renewables

An onboard renewable-energy generator may power an electric vehicle or at least assist it.

3.2.5.1 Solar-Powered Vehicles

A practical solar-powered passenger car may always remain in the realm of science fiction, not because of technological limitations, but due to the limited amount of energy in the sunshine that hits a passenger car.

I calculate the maximum range that a solar car may achieve in the future to be about 50 km/day[6]. This distance exceeds the 35 km of the average daily travel by car in Europe. If the technology advances to the point that solar panels are close 70% efficient, and we devise a technology that allows 100% of the sun that hits the car to fall entirely on these panels, and the efficiency of electric cars doubles to 100 Wh/km, and that people drive less than 50 km per day, then, yes, one day a solar passenger

---

3. Statistically, they don't. It's just that we are more likely to notice reports of cars stopped on tracks than of cars stopped elsewhere. People panic, so they don't manage to restart the car. It is true that low-clearance cars to get stuck on elevated train crossings. "Why do cars stall so often on railroad tracks?" The straight dope, www.straightdope.com/columns/read/2770/.

4. The designer of a hovercraft in Namibia.

5. Discharge current limit. See Volume 1, Section 4.6.1.

6. A passenger car has an area of about 3 square meters. The sun generates on the order of 5 kWh/m/day (the direct insolation). The maximum efficiency of solar panels may get up to 70%. Therefore, the theoretical maximum energy/day of the sun on a car is about 10 kWh. A passenger EV uses about 200 Wh/km. Therefore, the theoretical maximum daily range of a solar car is about 50 km.

car may be possible. I don't believe that we can achieve each of these goals, that we'll never have a practical solar-powered passenger car. Instead, I see a future with few personally owned passenger cars. Instead, people will use shared vehicles (see Section 3.11.1) operating continuously, powered by land-based renewable energy. At best, solar panels on the vehicle will contribute a small portion of the vehicle's energy consumption.

Having said this, onboard solar-powered vehicles (not passenger cars) do exist, including solar race vehicles (see Section 3.10.5) and vehicles that recharge while parked for a long time (see Section 3.7.4.3).

3.2.5.2   Wind-Powered Vehicles

An EV powered by a wind generator may travel faster than the downwind speed! Your visceral reaction should be: "Yeah right!"[7]. Indeed, no vehicle can power itself by harnessing the airflow generated by its forward motion (see Section 3.2.5.3). However, counterintuitively, naturally occurring wind can be harnessed to propel or assist an EV [1].

Let me walk you through a thought process to see how this is possible. Imagine a parked EV with a wind generator on top spinning in the wind, charging its battery. After some time, the EV stows the generator and drives away, powered by the energy in the battery. So far, so good? Now let the EV drive with the generator up, as the wind charges its battery, and as the motor discharges the battery (the EV drives in one direction, and the wind generator points towards the wind). Still with me? Now assume that the wind is so strong that it charges the battery as fast as the motor discharges it. As the net battery current is zero, we can disconnect the battery. The wind generator powers the EV.

This technology does not violate any law of physics. A wind generator can increase the range of an EV, if it's windy, just like a boat can go faster if it uses both a sail and its engine. The vehicle harnesses the energy from the natural wind, not the wind from the car moving through the air.

Even more counterintuitively, wind can propel the vehicle faster than the down-wind speed[8]. Rick Cavallaro built an EV that traveled 2.86 times faster than the wind [2]! The wind is not pushing the vehicle, the way a sail would, which would move the vehicle slower than the wind, and rotate the wind generator in one direction. Instead, the wheels move the vehicle faster than the wind speed, and the wind generator spins in the opposite direction, generating so much energy, enough to power the wheels at the higher speed [3]. The extra energy to propel the vehicle derives from the size of the wind generator, which is larger than the vehicle. This technology is similar to using mirrors to concentrate sunlight onto a small solar panel.

The Ventomobile, the Spirit of Amsterdam, and the Blackbird are just three among the many vehicles that use a rotary wind generator to propel themselves, some faster than the wind speed. Various vehicles use a sail of some sort: the Mercedes Benz Formula Zero uses a large tail as a sail, which may be rotated in the appropriate direction [4].

---

7.   Teacher says: "In English, two negatives make a positive. However, two positives do not make a negative." Smart Aleck at the back of the room quips: "Yeah, right!"

8.   Downwind, faster than the wind (DWFTTW).

3.2.5.3 Over-Unity, Perpetual Motion

No, you cannot mount a generator on your electric car to recharge the battery from the motion of the vehicle itself, allowing the vehicle to travel forever: this would be an attempt at creating a perpetual motion machine, which is impossible[9]. Quoting the first law of thermodynamics to a skeptic, the total energy of an isolated system is constant, is ineffective. Stating that you can't create energy out of nothing does not convince a skeptic. Therefore, it's better to discuss the inefficiency of machines in an example application.

You can easily use your hands to spin a generator that is not connected to anything. Once you connect this generator to a battery to charge it, it becomes much harder to spin the generator; you expend more energy turning the generator shaft than the energy reaching the battery. This is because the generator is not 100% efficient. The difference in energy is wasted in heat in the generator. Due to this inefficiency, a self-powered car doesn't work.

Consider a car that is already moving (for example, because it just came down a hill and is now coasting on a flat highway) (Figure 3.4). The front wheel extracts 10 kW from the car's forward motion and passes it to the generator. The generator is 90% efficient and generates 9 kW, which it passes to the motor. The motor is 90% efficient and generates 8.1 kW to drive the rear wheel. The car uses 50% of this power to overcome the friction of the car moving through the air, leaving the remainder, 4 kW, to spin the front wheel. Now we have only 4 kW to spin the front wheel, not 10 kW. The next time around the loop, only 3.6 kW leaves the generator, only 3.2 kW leaves the motor, and only 1.6 kW reaches the front wheel. By the tenth time around the loop, the energy to move the car has dropped to practically zero, and the car stops.

A car without a motor and generator also coasts to a stop; yet it takes longer to do so, because it's not wasting energy heating the motor and generator. Therefore, adding a generator reduces the range of an EV, it does not increase it.

3.2.6 Designing for Second Use

Much has been said about the beauty of reusing traction batteries after they are no longer suitable for vehicle use in stationary applications. So far, that has not materialized, due to economics:

- It costs too much to recover cells from a traction battery, qualify them, and reassemble them into a new battery (see Section 2.12.6).
- The cost of new cells is dropping fast, reducing the value of used cells.

FIGURE 3.4
A self-powered car coasts to a stop.

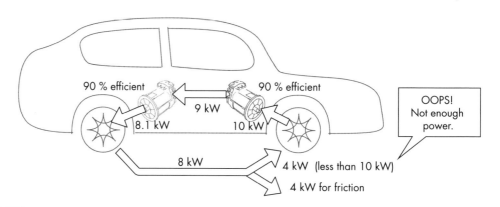

---

9.   And, no, I am not a shill for "them" when I tell you this (which, come to think of it, is precisely what a shill would say).

- In most cases, other than the cells, little else can be reused, especially when converting a 350V battery to 48V.
- Even if, technically, the BMS could be reused, that may be difficult because its communication link uses proprietary messages.
- Few people are willing to pay a premium for the bragging rights of having a recycled battery in their home.

A few steps may help reuse:

- Select a standard voltage: 48V for a small vehicle, 350V for a passenger car.
- In addition to proprietary messages on a CAN bus, include a publicly shared message that gives a second user access to the battery; I recommend the single message that includes all the essential data in just 8 bytes (see Volume 1, Section 5.10.3.9).
- Specify connectors that are commercially available to general users.
- Design a stand-alone, self-protected battery, with no need for support from an external system, to prevent the second user from abusing it.

3.2.7   Indoor EV Charging

The first thing that comes to mind when considering EV charging is a passenger EV in a parking spot plugged into a charging station. But, a plug-in electric vehicle may be charged indoors. That raises safety issues because a fire can have worse consequences indoors than outdoors. Li-ion cells are most likely to go into thermal runaway while charging.

The best insurance against such an event is a well-designed battery with a properly installed and functioning BMS. Still, accidents happen. A BMS is powerless against external ignition sources or damage within a cell. Therefore, it is wise to implement specific measures to contain a fire, should it occur, and have an action plan in place.

Indoor charging of flooded lead-acid batteries is a concern because they vent hydrogen, which can ignite in high concentrations. Li-ion batteries do not vent gases as part of their operation; yet large prismatic cells slowly release some solvents from the electrolyte, leaving an unpleasant odor when concentrated in a closed environment. Therefore, fresh air circulation is required not only for lead-acid batteries but also for Li-ion batteries.

The following measures must be of a level that is appropriate for the application (the type of vehicle and the type of building); charging multiple forklifts in a warehouse commands a larger investment in infrastructure than charging a hoverboard in a home. They must also be appropriate for the statistical risk; a homemade e-bike battery is far more likely to catch fire than a commercial one from a reputable brand.

- Charge a hoverboard or an e-bike battery next to a window, so you can quickly throw it outside.
- Personal transporters (or just their batteries) are likely to be charged at home or in an office, where there are no safety measures:
  - Place the vehicle or battery on tile or concrete, not carpet or wood floors.
  - Don't charge when no one is around.
  - Charge near a door to the outdoors (not to a hallway) or near a window that can be opened.

- Small electric vehicles or passenger vehicles in residential garages:
  - Do not store flammables near the EV (e.g., lawnmower and its gas tank).
  - Install a smoke detector.
  - Replace the door between the garage and the rest of the house with a fireproof one.
- Multiple EVs in a commercial or industrial environment (motorcycles, go-carts, golf carts, forklifts, cars):
  - For high power charging, wire one electrical circuit for each vehicle with individual circuit breakers, to avoid electrical fires from overheated wiring
  - Minimize propagation from one burning vehicle to the others: build a separate charging bays for each vehicle, with a concrete wall between adjacent vehicles to block flames, heat, and shrapnel; that also reduces smoke and fire-fighting water damage to the other vehicles
  - Install a fire detector in each bay and implement 24-hour surveillance
  - Install a fire hose and train staff to use it while waiting for the fire department to arrive
  - Install high volume ventilation to remove fumes and gases rapidly
- Marine winter storage:
  - Install a fire detector for each vessel.
  - Install a fire hose.
- EV development:
  - Work in a fireproof room or, at least, place the vehicle next to a door to the outdoors that can be opened up immediately.
  - Train staff on proper procedures while working on the vehicle and in case of fire.
  - Store other batteries away from the vehicle.

When the effort required to implement safety measures is not warranted, an action plan is the next best thing.

3.2.7.1 Action Plan

Consider how you would react if witnessing a Li-ion-powered vehicle or battery begins thermal runaway. An online video shows a traction battery burning up while charging in a store in Wenzhou, Zhejiang Province, China. The battery starts with a pop, which should have given time to throw the battery outdoors. Yet people were unprepared; all they did was to unplug the battery [5].

If you hear a pop and start to see white smoke, you might have time to move the vehicle or battery outdoors, before flames start. If it is too late, or if you can't move the vehicle, then move the flammables and valuables away, while telling someone else to call the emergency phone number. If an electric car is on fire in a garage, you do have time to get family members, pets, documents, laptop computers, and irreplaceable mementos out of the house. Do not waste time with a fire extinguisher; it is powerless against a Li-ion fire. Fighting an EV fire requires 10,000 liters of water [6]. Save the extinguisher to put out other items on fire, such as a swing door to the rest of the house. Do not bother to unplug the vehicle; it's too late for that. After the fire is put out, move the vehicle outdoors. Li-ion fires have the nasty tendency to restart later.

3.3 DEVICES

To discuss traction battery applications, we need to understand the devices that are likely to be connected to the traction battery. This section is just an overview of these devices (Figure 3.5); later in this chapter, we will go into the details of interfacing with them (see Section 3.1.5).

3.3.1 VCUs

Production passenger vehicles and larger vehicles include a vehicle control unit (VCU), which is an electronic control unit (ECU) dedicated to overseeing the various units in the vehicle (sensors, actuators, motor driver, battery, other ECUs) (Figure 3.6).

3.3.1.1 BMS Communications to the VCU

The BMS is expected to communicate with the VCU to report the state of the battery. An off-the-shelf BMS is not normally expected to be natively compatible with all VCUs; instead, it is more likely that the code in the VCU is customized to interface with all the devices in the vehicle, including the BMS, to achieve the desired performance.

At the very least, the BMS must report to the VCU this essential data:

· The charging and discharging current limit;
· Flags for the state of the battery.

The BMS may also report

· *State of charge*: so that it may be displayed to the user; in some cases, the VCU makes decisions based on the SoC[10].
· *Battery measurements*: voltage, current, and temperature.

FIGURE 3.5
Devices: (a) VCU (courtesy of New Eagle), (b) traction battery, (c) motor driver, and (d) charger (courtesy of Brusa).

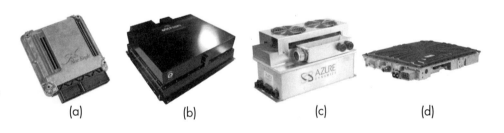

(a)   (b)   (c)   (d)

FIGURE 3.6
VCU in a vehicle, managing a throttle, a charger, a traction battery, and a motor driver through a CAN bus.

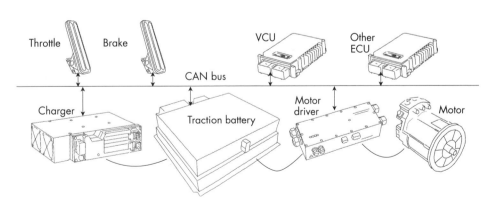

---

10. For example, in a hybrid car, so that the SoC may be maintained within the desired range.

- *Evaluated values beyond the SoC*: state of health, state of power (see Volume 1, Section 1.6.5.3), battery isolation, and resistance.

VCU development is done most effectively through the services of a company that offers VCU hardware and an associated Integrated Development Environment (IDE) [7].

3.3.1.2   VCU Acting as a BMS

A VCU that requires more than the essential data (e.g., it requires cell voltages) raises a red flag. It could be an indication that the VCU is attempting to manage the battery, which is a role that is reserved for the BMS. The only reason for a VCU to step into that role is if the BMS is deemed not trustworthy and is relegated to the simple role of a monitor. Given that most BMSs are designed by people who have more experience in managing batteries than the people who program VCUs, do not succumb to the temptation of letting the VCU act like a BMS.

3.3.2   Chargers

Generally, plug-in vehicles (BEV or PHEV) include an AC power inlet and an onboard charger, unless

- The vehicle operates in a limited area (e.g., a forklift, (see Section 3.8.1), a submarine (see Section 3.13.2)).
- The vehicle must be as light as possible (e.g., aircraft, including UAVs (see Section 3.5.2)).
- A small vehicle is charged at home (e.g., a hoverboard (see Section 3.6.1)).
- A fast-charging station is used (e.g., DC level charging for passenger vehicles).

If so, they have a DC inlet for connection to an external charger.

Note that a charging station is not a charger, although it may include one (see Section 3.9.4).

Some passenger vehicles include two chargers: a low-power convenience charger powered by a house-hold AC outlet, and a fast charger powered by an industrial AC outlet. The heat generated by the charger must be routed away from the cells, especially if it would affect some cells more than others (see Sections 3.2.8.1 and 5.17).

3.3.2.1   Charger Selection

Chargers are generally powered by AC power and isolate the AC input from the DC output[11]. The output voltage ranges from 12V for smaller vehicles to about 350V for passenger vehicles; 700V chargers for industrial vehicles are rare. Old chargers use heavy, large transformers; today's chargers use high-frequency switching electronics, requiring a much smaller and lighter transformer.

There are many off-the-shelf onboard chargers (see Section B.5) with power ranging from 1.5 kW to 10 kW. Two or more chargers (see Volume 1, Section 5.15.4) may be used to increase the charging power, to even out the load on a split or three-phase AC power, or to increase the voltage. A few chargers are powered by three-phase AC; most are air-cooled, while a few require liquid cooling.

---

11.   The Manzanita Micro PFC series is an exception. It is high power, but its lack of isolation is potentially dangerous.

3.3.2.2   Charger Connection

If the BMS can control and turn off an onboard charger, the string of cells could be connected permanently to the charger, before the protection switch (Figure 3.7(a)). Doing so avoids the need for a precharge circuit. The charger must draw negligible current from the cells when it is turned off; otherwise, it would overdischarge the battery over a long time of disuse.

The charger's output capacitance must be precharged at the factory, either through a resistor or by turning on the charger before carefully connecting it to the string of cells. Even if the charger gives the BMS a way to control the output, it's good practice to include a way for the BMS to disconnect the AC power to the charger as

FIGURE 3.7
Charger connection: (a) directly to the string of cells, (b) to motor driver bus, (c) through separate precharge and contactors, and (d) external charger.

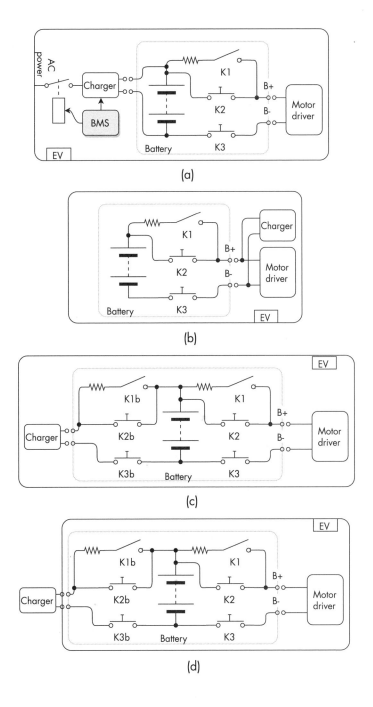

a last resort, in case the BMS is not able to control the charger for some reason (e.g., a disconnected communication cable).

If the BMS cannot control the charger, the charger must be connected on the far side of the battery's protection switch so that the BMS can disable discharging by opening the protection switch. This is particularly the case with external chargers. A precharge circuit is probably required across the protection switch (see Volume 1, Section 5.13.1) to charge the capacitor on the output of the charger (see Volume 1, Section 5.10.1). If the charger is connected directly to the motor driver, it could use the same precharge circuit that is used for the motor driver (Figure 3.7(b)). Otherwise, a separate precharge circuit is required (Figure 3.7(c)), especially if the charger is external (Figure 3.7(d)).

3.3.2.3   Charger Control

To protect the battery from overcharge, the BMU needs to control the charger, at the very least, to turn it off when charging is disabled, although ideally also to set its current and voltage (see Volume 1, Section 5.1.5.1).

From a control standpoint, there are three types of chargers for traction batteries:

- *Dumb*: AC in, DC out, CCCV settings are either fixed or adjusted with knobs.

- *Remote control*: same as above, but with an input that lets the BMS turn on and off the output.

- *CAN*: configured and controlled through the CAN bus (used in professional applications).

Passenger cars and larger EVs use CAN chargers. Since the VCU is probably off while charging, it falls to the BMS to control a CAN charger. Few BMUs are capable of interfacing directly with a CAN charger. A separate gateway may be required to translate between the two devices.

The BMS and the charger send messages to each other. The messages are

- BMS to charger:

  - The CCL (to limit the charging current);

  - An on/off flag;

  - Optionally, the maximum DC voltage, the maximum current from the AC inlet.

- Charger to BMS:

  - The measurement of the current[12].

No additional data is required for proper operation.

Ideally, the BMS should have a CAN bus port just for the charger, separate from the CAN bus port for the VCU. Doing so offers several advantages:

- The charger and the vehicle can use different CAN speeds.

- It doesn't matter if the VCU is off.

- A charger that cannot handle any messages on the CAN bus other than the ones it expects is not overburdened.[13]

---

12. That way, the BMS does not need a separate current sensor for the charging current.
13. ElCon charger.

3.3.3 Traction Motors

A traction motor[14] converts electrical power to mechanical power to propel a vehicle (see Section B.3). In contrast, an engine converts fuel to mechanical power[15]. The power comes from a battery, some other DC source (e.g., rails that power an electric train), or an AC source (e.g., the grid or an inverter).

3.3.3.1 Types of Motors

You may have learned that motors can be either AC or DC. This simple classification ignores the fact that many motors are neither (see Section B.3.1). Motors may be designed to be driven by one of these waveforms (see Section B.3.1.1):

- Straight DC (Figure 3.8(a));
- Sinusoidal AC (Figure 3.8(b));
- Trapezoidal waveform (Figure 3.8(c));
- Rectangular, unipolar (Figure 3.8(d)) or bipolar (Figure 3.8(e)).

Regardless of their classification, at their core, all motors are AC motors (see Section B.3.1.2). In a rotary electromagnetic motor, an AC voltage is required to create a rotating magnetic field to make the rotor turn. With an AC motor, this AC is generated externally and supplied to the motor. With a DC motor, a device inside the motor converts the DC at its terminals to the AC that it needs to run the AC motor within it.

Among the great variety of motors, only a few are likely to be used in an EV (Table 3.3)[16].

Some interesting exceptions are

- Maglev trains, some Japanese subways, and some airport people movers use linear motors, where the track functions as the stator of the motor.

FIGURE 3.8
Motor drive: (a) straight DC, (b) sinusoidal AC, (c) trapezoidal, (d) unipolar, and (e) bipolar.

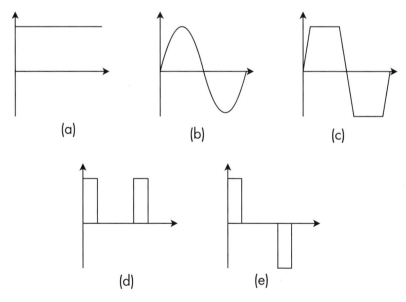

(a)   (b)   (c)

(d)   (e)

---

14. See Section B.3 for more details and directories.
15. Only a few languages (English, Greek, Japanese, Welsh) use two different words for a fuel-powered engine and an electric motor; most other languages use the same word for both.
16. Some vehicles use more than one motor: a Toyota Prius has two. A high-performance car may have a motor per wheel. A yacht may have one motor per screw propeller. An airplane may have one motor per propeller.

TABLE 3.3
Motors for EVs

Type	Drive	Quadrants
Induction	Three-phase AC sinusoidal	4
IPM (interior permanent magnet)/BLAC	Three-phase AC sinusoidal	4
Hysteresis/synchronous reluctance	Three-phase AC sinusoidal	4
External controller BLDC	Three-phase trapezoidal	4
Reluctance stepper	Rectangular	2: no braking
Permanent magnet stepper	Rectangular	2: no braking
PM (permanent magnet)	Straight DC	4
Series-wound or shunt-wound, brushed motor	Straight DC	2: no braking
SepEx	Straight DC	4

- Bumper cars (the ones in fairs) may use a universal motor.

3.3.3.2   Traction Quadrants

At any given time, a motor may operate in one of four quadrants, based on the direction of speed and torque (Figure 3.9(a)).

By convention, the quadrants in a coordinate plane are numbered in a counterclockwise direction, starting from the top-right quadrant[17]:

- I. Forward motion, acceleration;
- II. Forward motion, braking;
- III. Reverse motion, acceleration;
- IV. Reverse motion, braking.

Not all motors can operate in all four quadrants:

- Most motors operate in all four quadrants (e.g., induction AC motor, brushed permanent magnet DC motor).
- Some motors operate only in only the top two quadrants because they do not support regenerative braking (see Section 3.2.2) (e.g., some brushless DC motors).
- Some motors operate only in the right two quadrants: they work only in one direction (e.g., single-phase AC motors).
- Very few motors are limited to only one quadrant (e.g., universal motors).

For a brushed permanent magnet DC motor, the torque is proportional to the motor current (regardless of speed) (Figure 3.9(b)). A positive current produces forward torque, and a negative current produces reverse torque. Its speed at no load is proportional to the applied voltage (Figure 3.9(c)). A positive voltage produces forward speed, and a negative voltage produces reverse speed; loading the motor reduces its speed for a given voltage. This relationship holds roughly for other types of motors, although it is more complicated.

When powered by a fixed DC voltage, a brushed permanent magnet DC motor may operate at any point along the diagonal line (Figure 3.10(a)). The speed changes

---

17. Careful: A few people exchange the speed and torque axes, which results in the definitions for Quadrants II and IV being exchanged. There is no standard for speed versus torque, but there is one for voltage versus current: voltage on the x-axis. I place speed on the x-axis because torque is related to current, and speed is related to voltage.

FIGURE 3.9
(a) Traction quadrants,
(b) DC-motor torque
versus current, and
(c) DC-motor unloaded
speed versus voltage.

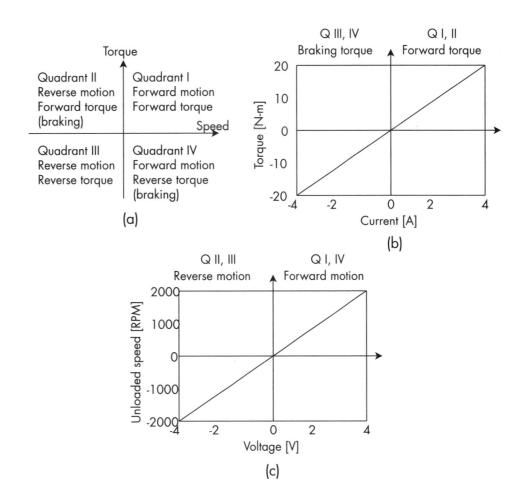

(a)

(b)

(c)

linearly as torque changes. The motor slows down when it generates torque and speeds up if torque is applied to it. This curve shows that, when powered at +20 Vdc, this motor can operate at 2,000 RPM at no load, or 1,000 RPM and 10 N-m (and draw 2A), or 20 N-m at stall (drawing 4A).

If the polarity of the applied voltage is reversed, the motor operates in the opposite direction (Figure 3.10(b)). At −20V, the operation ranges from −2,000 RPM at no load to −20 N-m at stall (drawing −4A).

The operating diagonal line is shifted by varying the voltage (Figure 3.10(c)): increasing the voltage moves the line towards the top right corner, decreasing the voltage moves it towards the bottom left corner. The straight diagonal line in the curves above is typical for a brushed permanent magnet DC motor by itself. Other motors have nonlinear curves.

By varying the voltage applied to the motor, we can allow the motor to operate anywhere within the gray area (Figure 3.10(d)) bounded by

- The maximum motor speed (left and right edges);
- The maximum motor torque or the maximum current from the variable DC source, whichever is more limiting (top and bottom edges);
- The battery voltage (diagonal lines at the top right and bottom left corners).

A DC motor controller provides this variable DC voltage.

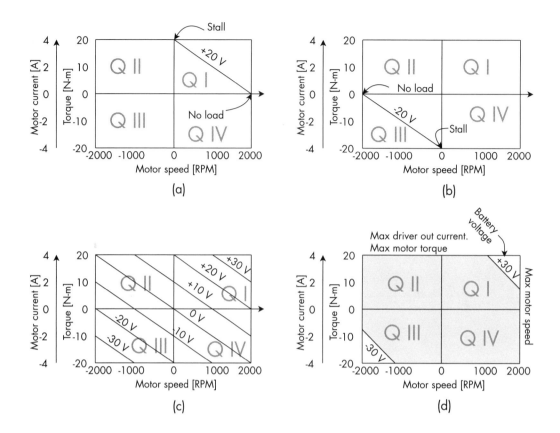

FIGURE 3.10   Operating range of a DC motor: (a) powered directly by a battery, (b) powered by a battery with the opposite polarity, (c) powered by various fixed voltages, and (d) powered by a motor driver, within the gray area.

3.3.4   Motor Drivers

In a toy, the DC source powers the DC traction motor directly (Figure 3.11(a)). Similarly, in a fair ride, the grid powers the AC traction motor directly (Figure 3.11(b)). In all other applications, a motor driver[18] sits between the DC source and the traction motor (Figure 3.11(c, d)) to regulate the power to the motor.

The motor driver converts the voltage from the DC source to the voltage required to drive the motor at the desired speed and torque. This voltage could be a fixed DC voltage or a varying voltage whose frequency and waveform are appropriate for the motor (see Section 3.3.3).

Almost universally, a motor driver reduces the DC bus voltage down to the voltage required by the motor. The motor driver also increases the current (see Section B.4.1); the motor current is higher than the battery current (Figure 3.11(c)). Keep that in mind when discussing current in a traction system: are we talking about the battery current or the motor current?

3.3.4.1   Motor Driver Quadrants

Regardless of the number of quadrants in which a motor can operate, the motor driver may limit operation to fewer quadrants. For example, a simple DC motor

---

18. See Section B.3 for more details and directories.

FIGURE 3.11
Motor drive: (a) toys:
direct connection to
battery, (b) amusement
park ride: direct
connection to AC,
(c) DC motor applications:
through a DC motor
controller, (d) AC motor
applications, through an
AC motor inverter, and (e)
a motor driver decreases
the voltage and increases
the current; the power
stays nearly the same.

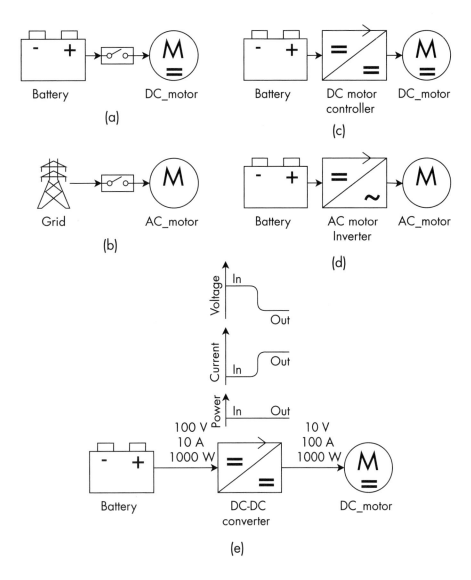

FIGURE 3.11
Motor drive: (a) toys: direct connection to battery, (b) amusement park ride: direct connection to AC, (c) DC motor applications: through a DC motor controller, (d) AC motor applications, through an AC motor inverter, and (e) a motor driver decreases the voltage and increases the current; the power stays nearly the same.

controller limits a brushed permanent magnet DC motor, which can operate in all four quadrants, to only one quadrant.

Regardless of the type of motor, the motor driver can modify the natural curve of a motor into any shape, as long as the entire curve is within the gray area shown in the graph (Figure 3.10(d)). The motor is free to operate at any point along the curve defined by the motor drive.

The following curves are likely in an EV (Figure 3.12). They assume that both the motor and the driver may operate in all four quadrants.

Forward (Figure 3.12(a)).

- *Quadrant I*: normal operation. The motor controller adjusts the voltage to maintain the desired speed,—vertical line; beyond a given torque (e.g., going a steep hill), the motor controller limits the current instead, and the speed drops—top horizontal line; note how this is the same shape as a CCCV charger: constant current at the top, constant voltage on the right.

- *Quadrant II*: It's possible that the vehicle cannot move forward with this maximum torque, and the speed actually reverses (e.g., the vehicle starts rolling back

FIGURE 3.12
Four quadrant operation
of a motor driver: (a)
forward, (b) stopped,
and (c) reverse.

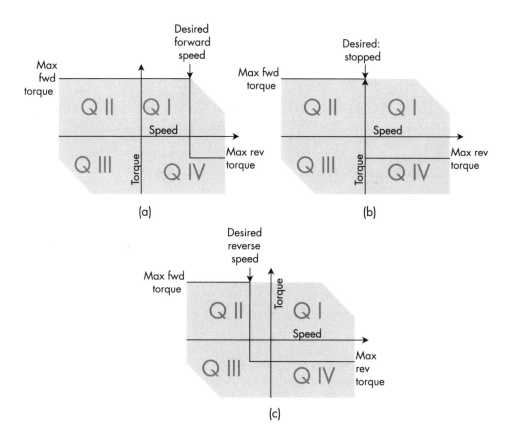

on a very steep hill); if so, the motor controller brakes the vehicle so that it does
not roll back too fast by applying the maximum forward torque to the motor.

- *Quadrant IV*: An external force may be attempting to move the vehicle forward
  more than the motor (e.g., going down a steep hill); if so, the motor applies
  reverse torque (braking) to maintain the speed—vertical line; beyond a given
  braking torque (e.g., going a very steep hill), the motor controller limits the
  current instead and the speed increases—bottom horizontal line; note that the
  value of the forward torque can be different from the value of reverse torque—
  the top horizontal line is farther from the X-axis than the lower horizontal line.

Stopped (Figure 3.12(b)) (e.g., a Segway stopped on a hill, balancing the user):

- *Quadrant IV*: If there's an external forward force (e.g., the vehicle faces down-
  hill), the motor applies reverse torque to try to stop the vehicle.
- *Quadrant II*: If there's an external reverse force (e.g., the vehicle faces uphill), the
  motor applies a forward torque to try to stop the vehicle.

Reverse (Figure 3.12(c)):

- *Quadrant III*: driving in reverse. The motor controller adjusts the voltage to
  maintain the desired speed,—vertical line; beyond a certain torque (e.g., going
  backward up a steep hill), the motor controller limits the current instead and the
  speed drops—bottom horizontal line.[19]

---

19. That is because of power conservation: power = voltage × current. The output power is almost the same as the input power. As the
    output voltage is lowered, the output current must increase to make up for it and conserve the power.

- *Quadrant II*: Conversely, an external force may be attempting to move the vehicle backward more than the motor (e.g., going backward down a steep hill); if so, the motor applies reverse torque to brake and maintain the speed—vertical line; beyond a given braking torque (e.g., a very steep hill), the motor controller limits the current instead, and the speed increases—top horizontal line.

- *Quadrant IV*: It's possible that the vehicle cannot move backward with this maximum torque and the speed actually reverses (e.g., the vehicle starts rolling back down on a very steep hill); if so, the motor controller brakes the vehicle so that it does not roll back too fast by applying the maximum reverse torque to the motor.

Many more curves are possible. For example:

- The curve for a torque motor[20] has a single horizontal line with maximum speed limits at both ends (Figure 3.13(a)).
- The curve may use sloped lines to emulate an ICE vehicle (Figure 3.13(b)).

In all cases, the curve is contained within the gray area that is limited by both the motor and the motor driver. To reshape a motor's natural curve to the desired curve, the motor driver dynamically changes its output based on feedback from the motor speed and torque.

For example, the motor driver may implement a servo loop function[21] loop to control the motor speed, regardless of torque. The servo loop monitors the actual speed, compares it to the desired speed, and varies the motor voltage until the motor reaches the desired speed (Figure 3.14(a)). The motor driver does so up to a maximum current (actually two currents, one for the positive current/forward torque, one for the negative current/reverse torque).

The motor driver may use a servo loop to set the motor current at the desired value regardless of speed (Figure 3.14(b)). Similarly, it may control the torque using a torque sensor (Figure 3.14(c)).

3.3.4.2 Types of Motor Drivers

Generally speaking, motor drivers may be

FIGURE 3.13
Four quadrant operation:
(a) torque motor, and
(b) ICE emulation.

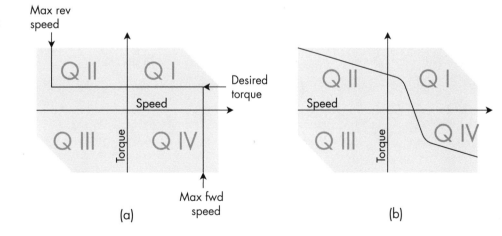

(a)   (b)

---

20. Used in the industry, for example, to keep tape at a constant tension.
21. Be aware that the hobby industry misappropriated the term "servo" to mean a small DC motor.

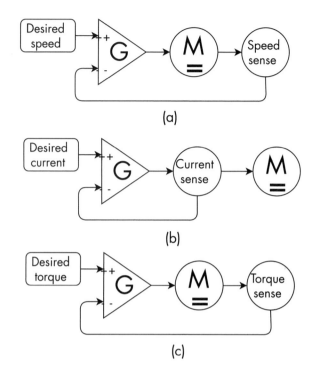

FIGURE 3.14
DC motor driver servos:
(a) speed, (b) current,
and (c) torque.

- *DC*: outputs a flat DC of variable voltage (DC motor controller);
- *AC*: outputs sine waves of variable voltage and frequency (AC motor inverter);
- *Trapezoidal*: outputs trapezoidal waves of variable voltage and frequency (e.g., BLDC controller);
- *Switched DC*: outputs a rectangular voltage of variable frequency, either unipolar or bipolar (e.g., stepper driver).

3.3.4.3   DC Motor Controllers

DC motor controllers reduce the battery voltage to a straight DC to power a DC motor. The typical DC motor controller works with almost any DC motor: permanent magnet (PM), wound stator, integral controller BLDC, and universal. The exception is a separate excitation (SepEx) motor, which requires a SepEx motor driver (it provides an additional output for the motor's field winding).

Off-the-shelf DC motor controllers for traction may be roughly divided into classes based on battery voltage and motor power (Table 3.4).

Many of these DC motor controllers are for low-voltage applications (12V, 24V, or 48V), such as golf carts. Typically they are not isolated.

3.3.4.4   DC Motor Speed Control

With a dumb DC motor controller, the driver controls the speed with a throttle (accelerator pedal). This throttle is coupled to a 5 kΩ potentiometer with one end to ground. The resistance ranges from 0Ω at rest, to 5 kΩ at maximum drive ("pedal to the metal") (Figure 3.15(a)).

The battery may need to limit the current by keeping the driver from driving too hard. Since these motor controllers do not provide a current limiting input, a workaround is to shunt the throttle pot with a variable resistor controlled by the BMS. The only BMS that could do this is no longer in production, but here is how it

Type	Voltage [V]	Power [kW]	Applications	Notes
Low voltage, dumb	12~48 maximum	1~20	Golf carts, bikes.	Throttle only (no CAN): not isolated
Low-mid-voltage, CAN	60~96 maximum	3~50	Small EVs	CAN bus: isolated
Mid-voltage, dumb	56~120 maximum	3~90	Small EVs	Throttle only (no CAN): not isolated
Mid-high voltage, dumb	140~156 maximum	30~200	Medium-size EVs, race motorcycles	Throttle only (no CAN): isolated
High voltage	~350	70~750	Passenger vehicles, race cars	Either CAN bus or throttle pot: no isolation

TABLE 3.4   Classes of DC Motor Controllers

FIGURE 3.15
Dumb DC motor
controller: (a) throttle,
and (b) throttle
override circuit.

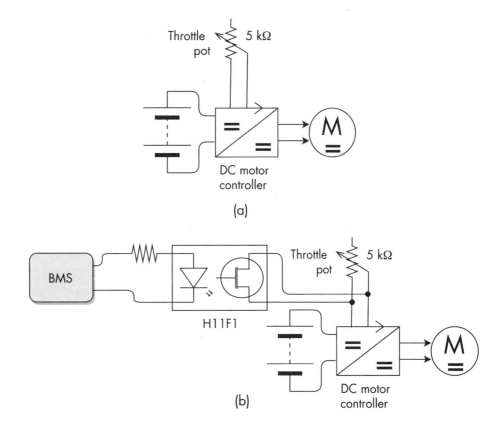

did it (Figure 3.15(b)). You can replicate it with a few electronic components, as long as your BMS has a 0~5V output for DCL[22].

More sophisticated motor drivers include a digital link (typically a CAN bus) to let the VCU set the desired speed and torque, based on the data from the battery's BMS (the Discharge Current Limit), the throttle and the brake (Figure 3.16(a)). The VCU sends messages to the motor driver with the desired speed and torque, and the motor driver replies with the actual speed and torque.

In case there is no VCU, and in case the motor driver supports BMS messages, then the BMS can communicate the DCL directly to the motor driver (Figure 3.16(b)). Today, few motor drivers can operate just with a BMS and no VCU[23].

---

22. Discharge current limit.
23. At the time of this writing, only CEBI/MES TIM is. The TM4 motor inverter is compatible through their VCU: TM4 Neuro/TM4 motor inverter; and Curtis inverters are compatible through a VCU from HPEVS.

FIGURE 3.16
Traction system: (a)
with VCU, and (b)
without VCU.

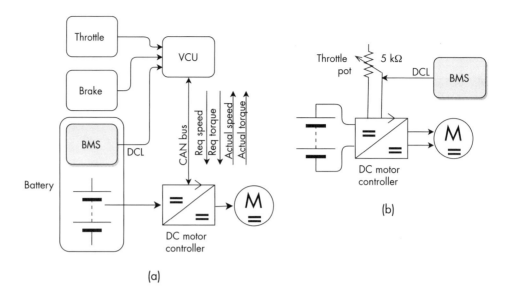

Motor drivers that require a value for the maximum motor speed (or voltage) and torque (or current) are not directly compatible with a BMS because the motor driver is concerned with the motor current while the BMS is concerned with the battery current. The motor driver has enough information to translate the current limits from the BMS to the corresponding motor current. If the motor driver doesn't do so, then the only other device that may be able to do so is the VCU. The BMS can't because it doesn't know the motor voltage and current.

3.3.4.5   DC Motor Reverse

Many motor controllers for small DC motors support four quadrants[24]. Conversely, many high-power DC motor controllers operate in fewer quadrants even though the DC motor itself can operate in all four quadrants:

- One quadrant[25]: forward only, no regenerative braking;
- Two quadrants[26]: forward only, with regenerative braking;
- Two quadrants[27]: forward and reverse, no regenerative braking.

An EV with a simple DC motor controller requires a way to drive in reverse[28]. This is done by reversing the motor connections

- With a PM motor, a reversing contactor is normally used for this purpose (Figure 3.17(a)).
- With a series-wound motor, only a single winding must be reversed (Figure 3.17(b)).
- With a shunt-wound or SepEx motor, only the field winding needs to be reversed, which is easier to do because the field winding operates at a relatively low current (Figure 3.17(c)); if both windings are reversed, the direction won't change (Figure 3.17(d)).

---

24.   They use a full-bridge driver circuit with four transistors.
25.   It uses a half-bridge driver circuit with a single transistor and a single diode.
26.   It uses a half-bridge circuit with two transistors.
27.   It uses a full-bridge circuit with four transistors.
28.   An EV conversion may retain the original transmission, in which case there is no need to reverse the motor direction.

FIGURE 3.17
Reversing direction for
a DC motor: (a) PM
motor, (b) series-wound
motor, (c) shunt-wound
or SepEx motor, and
(d) incorrect wiring for
shunt or SepEx motor.

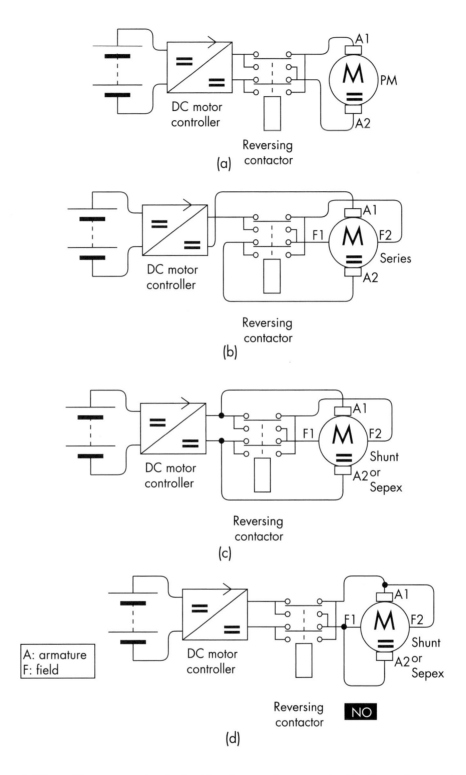

Adding this reverse circuit doubles the quadrants attainable from the traction system:

- One quadrant traction system: converts to two quadrants, adds reverse, still no braking;
- Two quadrants traction system: converts to four quadrants, adds reverse, with braking.

3.3.4.6 BLDC Motor Controllers

BLDC motor controllers (see Section B.4.3) (also known as BLDC motor drivers) generate 3 or more phases of switched DC for brushless synchronous motor: a permanent magnet stepper motor or a reluctance stepper motor.

Off-the-shelf BLDC motor controllers for traction may be roughly divided into classes based on battery voltage and motor power (Table 3.5).

3.3.4.7 AC Motor Inverters

An AC motor inverter (see Section B.4.2.1) generates a 3-phase sinusoidal AC at a variable voltage and frequency to drive an AC motor: Induction, IPM (Interior Permanent Magnet)/BLAC, or Hysteresis/synchronous reluctance.

These inverters are powered by DC, unlike AC motor inverters for industrial use, which are AC-powered. They are also different from grid inverters, which operate at 50 or 60 Hz only.

Off-the-shelf BLDC AC motor inverters for traction may be roughly divided into classes based on battery voltage and motor power (Table 3.6).

Most AC motor inverters are controlled through CAN bus messages to set speed and torque. They are isolated.

3.3.5 DC-DC Converters

Electric vehicles are likely to include one or more DC-DC converters for a variety of functions, such as to transfer power between two DC buses at different voltages.

Vehicles other than smalls EVs (e.g., bikes, golf carts) have two voltage buses: high voltage from the traction battery and low voltage, typically 12V.

The low-voltage system may be from

- A low-voltage SLI battery (see Section 2.9) (possibly lead-acid), which is charged by:
  - A charger, powered when the vehicle is plugged into an outlet;
  - A low-power DC-DC converter, powered by the traction battery, that can power the average load on the low-voltage system.
- A medium-power DC-DC converter (if there is no battery) that can power the peak load on the low-voltage system;
- An alternator in a hybrid vehicle.

TABLE 3.5
Classes of BLDC Motor
Inverters

Type	Voltage [V]	Power [kW]	Applications	Notes
Low voltage	24~48	3~10	Golf carts, bikes.	Not isolated
Mid-voltage	72~144 Max	30~150	Small EVs	Not isolated

TABLE 3.6
Classes of AC Motor
Inverters

Type	Voltage [V]	Power [kW]	Applications	Notes
Low voltage	24~48	4~17	Golf carts, bikes.	
Mid-voltage	80~160 Max	10~100	Small EVs	
Passenger	~350	30~200	Passenger vehicles, race cars	
Industrial	~700	70~750	Industrial vehicles, marine	May be a split battery

The BMS in the traction battery is likely to require a low voltage power supply. This supply can come from

- The cells, directly;
- The vehicle's low–voltage system;
- A small, dedicated DC-DC converter, powered by the traction battery.

Some specialized hybrid vehicles include a power source of some kind (e.g., a fuel cell) whose output voltage is different from the traction battery voltage. A high power DC-DC converter is used to transfer power between the two. In some applications, the DC-DC converter needs to be bidirectional.

We saw four types of DC-DC converter:

1. Very low power (~5W), for just the BMS, CV output (Figure 3.18(a));
2. Low power (~50W), to charge a 12V battery, CCCV output (Figure 3.18(b));
3. Medium power (~500W), to power the entire 12V system without a 12 battery (Figure 3.18(c));
4. High power (~5 kW) to very high power (~50 kW), between a power source and the traction battery (Figure 3.18(d)).

In the first three cases, the DC-DC converter powered by the traction battery and is isolated. In the fourth case, the DC-DC converter is probably not isolated (both sides are in the high voltage system) and may be bidirectional (if the motor driver is on the same side as the power source).

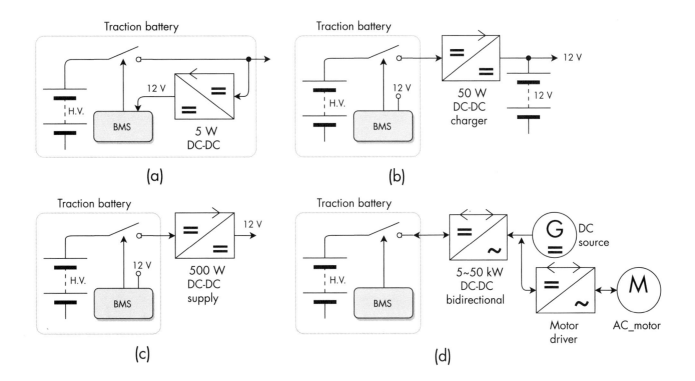

FIGURE 3.18   DC-DC converters used in EVs: (a) very low power to power the BMS, (b) low power to charge the 12V battery, (c) mid-power to power the 12V systems, and (d) high and very high power for traction.

A 5 W DC-DC converter is easy to find. A 50-W charger and a 500-W power supply not so much (see Section B.8 for a directory). For higher power levels, you probably need a custom product designed just for your application.

3.3.6 Displays

Usually, the VCU controls any displays, because the VCU has access to the complete status of the EV. The BMS may drive a display, though it is only able to display the limited information that the BMS has access to: the SoC. Therefore, it is more likely that the BMS tells the VCU the SoC, and the VCU makes sure that the display handles it somehow.

In a simple EV, the BMS may drive a fuel gauge directly. For example:

- Bikes and mopeds with an analog SoC meter;
- The same, but with a simple alphanumeric LCD[29];
- EV conversion of old cars with an analog fuel gauge originally using a sender unit in the fuel tank.

3.3.6.1 *Analog SoC Meter*

EV mopeds have a simple analog meter with a 0 to 100% dial. The meter is powered by 0 to 5V. Any BMS with a 0~5V output proportional to SoC can be used to drive this meter (Figure 3.19(a)).

3.3.6.2 *Alphanumeric LCD*

A simple display, such as a 2-line, 16-character LCD[30], may be used to display SoC. The BMS sends a 1-wire serial stream to the display, either as RS232 or simple UART Tx TTL (Figure 3.19(b)).

3.3.6.3 *Sender Emulation*

In old cars, a sender unit is a variable resistor controlled by a float in the fuel tank. Its resistance is quite low and may either increase or decrease as the fuel level drops. The sender feeds a peculiar meter in the dashboard that uses a heater on a bimetal leaf to move the fuel gauge needle.

To reuse this fuel gauge to show the SoC level of the traction battery, the BMS must emulate the sender. Only one BMS ever did this, but it is no longer in production. If your BMS has a 0~5V output proportional to SoC, you can build your own with a few electronic components (Figure 3.19(c)).

3.4 TRACTION TOPOLOGIES

Virtually all electric vehicles use one of the following topologies:

- *Batteryless EV:* EV without a battery;
- *Batteryless HEV:* hybrid without a battery;
- *BEV:* battery EV;
- *HEV:* hybrid EV (series or parallel);

---

29. Liquid crystal display.
30. Liquid crystal display.

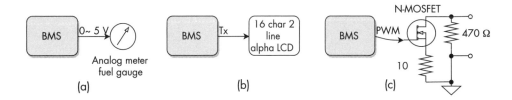

FIGURE 3.19
Fuel gauge circuits:
(a) analog SoC meter,
(b) LCD, and (c)
sender emulation.

• *PHEV:* plug-in hybrid EV.

3.4.1 Batteryless Electric Vehicles

Batteryless electric vehicles are off-topic for a book on batteries; still let's take a quick look at them for the sake of completeness.

3.4.1.1 Batteryless EV

Many batteryless EVs receive electrical power along their travel, such as from overhead lines or power rails. They include electric trains, light-rail trains, and trolleybuses. The vehicle contains just a motor controller and a motor (Figure 3.20).

It could be argued that any conveyance that is propelled by electricity is a batteryless EV. If so, that would include a wide range of people and freight movers:

- The vehicle is moved by cables or hydraulics powered by an external, stationary electric motor: elevators/lifts, ski lifts, funiculars, roller coasters, and gondola lifts; if you consider skis a vehicle, then also surface ski lifts (skiers hold onto a cable that pulls them up the hill).
- The vehicle stays in place as it moves people or freight: escalators, Ferris wheels, moving sidewalks, and conveyor belts.
- Half the motor is in the vehicle, half is stationary: people-movers with no active components, propelled by a linear motor in the tracks[31].

3.4.1.2 Batteryless HEV

Most diesel train locomotives are batteryless hybrid vehicles. Specifically, they are series hybrids (see Section 3.4.3.1): a diesel-powered generator provides electrical power to the traction motor through a motor driver (Figure 3.21).

Compared to a standard diesel truck, a diesel train locomotive has many advantages:

- *Power at stall:* the electric motor has the highest torque when the train first starts.

FIGURE 3.20
Topology of a
batteryless EV.

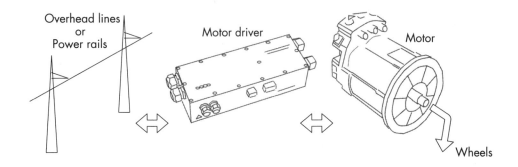

---

31. The Subway at Bush Airport in Houston Texas, and the Tomorrowland Transit Authority People Mover at the Walt Disney World Resort in Florida. See Section B.3.2.

FIGURE 3.21
Topology of a batteryless
series HEV.

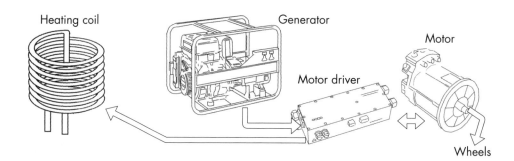

- *Energy:* the diesel-powered generator is operated in a region where its efficiency is highest.
- *No clutch:* the diesel engine does not drive the wheels directly.
- *Single gear ratio:* the motor can run from stall to way faster than 100 km/h.

However, as it has no battery, this locomotive has some disadvantages:

- When braking, energy from regen braking is wasted in heaters.
- When accelerating, the peak power must come from the generator; therefore, the generator is larger than it would be in a battery HEV, in which the battery provides the peak power, and the generator only needs to provide the average power.

3.4.2   Battery Electric Vehicle (BEV)

Fundamentally, a BEV includes a battery, a motor driver, and a motor (Figure 3.22). A charger powered by the AC power grid may be onboard or stationary[32]. A BEV uses an energy battery, providing additional energy at all times. The power handling capability of the battery is secondary to its capacity.

3.4.3   Hybrid Electric Vehicle (HEV)

An HEV includes an electric motor, a fuel-powered engine or power source of some form (e.g., a genset, a fuel cell), and a battery. There is no AC-powered charger.

The buffer battery (see Volume 1, Section 5.1.4) powers the motor when accelerating and receives regen power from the motor when braking. The engine or other power source provides the power to maintain the battery charge at around 50% SoC. The battery buffers the power in an HEV in the same way that a flywheel does so in an engine; it provides additional power when needed, and it absorbs extra power when available. Therefore, the power source (typically an engine) can operate

FIGURE 3.22
Topology of a battery
electric vehicle.

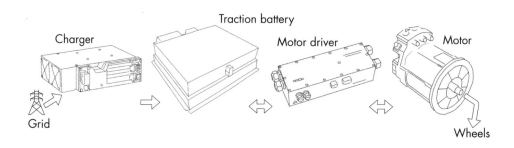

---

32.   Alan Cocconi invented a motor driver that doubles a charger repurposing the motor as an inductor.

within a range of maximum efficiency, regardless of the demand from the vehicle at a given moment.

Compared to a batteryless HEV, a traction battery in an HEV provides three advantages:

- Regen energy is stored in the battery (rather than wasted in heat).
- The HEV may use a low-power source, which only needs to provide the average power (not the peak power); this is because the battery acts as a buffer, sourcing the peak power when needed, and absorbing excess power when available.
- The HEV may use a power source whose output power cannot be changed rapidly (e.g., a hydrogen-powered fuel cell[33]): a buffer battery provides the power rapidly when needed, and absorbs regen power rapidly.

The traction battery is a buffer battery, meaning that its power handling capability is more important than its capacity. The battery is operated around 50% SoC (or some other mid-SoC level) so that it has headroom in either direction.

An HEV can be series or parallel.

3.4.3.1 Series HEV

The most common example of a series HEVs is a fuel cell electric vehicle (FCEV)[34]. There are two possible topologies, depending on which device is connected to the DC bus, that powers the motor driver:

- *Power source on the DC bus* (Figure 3.23(a)): The battery feeds the motor driver through the DC-DC converter; this solution is preferable if the peak power is less than twice the average power, meaning that most of the peak power comes from the power source, since the DC-DC converter needs to supply only the small difference between the peak and average power.

FIGURE 3.23
Series battery HEV topology: (a) motor driver connected to the generator, and (b) motor driver connected to the traction battery.

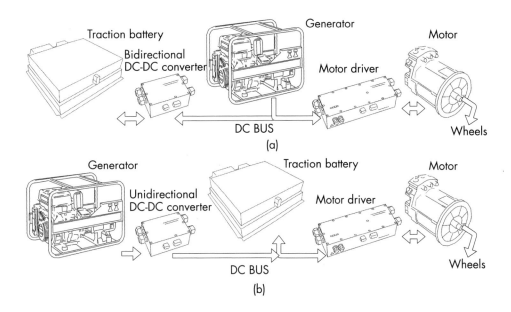

---

33. Heck, a small nuclear power plant could be used as a generator.
34. Toyota Mirai passenger vehicle, for example.

- *Battery on the DC bus* (Figure 3.23(b)): The power source feeds the motor driver through the DC-DC converter; this solution is preferable if the peak power is more than twice the average power, meaning that most of the peak power comes from the battery, since the DC-DC converter needs to supply only the average power; also, this solution uses a unidirectional DC-DC converter, which is cheaper.

A series hybrid has three different voltages:

- *Traction battery:* DC, varies with SoC and load;
- *Power source (e.g., generator):* DC, varies with operating point;
- *Motor:* likely three-phase AC, varies with speed.

Two power devices are required to convert between these three voltages:

- A DC-DC converter converts between the traction battery and the power source; in the first topology, this converter is bidirectional; in the second one, it isn't. For the highest efficiency of the DC-DC converter, their voltages should be close to each other; yet they are different because the battery voltage changes with SoC level and the load, while the voltage of the power source changes depending on its operating point. The CCCV output of this converter also limits the current into the battery.
- The motor driver converts a DC bus voltage to the motor voltage.

To prevent overcharge when the battery is nearly full, the BMS must be able to shut down the power source and disable regen braking. Therefore, the vehicle must include mechanical brakes as well to dissipate the braking energy when the battery is full.

To prevent overdischarge when the battery is nearly empty, the BMS must be able to stop the battery discharge. The power source limits the vehicle's power. Therefore, a power source capable of more than just the average power may allow the vehicle to climb up a long steep hill.

3.4.3.2 Parallel HEV

In a parallel HEV, both an electric motor and an engine provide the traction power (Figure 3.24). The point of a parallel HEV is for the electric motor to complement the limited power of a small engine operating in a narrow range of speeds for peak efficiency. A well-known example of a parallel HEV is the Honda Insight.

Retrofitting a truck to make it an HEV involves placing a motor in line with the driveshaft between the transmission and the differential. An ECU monitors the truck's acceleration to determine when to add power to the drivetrain. It monitors braking to determine when to absorb power from the drivetrain.

FIGURE 3.24
Parallel HEV topology.

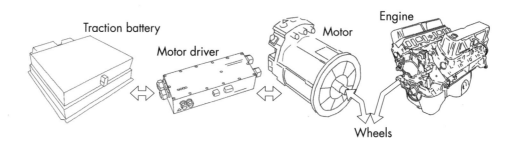

3.4.3.3 Series/Parallel HEV

A series/parallel HEV can operate in both modes. The best-known example is Toyota's trademarked Hybrid Synergy Drive. The Toyota Prius, Camry, and Highlander use this technology, as does the Ford Escape. I describe this technology in the appendix (see Section B.8).

3.4.4 Micro Hybrid (MHEV)

In a micro-hybrid, the motor is simply a starter motor, too small to propel the vehicle. A micro-hybrid uses a small 48V battery with a storage capacity of up to 1 kWh to power the starter motor when the car starts moving again. A standard alternator recharges the battery (Figure 3.25(a)).

Making the starter motor double as a generator (Figure 3.25(b)) saves one component but complicates the design. This is because the alternator is designed to remain connected to the engine and to operate over the full range of speeds of the engine, while the starter motor is designed to be engaged to the engine only while starting it, at low speed.

FIGURE 3.25
Micro-hybrid topology: (a) with an alternator, and (b) with a motor/generator.

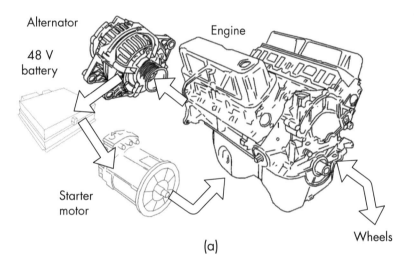

(a)

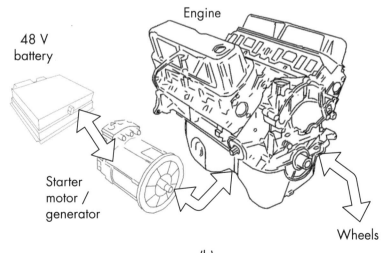

(b)

3.4.5 Plug-In Hybrid (PHEV)

A PHEV is like an HEV, except that it includes an onboard charger and an AC inlet (or at least a DC charging port for an external charger). Just like an HEV, a PHEV can be:

- Series:
  - Power source (generator) on DC bus (Figure 3.26(a));
  - Traction battery on DC bus (Figure 3.26(b)).
- Parallel (Figure 3.26(c)).

When the battery is charged, a PHEV operates mostly as a BEV[35], in the Charge Depleting mode. After the battery is nearly depleted, the PHEV switches to the

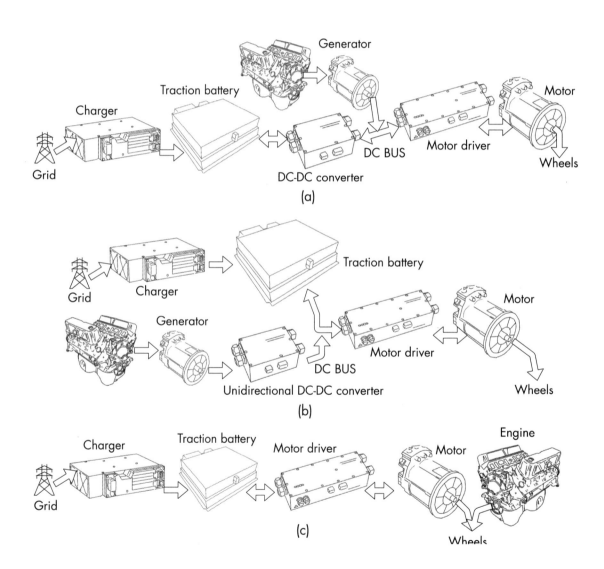

FIGURE 3.26   PHEV topologies: (a) series, generator on DC bus, (b) series, battery on DC bus, and (c) parallel.

---

35. If the motor is undersized, the engine may be turned on at times of peak demand.

Charge Sustaining mode and operates as an HEV, at about 10% SoC. If the headroom is insufficient, the engine can be used to raise the SoC slightly to keep cells from "bottoming out" at 0% SoC.

For optimal energy storage, the battery is top-balanced. Mid-balancing the battery in the Charger Sustain mode is counterproductive as it conflicts with the top-balancing that occurs after the vehicle is fully charged. Also, the energy lost during balancing at 10% SoC may not be acceptable.

3.5 UNMANNED APPLICATIONS

Having explored the technology in electric vehicles, let's now look at classes of applications, starting from vehicles that do not carry people[36].

3.5.1 Robots

Autonomous vehicles can replace humans in dangerous or stressful jobs.

3.5.1.1 Warehouse Robots

The largest online retailer in the world employs thousands of warehouse team members working under extreme conditions; the company is unlikely to let its employees unionize and to improve the working conditions. The solution is to let robots do those jobs. Warehouse robots work at least 80% of the time, day and night.

Warehouse robots use a small traction battery because they can be recharged often since they're always close to a charging station. This reduces their weight. A typical cycle would consist of 10 minutes charging and 50 minutes operating. LTO is the only Li-ion chemistry that can be charged safely in 10 minutes. The energy density of LTO is unimpressive, but the fast charge times make the use of these cells worth it.

Given that the traction battery is charged much faster than it is discharged, a two-port BMS that has a high current charging port and a low current discharging port is most effective. Such BMS is not readily available off-the-shelf. A custom BMS is required.

3.5.1.2 Hazard Robots

Who but a robot can reach the "elephant foot" under reactor #4 in Chernobyl or the flooded depths of the Fukushima Daiichi power plant?

The electronics in a traction battery in a robot for nuclear accidents must be radiation-hardened. Ionization scrambles digital memory, making a BMS unreliable. An analog BMS, using older integrated circuits (with large silicon dies), is less likely to misbehave.

Indeed, it may be best to forgo a BMS altogether. The robot can use a top-balanced traction battery that is charged with a balancing charger; the risk of an overdischarge is reduced by using the robot for relatively short runs. The cost of the potential damage to the battery is minor compared to the inconvenience of a robot shutting down and blocking the bowels of a nuclear power plant due to a radiation-damaged BMS.

---

36. I haven't found a gender-neutral term in place of "unmanned" that is technically correct.

3.5.2 Unmanned Aerial Vehicles (UAVs)

The popularity of UAVs3[37] for civilian use has exploded (Figure 3.27(a)). These products are infamous for their abuse of their Li-ion batteries (Figure 3.27(b)). These are not just power applications; they are "insane power applications."

Outlandish claims from unscrupulous battery vendors, combined with the users' feverish need for performance, lead to extreme abuse of the cells, well outside their safe operating area (see Volume 1, Section 2.3.2):

- Discharged in 10 minutes;
- No BMS;
- No restraint for pouch cell expansion;
- Chintzy power connectors, undersized wires;
- Batteries rated for "60 C" operation[38].

"Battery life be damned, I need performance now!"

This abuse is stomach-churning for those of us whose professional goal is to ensure the safe and prolonged use of Li-ion. You're satisfied with your UAV's performance, and Hobby King is happy to sell you a new battery in a month. These batteries are commonly charged with a balance charger (see Volume 1, Section 6.5.4); this is safe, even without a BMS.

During discharge, there is no BMS to protect the cells from overdischarge. While normally I would scream: "You MUST use a BMS!", in this case, I just sigh and say: "Whatever. If you're going to ruin your cells anyway by discharging them too fast, who cares if you also overdischarge them?"

I must accept that this industry's requirements are not negotiable:

- No protector BMS because a PCM capable of such high current is heavy and bulky; it wastes precious power.
- Cell longevity is not a concern; performance is the only concern. However, we don't want the UAV to go up in flames.
- Having one battery in the air and one fast charging on the ground is ideal.

Having understood these requirements, we drop two bismuth subsalicylate tablets in our glass of water and offer the following suggestions to the UAV designer.

FIGURE 3.27
UAV: (a) in flight,
and (b) battery.

(a)   (b)

---

37. Small quadcopters and multirotor copters are commonly, though incorrectly, called drones.
38.   Yeah right! 60 C means full discharge in 1 minute, something that a capacitor can do, but not a Li-ion cell if you want to get any useful life and power out of it.

3.5.2.1 BMS

The following approach performs the same function as a protector BMS without the extra weight and cost. It has the advantages of providing advanced notice when the battery is empty, and of avoiding operation past the maximum power point, an area of diminishing returns.

Have the main controller in the UAV monitor the following:

· Tap voltages: provide a connector in the UAV to mate with the battery's balance connector;
· Battery current: use a Hall effect sensor as it is more efficient than a shunt;
· Battery temperature.

Let the UAV do the following:

· Calculate the terminal voltage of each cell.
· Given the cell resistance, estimate the OCV of each cell.
· If any cell OCV[39] drops below one-half of the cell terminal voltage, reduce the motor drive to operate below the maximum power point; this creates an MPPT function[40].
· If an OCV is starting to get low, limit the drive to the motors and warn the pilot; this may be enough to let the OCV recover a bit.
· If an OCV approaches the absolute minimum voltage, land, overruling the pilot; then, when an OCV reaches the minimum, shut off the motors to effectively stop the battery current.
· Do the all the above also based on cell temperature.

Given that there is no protector switch and no shunt, this implementation of the BMS functions does not reduce the efficiency of the UAV.

3.5.2.2 Battery

Pick cells with a low MPT (see Volume 1, Section 1.5.2). Even if their capacity is lower than energy cells of the same physical size, they may last longer. The energy cells may have more charge, but, when discharged so fast, they waste so much energy in heating themselves that they won't last as long as more efficient power cells.

Do not believe the C-rating claims of the battery vendor (see Volume 1, Section 2.8.1). Instead, compare cells based on their maximum power time; before buying the battery, calculate the MPT from the discharge curves provided by the cell manufacturer: not the battery vendor; the cell manufacturer. Then, after receiving the battery, confirm the MPT based on data from actual discharge tests.

Pouch cells are lighter than other cell formats because they do not have a hard case. However, they must be placed in a hard case to contain expansion (see Volume 1, Section 5.6.2). It is better to add this constraint to the battery, rather than to the UAV so that the battery is constrained while charging. Place a thermistor in close contact with the cells, so that it may be used to sense the battery temperature while charging and while in use in the UAV.

---

39. Open circuit voltage.
40. Maximum power point tracker, same as a in a solar change controller used to optimize the power from solar panels.

3.5.2.3   Charging

If no BMS is present while charging, cells should be charged individually, either with individual chargers or with a balance charger (see Volume 1, Section 6.5.4). In either case, the battery requires a balancing connector.

Don't buy one fast charger and two batteries, charging them at 6 C (10 minutes)[41]. Instead, buy four standard speed chargers and five batteries. At any given time, one battery is in use in the UAV and the other four are charged at 1 C (taking about 90 minutes). In the long run, it's cheaper to buy five batteries and charge them slowly rather than buying only two batteries and abusing them with fast charging.

If you care for the life of your battery, charge at a slightly lower voltage than rated (say, 4.1V per cell). You won't lose much range when the battery is new, and you'll retain a lot of the range when the battery starts getting old, compared to a battery that is charged to the full voltage and ages faster. Do not let the battery remain in the charger after it's full (unless the charger shuts off automatically).

3.6   LIGHT EV APPLICATIONS

Light electric vehicles (LEVs) include electric skateboards, electric bicycles, and go-carts. The user can easily lift them as they weigh less than 50 kg. A light EV can have as few as a single wheel. The traction battery has a voltage of 12V to 48V and uses a PCM (see Volume 1, Section 4.3).

3.6.1   Personal Transporters

Personal transporters are among the smallest EVs They include self-balancing scooters (also known as hoverboards) (Figure 3.28(a)), self-balancing personal transporters (Segway PT), kick scooters, skateboards, unicycles, and golf course personal transporters (Figure 3.28(b)).

They use a tiny traction battery. The typical battery for a hoverboard consists of ten 18650 cells in series (36V, ~2.5 Ah) and a 30-A protector BMS mounted directly on the cells (see Volume 1, Section 4.2.2.5). The cells must have a low MPT (see Volume 1, Section 1.5.2) because the vehicle has high-power demands for its size, yet a decent energy density.

Fires in self-balancing scooters have damaged the reputation of the industry. The causes included

- Impacts against the traction battery resulted in penetration into the cells.

- Use of energy cells (instead of power cells) resulted in overheated cells and thermal runaway.

FIGURE 3.28
Light EVs: (a) self-balancing scooter, and (b) GolfBoard.

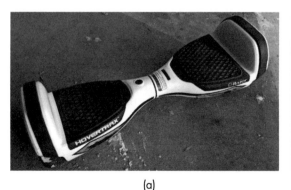

(a)   (b)

---

41.   You could charge an LTO battery in 10 minutes, but the energy density of LTO cells is too low for UAVs.

- Ineffective protector BMSs not shutting down the battery in case of overcurrent.
- BMS not preventing charging below freezing, cell overcharge, or charge after an overdischarge.

A fire while charging is particularly dangerous because these transporters are often charged inside the home.

Despite its small size and lightness, the traction battery must be rugged, to protect the cells from mechanical intrusion in case of an accident and to keep the bus bars from being pushed against each other and creating a short circuit.

A mid-battery fuse is a good idea. When it blows, it damages the BMS, yes, but so what: the battery is sealed and not repairable.

The cell arrangement must be well thought out to minimize the voltage between physically adjacent cells (see Volume 1, Section 5.4). The power straps between cells must be constrained to prevent them from touching each other in case the battery is deformed.

The protector BMS must be able to shut off the battery current. The battery usually has only two contacts, which is unfortunate because the BMS has no way to warn the vehicle that it's about to shut down the battery. If the battery had a communication link, the vehicle could warn the user to step off before shutting down the battery current.

The BMS should prevent charging below 0°C, as doing so damages the cells. However, personal transporters are likely to be charged inside the home, where the temperature is well above freezing.

3.6.2 E-Bikes

Small electric bikes and mopeds use a small, medium-voltage traction battery. Often the battery can be removed from the bike for charging at home. A simple fuel gauge to displays the state of charge is often just a voltmeter with three to five LEDs.

Full-size e-bikes use a somewhat larger battery, too large to carry inside the house for recharging. E-bike manufacturers typically buy a complete, ready-made traction battery from a Chinese manufacturer. E-bikes usually use an off-the-shelf PCM: those are inexpensive and readily available, giving little reason to develop a custom BMS. The typical voltages are 24V to 96V.

A simple fuel gauge in the bike displays the state of charge to the user. The BMS should support this display with an analog signal or a simple serial stream.

3.6.3 Go-Carts

The e-go-cart is likely to use a 48V battery and a 600-W brushless motor. It is likely to use a relatively small battery (~300 Wh) with power cells, as the go-cart is used at high acceleration and for a short-duration. It would probably use pouch cells. A cheap, 48V, 20-A, Li-ion PCM protects the battery.

3.7 SMALL PASSENGER EV APPLICATIONS

Small passenger EVs are smaller than a passenger car. They include electric motorcycles, snowmobiles, golf carts, micro commuter cars, and auto-rickshaws (tuk-tuks). The traction battery has a voltage of 48V to 156V.

You can find unbranded PCMs on eBay, AliExpress, and AliBaba. Table 3.7 lists analog PCMs explicitly designed for small EVs (Table 3.7). When two values for current are listed, the first one is for charging.

TABLE 3.7
Analog Protector BMSs for
Small EVs

Brand	Model	Cells	Communication	Notes
Ayaa	BMS-LB16S150-1501	16	100	e-bikes
Ayaa	PCM-L26100-313	30	100	e-bikes
Leadyo	PCM-L24S60-622 (24S)	16~24	20/60	
Leadyo	PCM-L13S25-B24	13	5/20	e-bikes
Shenzhen Li-ion Bodyguard	?	20	60	
SuPower Battery	86V 24S xxA Lion LiPo	24	30~60	
Qwawin	PCM-L16Sxxx-xxx	16	12~350	25 models
Qwawin	PCM-L20Sxxx-xxx	20	40~150	2 models
Qwawin	PCM-L24Sxxx-xxx	24	12~100	8 models
Qwawin	PCM-L26Sxxx-xxx	26	12~60	5 models
Qwawin	PCM-L32Sxxx-xxx	32	16~100	2 models

Table 3.8 lists digital PCMs were also designed for small EVs.

Larger batteries use a BMU (see Volume 1, Section 4.4.2) plus a contactor. Table 3.9 lists digital BMUs that are designed for small EVs.

Let us look at specific examples of small EVs.

3.7.1 Motorcycle

Electric motorcycles[42] are blessed by high torque, while also being refreshingly quiet (Figure 3.29(a))[43]. High demands are placed on batteries for full-size electric motorcycles, yet there is not enough space for a full-size traction battery. Even though the battery is small, a motorcycle's range is comparable to the range of modern EV cars because the bike is so light[44].

In the last decade, many full-size EV motorcycle companies have come and gone, proportionally more than EV car companies; none have cracked open any market outside of China. Harley Davidson just started selling their LiveWire EV bike.

TABLE 3.8
Digital Protector BMSs for
Small EVs

Brand	Model	Cells	Current	Communication	Notes
Ayaa	AY-EK01	32	30		Motorcycles
BestTechPower	Many	16~24	up to 200	CAN, Bluetooth	e-bikes
JTT	P-Series	5~13	25	—	e-bikes
BMS PowerSafe (Ventec)	iBMS	10	50	CAN, RS232	e-bikes
BMS PowerSafe (Ventec)	iBMS	18	350	CAN, RS232	e-bikes

TABLE 3.9
BMUs for Small EVs

Brand	Model	Cells	Additional Features
Micro Vehicle Lab	?	5~120	CAN, RS 232 C, RS 485
Shenzhen Klclear Technology	BMS05-48S8T-IG-12V-200V	50	CAN

42. Thanks to Travis Gintz for help with this section.
43. In 2010, Matt Dieckman was killed in a collision with a car while riding this bike. Be careful out there.
44. Zero claims that their Zero SR ZF14.4 + Power Tank model has a range of 350 km in the city.

FIGURE 3.29
(a) electric motorcyle
(Courtesy of Travis Gintz),
and (b) snowmobile.

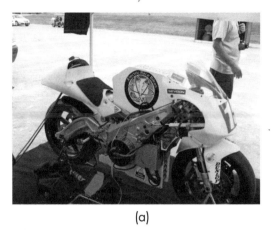

(a)   (b)

However, the traditional Harley riders prefer the shattering sound of the traditional Harleys, and most environmentalists cannot afford to pay $30,000 for a motorcycle. If you plan to start an EV bike company, you need to be aware of the odds.

The battery should use cells that offer a good compromise of energy and power. Prismatic cells perform poorly in a bike[45]; people who in the 2000s were misled by false claims from manufacturers of large cylindrical cells were sorely disappointed. The only cells you should consider are small cylindrical cells (see Volume 1, Sections 2.2.3 and 2.6.1.4), large pouch cells, and some of the new small prismatic cells for EV use. Small cylindrical cells allow you to fit the contour of the space normally taken up by the gas tank and the cylinders in a standard motorcycle. Pouch cells and small prismatic cells do not offer this flexibility, so it's harder to fit them in an EV conversion of an ICE motorcycle. They do work well in brand-new motorcycle designs.

A battery voltage of 48V to 96V is appropriate. The energy that fits in the bike ranges from 3 kWh to 18 kWh. The battery must be designed to withstand the level of vibrations in a motorcycle, which is more severe than in most other EVs (see Section 3.14.6.1). Mechanically, the battery must remain safe if the bike is dumped and lays on its side.

An EV motorcycle is too big for products appropriate for an e-bike and too small for products appropriate for a passenger EV. For example:

- The battery current in an EV motorcycle is rather high for a protector BMS with solid-state protector switches (for an e-bike) and rather low for a BMU that drives contactors (for a car).
- The battery voltage is around the threshold where a battery isolation detector should be used.

Table 3.8 lists BMSs that are marketed for motorcycle use, although any BMS for passenger vehicle use must be considered as well.

If using a permanent magnet DC motor, since motorcycles do not have reverse, there's no need for a reversing contactor or a four-quadrant motor driver. Regen braking requires a two-quadrant driver (see Section 3.3.4.1). Four-quadrant motors, such as AC and brushless motors, work well.

---

45. The GBS battery motorcycle package that you see advertised online uses large prismatic cells.

3.7.1.1 Conversions

When converting an ICE motorcycle to EV, attention must be paid to the change in mass; although Li–ion batteries are light, a large battery increases the total mass of the bike, so the suspension must be adjusted accordingly. The raised center of gravity is a bigger concern, as it affects handling.

Brushed DC motors are no longer used; today, people doing EV conversions use PMAC motors (including BLAC/IPM and externally excited BLDC motors) (see Section B.3), driven by drivers for such motors[46]. Rather than developing a design on your own, you may opt for buying a complete conversion kit[47].

3.7.2 Snowmobiles

You're cross-country skiing in the peaceful, white forest, when the obnoxious roar of a snowmobile ruins the silence and the stench of poorly burned fuel mars the fresh smell of the evergreen (Figure 3.29(b)). Recently, electric snowmobiles have been introduced to allow skiers and snowmobile riders to coexist peacefully.

Compared to gas-powered snowmobiles, electric ones suffer from a shorter range and the effects of cold weather. While the range of ICE snowmobiles exceeds 150 km, electric snowmobiles claim a range of up to 90 km. Do consider that the typical outing lasts only 10 to 30 km.

In cold weather, it may be harder to start a gas snowmobile; yet, once it starts, it doesn't mind the cold. On the contrary, the performance of electric snowmobiles worsens as the temperature drops. Additionally, the battery cannot be recharged below freezing.

Two solutions help electric snowmobile operate in cold weather: LTO cells and heating. The internal resistance of LTO cells does rise at low temperatures, although not nearly as much as other Li–ion chemistries. LTO cells are almost as good at $-20°C$ as at room temperature and are still usable at $-30°C$. However, LTO cells have the lowest energy density of any Li–ion chemistry.

If LTO cells are not acceptable due to their high cost and low energy density, then use an electric blanket. When plugged into AC power, the blanket raises the battery temperature to above $0°C$ so that charging may start. Insulation keeps the heat within the battery. Care is required to distribute the heating evenly and to prevent uncontrolled overheating (see Volume 1, Section 5.17.8). During use, the current flowing through the cells' internal resistance heats them somewhat. The battery may power the electric blanket to raise its temperature to $0°C$.

The BMS must communicate with the motor driver to gradually limit the battery current if the battery is too cold or the charge is too low. The snowmobile becomes sluggish, keeping the driver from gunning it and damaging the battery; yet the driver should be able to return home.

The battery must be designed to withstand the intense shaking that snowmobiles experience (more intense than in a passenger vehicle). The interconnections in a normal battery would break after a while. I recommend

- Sealing the traction battery against moisture penetration.
- Following the section on battery design for high vibration (see Section 3.14.6.1).

---

46. Sevcon or Motenergy. See Section B.3.
47. Such as from HPEV.

• Securing the battery so that it remains in place if the snowmobile is turned on its side or upside down.

3.7.3   Golf Carts

ICE golf carts are ubiquitous micro-hybrids (see Section 3.4.4), making them the first HEV widely experienced by the public (Figure 3.30(a)). They use a standard SLI battery (see Section 2.9) to start the engine. Similarly, electric golf carts have been the most widely available and best-known electric vehicle. The rest of this section is specific to electric golf carts.

3.7.3.1   Battery

The battery in an electric golf-cart is either 36V or 48V. For lead acid, this consists of

• 36V: six 6V batteries in series;
• 48V: either six 8V batteries, or four 12V batteries.

The capacity is between 95 Ah and 225 Ah. The depth of discharge is limited to 50% for the sake of the life of the lead-acid battery. Therefore, the usable energy is in the range between 1.7 and 5.5 kWh. Li-ion batteries may be used instead of lead acid for better performance. Since a Li-ion battery can be discharged completely and safely, a 2-kWh to 5-kWh battery is sufficient.

There are three approaches to a Li-ion golf cart retrofit:

• Replace individual lead-acid batteries with Li-ion batteries of the same voltage and form factor.
• Replace the entire 36V or 48V battery with a complete Li-ion replacement kit.
• Design a new Li-ion battery using individual cells and a BMS.

Ready-made, 12V Li-ion batteries that are a drop-in replacement for lead-acid batteries are on the market. They include protectors[48].

FIGURE 3.30
Small EVs: (a) golf cart, and (b) Polaris GEM NEV.

(a)   (b)

---

48.   Smart Battery (lithiumion-batteries.com): SB100 battery: 12V, 100 Ah, 100A protection. Nexgen (nexgenbattery.com): NG100, the same battery, just different branding. Note: Relion batteries are not protected and therefore unsafe in this application, which does not have a separate BMS.

You can buy ready-made Li-ion kits for golf carts, complete with BMS and charger, and drop them into an existing golf cart[49]. The battery form factor is an issue because each golf cart brand and model places the lead-acid batteries in different locations. A significant concern is whether the drop-in battery is mechanically stable and whether it presents a safety concern in case the golf cart is overturned.

Otherwise, you can design your own lead-acid compatible Li-ion batteries; your options are limited:

- 6V battery: two LFP cells in series.
- 8V battery: no combination of Li-ion cells of any chemistry works well for 8V batteries; the voltage of two 3.7V cells in series is a bit too low, and with three LFP cells in series, it is too high.
- 12V battery: four LFP cells in series.

Each battery requires its own protector BMS, which makes it expensive, especially if you use eight 6V batteries to reach 48V.

If designing a golf cart from scratch, the battery needs several large prismatic 100-Ah cells in series and a protector BMS capable of a given continuous and peak current (Table 3.10).

Energy cells work well because the discharge is not too fast.

3.7.3.2 Motor and Driver

Golf carts use DC motors[50] with a power of on the order of 3 kW continuous, 8 kW peak. They can be either series-wound or a SepEx[51]. Table 3.11 helps to identify the type of motor driver.

TABLE 3.10
Cells and Current in a
Li-Ion Golf Cart Battery

Voltage	Energy	Continuous Current	Peak Battery Current	Charger Current	LFP Cells (3.2V)	LCO, NMC Cells (3.6V)
36V	3.6 kWh	80A	220A	21A	12 in series	10 in series
48V	4.8 kWh	60A	160A	13A	16 in series	13 in series

TABLE 3.11
Golf Cart Motor Type
Identification

	Series	Shunt, SepEx/Regen
Motor terminals	4 large (A1, A2, S1, S2)	2 large (A1, A2), 2 small (F1, F2)
Stator windings	Wide, flat	Thin, round
Speed encoder	No	Probably
Run/maintenance or tow/ run switch	No	Yes
Forward/reverse switch	Large and clunky, high current, or drives four contactors (solenoids)	Small toggle, low current
Regenerative braking	No	Yes

---

49. HPEVS, Elite Power Solutions.

50. Conversion kits from DC to AC motors are available, which deliver higher power in the same form factor.

51. Which people in this industry may call a regen motor.

Older golf carts used a resistor and tap voltages to change the speed; today, they all use DC motor controllers[52], either a simple controller (for series-wound motors) or a SepEx controller (with two extra output terminals for the field). These controllers are not isolated and use a 5 kΩ throttle pot. They do not include precharge.

The Run/Maintenance or Tow/Run switch in a shunt-type golf cart turns off power to the field coil. It has two functions:

1. It releases the electrical braking.
2. It is the main power switch that disables all power when the golf cart is not in use.

3.7.3.3 Power Circuit

The traction battery uses just one contactor and a precharge resistor without a relay. When off, this circuit offers no galvanic isolation between the battery and the rest of the golf cart.

The Forward/Reverse switch on the dashboard reverses direction by reversing the current in the field winding. For a series-wound motor (Figure 3.31(a)), this is a large and rugged switch because it must handle the entire motor current. For a SepEx motor (Figure 3.31(b)), it's a small rocker switch because the field current is low.

The charger is off-board. When converting to Li-ion, the voltage of the original lead-acid charger may not be appropriate for Li-ion batteries, so it should be replaced by a properly adjusted Li-ion charger.

3.7.4 Commuter EV

The typical commuter drives alone and, on average, 9 km in the United States or 5 km in Europe and most other countries. Over the years, many companies have developed small commuter EVs, especially for this typical commuter.

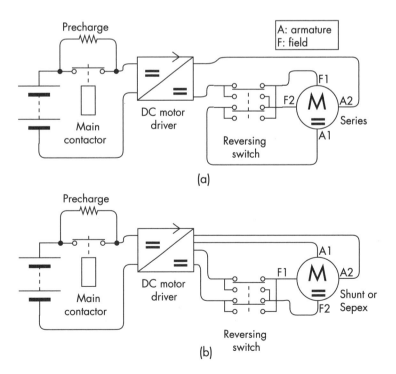

FIGURE 3.31
Typical traction circuit in a golf cart: (a) series-wound motor, and (b) shunt-wound or SepEx motor.

---

52. From Curtis, Alltrax, and many others.

This EV is typically an enclosed, three- or four-wheeled vehicle for one or two passengers. If limited to 55 km/h, it is called a NEV[53], which is truly a glorified golf cart (Figure 3.30(b)).

In the United States and Europe, no commuter EV has been adopted in the degree that would be required to make a dent in traffic. Commuter EVs are starting to have an impact in China[54]. Electric tuk-tuks are popular in South and East Asia.

3.7.4.1 Chances of Success

Most small companies that began the process of offering a small commuter EV did not take off; some failed even before the vehicle went into production. Aptera, Bug-E, Citicar, Corbin Sparrow, Gizmo, Saba Motors, Tango, Think, Weego Whip, Zap Zebra, Zenn, and Z-wheels come to mind.

Having worked directly or indirectly with most of these companies, I believe that the failure was particularly due to poor engineering, besides a lack of funds[55]:

- Corbin did not have a single engineer on its staff; the EV enthusiast who put together the electrical system purposely did not write down the schematic diagram for the sake of job security[56]; no two Sparrows are the same.

- I visited the Zenn headquarters in Toronto, and I was shocked to find that it was not a factory, but a two-bay garage.

If you believe that you are the entrepreneur who will finally succeed in bringing a small EV commuter car to the market, best wishes to you! However, please learn from history and the mistakes of others:

- Start from an existing vehicle chassis (see what's available from China) with working brakes, comfortable suspension, and no water leaks.

- Have a $1 million fund available before you even start; anything less than this, you might as well buy a boat instead, or go gamble in Macao or Las Vegas: you'll have more fun, and the result will be the same.

- Read the existing market research.

- Hire engineering consultants: mechanical, electrical, automotive, and industrial design; be ready to pay them for their services.

- Form partnerships with industry leaders; they don't have to be giants, but they do need to be an established company; this doesn't mean to ask for free product under the guise of "partnership." The moment you ask for free product, you're tagged as a losing proposition; partnership means issuing a press release in common, it means spending time at each other's facilities, it means signing a memorandum of understanding, it means having both company names on white papers, and it means working collaboratively to codevelop or adapt products, systems, or subsystems; it does not mean freeloading.

- Work with EV enthusiasts: yes, some turn out to be gadflies; but others have valuable experience to share.

---

53. Neighborhood electric vehicle.
54. Geely Zhi Dou D2 EV.
55. And, yes, management, of course.
56. Back then, I reverse-engineered the electrical schematic diagram of my Sparrow so I could service it. Today, the man who did not write down what he designed called me to ask where he could find my diagrams.

3.7.4.2 *Technology*

The traction system for a small EV is the same as for a golf cart, except

- The battery voltage is between 96V and 156V.
- The range is at least 40 km.
- The charger is onboard.
- The motor can be DC series-wound (Figure 3.32(a)), brushless DC, or AC (Figure 3.32(b)).
- With a DC motor, a reverse contactor is used instead of a reversing switch.

3.7.4.3 *Case Study: Solar Micro-Car*

A solar micro-car is a small EV with a small area of solar panels (because this is all you can fit on top of such a small vehicle) (Figure 3.33). I made myself one out of a Corbin Sparrow [8]. It's ridiculously impractical: it has to sit in the sun for 2 months to get its full range of 100 km from its 100-W panel. The second problem is that the BMS is unable to evaluate the SoC very well because the charging current is about 100 mA,

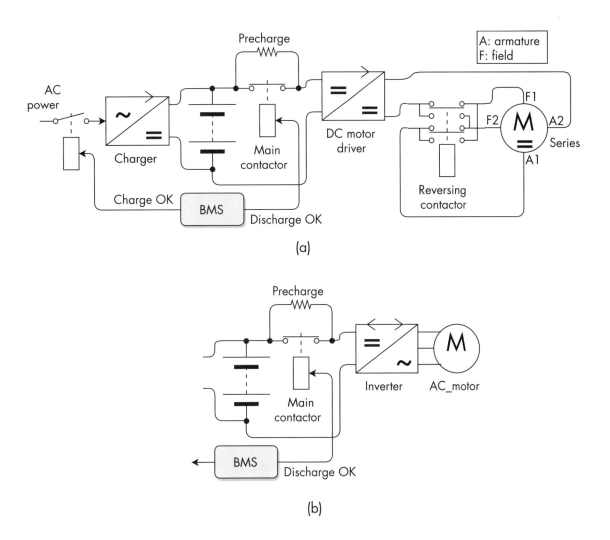

FIGURE 3.32   Typical traction circuit in a small EV: (a) DC motor, and (b) AC motor.

FIGURE 3.33
My solar micro-car: a
Corbin Sparrow converted
to Li-ion and 100%
solar: (a) charging in the
sun, and (b) circuit.

(a)

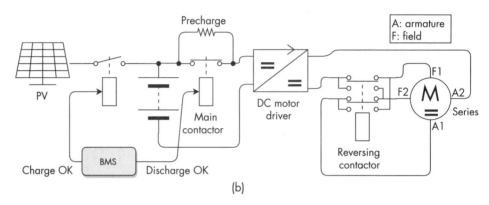

(b)

yet the load current is 200A. The current sensor offset dwarfs the 100-mA reading, so the SoC evaluation drifts terribly during the long charging period.

There is no solar charge controller; I selected the number of Li-ion cells in series for a total voltage that matches the maximum voltage from 11 solar panels. The twelfth panel charges a 12V lead-acid battery. The battery uses 104 each 40-Ah cells in a 2P54S arrangement, for a total of 13 kWh.

3.7.5 Auto-Rickshaws

Tuk-tuks are a ubiquitous taxi in developing countries, especially in Southeast Asia. This three-wheeled vehicle is based on the Italian Piaggio Vespa moped design. An added cargo area behind the driver seats two to 10 people. The two-stroke engines generate heavy pollution in cities.

The conversion of tuk-tuks to EV dramatically benefits the environment. Electric versions are being introduced, and older tuk-tuks are being converted, despite eye-watering costs for low-income rickshaw drivers. Grants from developed countries are offsetting some of these costs. In India, a proposal is that the rickshaw driver would rent a battery rather than buying one.

The battery is 48V to 96V and on the order of 5 to 10 kWh. It's divided into two sections, one under each bench, and one along the starboard side and one along the port side. Most of them use brushed DC motors, although some use AC or BLDC motors. Other than this, the technology is the same as for golf carts.

3.7.5.1   Case Study: Kathmandu

In Nepal, auto-rickshaws are called Tempo, and the electric version is called Safa Tempo (clean tuk-tuk). There are more than 600 Safa Tempos in the Kathmandu valley alone (Figure 3.34) [9].

A Safa Tempo uses

- Four each, Trojan T-125 Deep-Cycle Flooded batteries, 240 Ah, for a total of 48V, 11 kWh;
- A Curtis DC motor controller;
- An Advanced DC series-wound motor;
- An off-brand charger.

Swiss engineers Markus Eisenring and Thomas Kuster, under a grant from REPIC[57], developed an open-source conversion from lead-acid to Li-ion EV. It uses 24 large prismatic cells in series and a Lithiumate HD BMS. At the time of this writing, just 25 of the Sofa Tempos have been converted to Li-ion.

3.8 SMALL INDUSTRIAL APPLICATIONS

EVs are preferred in industrial applications because they do not pollute and are quiet, powerful, and economical.

3.8.1   Forklifts

Forklifts[58] for use inside warehouses use propane fuel because, in low doses, its emissions are considered safe. Electric propulsion using lead acid is even better, as it has fewer emissions, just some hydrogen at the end of charge. The upfront cost of an electric forklift is higher than for a propane one, but the significantly lower operating

FIGURE 3.34
Safa Tempo in
Kathmandu, Nepal.
(Made by Mr. Shravan
Chaulagain.)

---

57. Renewable Energy Promotional International Cooperation, repic.ch.
58. Thank you to   of EnSol Technology, Russia, Juan Francisco Otth of Royal America, Chile, and Sylwester Zawadzki of Emtor, Poland, for help with this section.

costs soon make an electric forklift more economical. Today, most forklifts for indoor use are electric (Figure 3.35(a)).

The energy in forklift batteries is on the order of 10 kWh to 40 kWh. The voltage ranges from 24V to 80V (Figure 3.35(b)). Typical batteries are

- 24V, 500 Ah, 10 kWh for indoor use[59];
- 48V, 1,000 Ah, 40 kWh for heavy-duty indoor use;
- 80V, 1,000 Ah, 66 kWh for outdoor use.

In the Americas, they often use Anderson Powerpole SB-175 (175A) or SB-350 (350A) hermaphroditic connectors[60]; various versions are keyed differently and color-coded to indicate the voltage: red for 24V, gray for 36V, blue for 48V, and black for 80V (Figure 3.35(d)).

The standard connector in Europe is based on the EN 1175-1 and DIN VDE 0623-589 standards[61]. It is designed for 80A, 160A, 320A, or 640A. The same connector can be keyed for different voltages and different battery types by inserting a hexagonal key in the plug (Table 3.12).

The hex key may be placed in one of six positions, to code the connector for 24V, 36V, 48V, 72V, 80V, and 96V. Two optional auxiliary contacts are available used for CAN bus communications.

Charging involves disconnecting the battery cable from the forklift connector, pulling it out, and connecting it to a long cable from the stationary charger (Figure 3.35(c)). Opportunity charging occurs during lunch breaks or other pauses. Fast charging (at high current) is effective, although it should be stopped at 80% SoC to reduce battery degradation. Fast charging of lead-acid is done at a C/4 or

FIGURE 3.35
Electric forklift: (a) battery side, (b) lead-acid battery close-up, (c) charger, and (d) Anderson Powerpole hermaphroditic connectors.

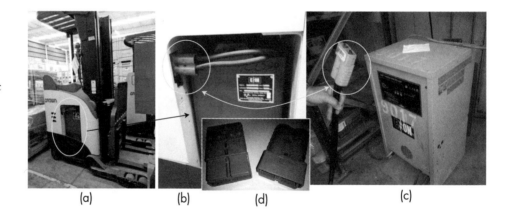

(a)   (b)   (d)   (c)

TABLE 3.12
DIN Forklift Connectors

		Charger	Battery	Forklift
Gender		Male, shroud	Female, insert	Male, shroud
Key insert color code		Red: wet, high current Gray: wet Green: dry, immobilized electrode Blue: dry		Yellow (universal)

---

59. These values are the energy available from a lead acid battery, which is not fully discharged.
60. Hermaphroditic means that the two mates are identical, which is a rare feature in connectors.
61. These connectors are made by REMA (DIN series), Schaltbau (Series LV), Eaxtron (FEM type), and others.

even C/2 rate, meaning that it still takes 2 to 4 hours to bring an empty battery to 80% SoC.

Electric forklifts have two motors: traction (~5 kW) and lift (~10 kW). Assuming that both motors are active simultaneously, the peak battery power is about 15 kW, which at 48V is about 300A.

The heavy lead-acid batteries are ideal as they provide the weight with a low center of gravity that a forklift requires. However, lead-acid batteries, have a short life, have relatively high internal resistance, the voltage sags under load, and the efficiency is low. In particular, flooded lead-acid batteries require regular maintenance, emit flammable hydrogen gases, and cause splashes of acid.

Most modern forklifts use regenerative braking when lowering a load. However, be careful when saying "regen" because, when we talk about regen, we mean using braking energy to recharge a battery; when forklift people talk about regen, they mean rejuvenating a lead-acid battery through desulfation.

In many industries, the battery charge is sufficient to power the forklift during a working day. Various solutions have been developed if the battery cannot be charged overnight:

- Battery swap;

- Hybrid forklift;

- Fast charging or opportunity charging;

- Li-ion battery.

Battery swap involves rolling out the depleted battery and rolling in a charged one. Batteries are charged in a battery room designed to deal with the chemicals and emissions associated with lead-acid charging.

A hybrid forklift is a series HEV (see Section 3.4.3.1) powered by hydrogen. Filling the hydrogen tank is as fast as filling a propane tank. A fuel cell converts the fuel to electricity, continuously at low power. Conveniently, the emissions of a fuel cell are just water steam, not hydrocarbon fumes. A buffer battery (see Volume 1, Section 5.1.4) supplies peak power to the motor and accepts regen power from the lift motor.

3.8.1.1 Li-Ion Forklifts

Recently, forklifts with Li-ion traction batteries have become available, offering many advantages:

- Higher reliability, no requirement for scheduled maintenance, longer life, higher energy efficiency, faster charging, zero emissions, and a smaller, lower-to-the-ground vehicle.

- With lead acid, companies own extra batteries and replace an empty battery with a charged one; with Li-ion, there's no need to exchange batteries, and there's no need for extra batteries.

- When building a new warehouse, it's better to start right off with all Li-ion forklifts because building a charging room for lead-acid batteries is expensive and has ventilation and safety measures.

- What is fast charging to lead acid is normal charging for Li-ion; depending on the cells, a Li-ion battery may be charged safely at 1C.

However, the forklift industry is conservative and prefers to stick to the tried and true: lead acid. Among a chorus on naysayers [10], change is coming slowly.

Rather than buying new forklifts, managers may prefer to convert their present fleet of lead-acid forklifts to Li-ion, either by replacing the lead-acid battery with a plug-and-play Li-ion battery designed for a particular forklift or by designing a custom battery.

3.8.1.2 Single-Port Battery Design

Lead-acid batteries have a single port that handles both charging and discharging: a single battery cable is moved from a plug in the forklift to the charger and back. A Li-ion battery designed to be a direct replacement to a lead-acid one would also need to have a single port. This is inconvenient because a single-port Li-ion battery is more complicated: the BMS must control charging and discharging independently through this single port. A single-port battery can be implemented with one or two contactors, depending on the BMS (see Volume 1, Section 5.12.2). Using one contactor (Figure 3.36(a)) is cheaper and more efficient but requires a sophisticated BMS that can read the terminal voltage and the battery current to determine whether the contactor should be on or off.

Otherwise, a two-contactor solution is used (Figure 3.36(b)). The contactors are wired in series, each with a rectifier diode, to control charging and discharging separately. Besides the additional components, this solution wastes power in a rectifier diode when initially charging an empty battery, until the cell voltage rises to the point

FIGURE 3.36
Forklift battery, single port: (a) single contactor, (b) dual contactor, and (c) with precharge.

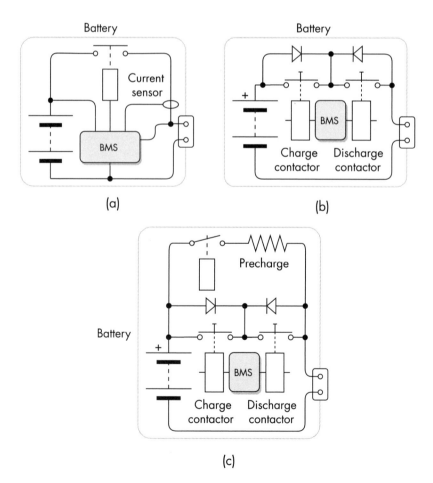

that the BMS reenables discharging. This is also an issue when initially discharging a full battery until the cell voltage drops to the point that the BMS reenables charging. Certain BMSs can turn on both contactors and bypass the diodes as appropriate, therefore avoiding the power loss in the rectifier diodes.

An engineer at a forklift company that tried making a single-port battery tells me that the first approach didn't work so well. He did not try the second approach.

A Li-ion battery for a forklift could use a centralized BMS or a distributed one. Because of the high current, a forklift would use a BMU that can drive EV contactors.

An advantage of a Li-ion battery is its low internal resistance, which means a low-voltage sag under load. However, this is also a disadvantage because of the high inrush current into the load capacitance and charger capacitance. Adding the typical precharge circuit inside the battery (Figure 3.36(c)) is pointless if the BMS has no idea when the battery will be connected to a charger, or when the forklift operator will turn on the ignition key (see Volume 1, Section 4.9.1). Instead, the BMS must be able to withstand a possible, short-duration reversal of cell voltage when the battery charges this capacitance without precharge (see Volume 1, Section 3.2.12). The battery handles regen power no differently than the power from a charger.

3.8.1.3   Dual-Port Battery Design

Forklifts are designed so that the operator disconnects the battery's connector from the machinery and plugs that very same connector into a stationary charger (Figure 3.37(a)). This design choice turned out to be lucky because it allows us to replace a single-port lead-acid battery with a dual-port Li-ion battery. Such a battery is simpler to implement than a single-port battery because the discharge port is semi-permanently connected to the forklift's machinery, while the charge port is either left disconnected or is connected to a charger (Figure 3.37(b)).

FIGURE 3.37
Forklift battery: (a)
single-port lead-acid, and
(b) dual-port Li-ion.

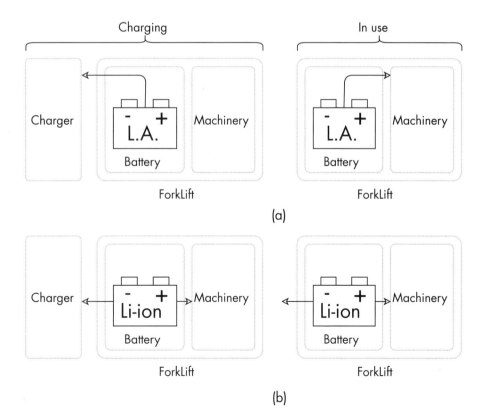

A dual-port solution is advantageous:

- The operator needs to make a single connection to the charger, instead of two connections: disconnect from the vehicle, and connect to the charger.
- The battery and its connection to the vehicle can be enclosed for safety and aesthetics; in a single-port solution, disconnecting the battery from the vehicle would entail opening some covers to reach the cable.
- The charging cable may be located on the forklift for a convenient connection to the charger.

To prevent the operator from driving off while the forklift is still connected to the charger, an interlock disables the forklift while it sees voltage or CAN bus communications from the charger.

This dual-port battery uses two contactors, one to control charging and one to control discharging (Figure 3.38(a)). The BMU can easily control the two paths independently.

Normally, regen would be a problem because, if the battery is full, the BMU opens the charging contactor, but not the discharging one; regen current comes in from the discharging port. However, this is generally not a significant problem with a forklift because most regenerated energy originally came from the battery itself, so it can't overcharge the battery. There could be a problem if a fully charged forklift is used to lower many heavy pallets: the potential energy in the pallets (from the top shelf to the ground) would go into an already full battery, and could overcharge it.

This solution does not include precharge because it would be pointless. The BMS has no way of knowing when a charger is connected, or when the operator turns on the ignition key.

Modern, high-frequency chargers[62] include a digital link, usually a CAN bus, to control the charging current. If so, and if the forklift supply company is allowed to modify the forklift to connect to the ignition line, precharge can be implemented (Figure 3.38(b))

- When the BMS hears a response from the charger, it knows that it is connected to it, and starts the charger side precharge circuit; then it turns on the charge contactor. Finally, it instructs the charger to start charging.
- When the BMS sees that the ignition is turned on, it knows that it is connected to the forklift machinery and starts the load side precharge circuit; then it turns on the discharge contactor.
- If the BMS sees that both the ignition is on and the charger is connected, it issues a plug and drive fault and shuts off the discharge side contactor to prevent the forklift from driving off while still plugged in.

A diode and a resistor can precharge the capacitance across the output of the charger be through a diode and a resistor (Figure 3.38(c)). A second diode in series with the charge contactor blocks the inrush current directly through the charge contactor. This solution works on the charger side because current flows in one direction during charging (power goes from the charger to the battery) and the other direction during precharge (power goes from the battery to the charger). This solution would not work on the load side because current flows in the same direction for both precharge and operation (power goes from the battery to the motor driver).

---

62. Switch-mode converters.

FIGURE 3.38
Dual-port forklift Li-
ion battery: (a) basic, (b)
with precharge, and (c)
same but no CAN bus.

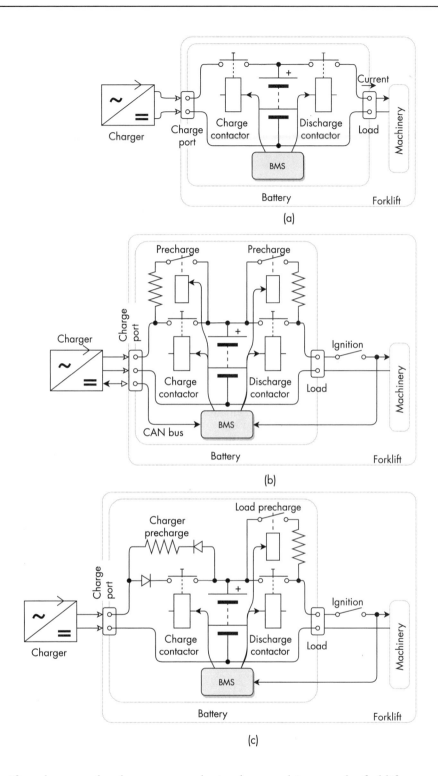

If precharge to the charger cannot be implemented, instruct the forklift operators to turn on the charger before connecting it to the battery. This way, the charger charges its output capacitor, and the battery doesn't need to do so.

If designing a Li-ion forklift from scratch, consider a higher voltage than 48V to reduce copper wiring and increase motor and driver efficiency. However, sticking with 48V allows the use of the present supply of forklift components, including the charging infrastructure.

3.8.2 Lawnmowers

The magic of owning a house on a golf course wanes when the lawnmowers wake you up at 7:00 on a Sunday morning. Riding lawn mowers with electric traction solve this problem. The traction battery ranges from 48V, 3.5 kWh for home use, to 120V, 20 kWh for professional grounds maintenance: golf courses, parks, housing developments. The current is not too high (a 2-hour discharge is typical), so a PCM may work well for a BMS.

3.8.3 Micro Pickup Trucks

These trucks are just like the auto-rickshaw described above but, instead of carrying people, they carry cargo. They are used in tight streets in Asian countries and inside buildings in industrial plants in developed countries.

A Li-ion traction battery would use large prismatic cells, a BMU, a contactor for the discharge switch, and a relay to control the charger (Figure 3.39(a)). As the vehicle may use its original transmission for reverse, there is no need for an electrical reverse.

For example, a micro truck could use solar power to help keep the battery charged (Figure 3.39(b)). There is no reversing contactor because the truck uses the transmission for reverse.

3.9 PASSENGER CAR APPLICATIONS

The electrification of passenger vehicles is having one of the most significant positive impacts on the environment. After decades during which EVs were a curiosity pushed

FIGURE 3.39
Micro EV truck: a Mitsubishi micro truck converted to Li-ion and solar: (a) traction battery assembly, and (b) circuit.

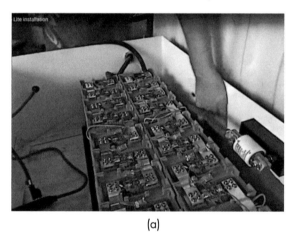

(a)

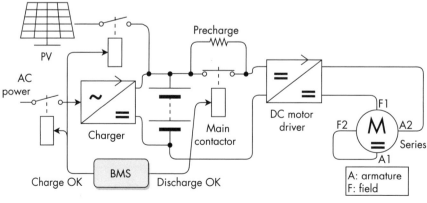

(b)

by "EVangelists," production passenger EVs are have finally entered the mainstream[63]. One by one, major car companies are announcing dates when they will stop manufacturing ICE cars.

As a compromise between a standard car and a BEV, hybrid cars are a stepping-stone towards a future when only BEVs will be produced and charging will be entirely from nonpolluting resources. Unfortunately, some jurisdictions classify gs-powered HEVs as alternative fuel vehicles, even though hybrids burn gas just like any other car and also pollute.

This section is directed to the designer of a traction battery for a low to mid-volume production or a mule used in the initial development of a high production vehicle. The design of a traction battery for a high-volume production EV (Figure 3.40) is quite involved. Regulations, long-term reliability, cost accounting, accident prevention, stress, thermal analysis, corporate alliances, and market analysis are just a few of the many aspects that go into the design of such a battery. Specialized books cover the subject in greater depth than this book can (see Section C.2.1.4).

3.9.1 Technical Considerations

Traction batteries for passenger EVs are subject to a wide array of regulations to ensure the safety of the passengers.

3.9.1.1 Traction Battery Specifications

The typical traction battery in a passenger electric vehicle has a voltage of about 350V, although, in some cases, it is as low as 96V. An HEV car uses a buffer battery (~20 kW). A plug-in electric car, BEV or PHEV, uses an energy battery (10 to 20 kWh) (see Volume 1, Section 5.1.4).

3.9.1.2 Cell Selection

The Verband der Automobilindustrie[64] (VDA) created a standard for pouch cells and small prismatic cells for the automotive industry, some optimized for HEVs, and some for plug-in cars (Table 3.13) [11].

FIGURE 3.40
Passenger EV
traction battery.

---

63. Some EV enthusiasts claim that the present availability of production EVs is a result of their tireless work; I compare this to the claims that Ronald Reagan was responsible for the fall of the Soviet Union.
64. German Association of the Automotive Industry.

Format	Class	Dimensions [mm]	Capacity [Ah]	Power [W]	Current [A]	Minimum Specific Energy [Wh/kg]	Minimum Energy Density [Wh/l]	Minimum Specific Power [kW/kg]	Minimum Power Density [kW/l]
Small prismatic	ISS-LV	150 × 150 × 15	20						
	HEV	85 × 125 × 12.5	5.5	700	225	90	160	3	5
	PHEV1	85 × 173 × 21	22	850	300	115	225	1.4	2.8
	PHEV2	91 × 148 × 26.5	24~28	850	300	115	225	1.4	2.8
	BEV1	115 × 173 × 32	40~44	1,000	300	120	270	0.5	1.1
	BEV2	115 × 173 × 45	60~66	1,200	300	120	270	0.5	1.1
Pouch	HEV	121 × 243	5.5	700	225	100	180	3.5	5.5
	PHEV	165 × 227	20~22	850	300	125	250	1.6	3
	BEV	162 × 330	50~54	1,100	300	140	300	0.7	1.5

TABLE 3.13   VDA Standard Cells

Samsung and CATL (Amperex) are among the first companies to manufacture cells that meet this standard. They are now becoming available to the general public.

Otherwise, for small runs, small cylindrical power cells are recommended for HEVs, and large prismatic or ready-made stacks of pouch cells are recommended for EVs. For high-volume production, pouch cells are recommended for any vehicle. EV conversions typically use LFP cells due to their perceived safety. Production vehicles use LCO or NMC cells, and occasionally some more exotic cell chemistries, such as LMO and NCA.

3.9.1.3   Thermal Management

A buffer battery in an HEV requires cooling to remove the internally generated heat. An energy battery should not require cooling, as the specific current is so low that little heat is generated. If that is not the case, then the choice of cells should be reconsidered; high-power cells are more efficient and should not require cooling.

Regardless of whether an automotive traction battery requires cooling to remove internally generated heat, it does require thermal management to handle the wide range of ambient temperatures (normally considered to be −40°C to 85°C); an HEV generates additional heat from its internal combustion engine, which may affect the battery, depending on their relative locations.

This range of temperatures is well beyond the safe range for Li-ion batteries. Thermal management can help, to some extent (see Volume 1, Section 5.17).

At the cold end, the cells are not degraded, but the performance is reduced. It may be possible to heat the cells when first turning on the car.

At the hot end, the cells are degraded. Yes, it is possible to use a heat pump to cool the cells while the car is in use; but the temperature can be the hottest when the car is parked, which is most of the time. When the car is parked and turned off, it is not acceptable to run the air conditioning.

Therefore, traction batteries for cars require a cell chemistry that can withstand prolonged high temperatures with minimal degradation. Infamously, Nissan chose to use a cell chemistry that included an LMO; that chemistry degraded quickly in hot climates; Nissan agreed to buy back cars from owners in the U.S. state of Arizona under that state's Lemon Law [12]. Nissan has long since changed to a different cell chemistry.

Nissan is often unfairly mocked for not implementing liquid cooling [13]. Such comments completely miss the mark, because the degradation was not due to internally generated heat during operation; it was due to exposure to hot ambient temperatures while the car is off. Whether or not the car uses air or liquid cooling is moot since cooling is off when the car is off.

3.9.1.4 Contactor Protector Switch, Precharge

In a high-voltage traction battery, the protector switch uses two contactors to provide galvanic isolation (see Volume 1, Section 5.12.3). Devices connected to the battery (motor drivers, chargers, DC-DC converters) include large filtering capacitors across the input. Connecting the battery directly to such devices without precharge with result in a massive inrush of current (see Volume 1, Section 5.13).

3.9.1.5 Contactor Control

There are various ways to control the contactors, depending on the preferences of the vehicle designer and on whether the BMS includes contactors drivers and logic (Table 3.14).

For safety, the contactors are always located inside the traction battery regardless of who controls them.

It is acceptable for the VCU to control the contactors as long as the VCU obeys the BMS when it commands that the current must be stopped (Figure 3.41).

3.9.1.6 HEV Power Circuit

The traction battery for a hybrid vehicle uses two contactors (one for the positive terminal and one for the negative one) and a relay and resistor in the precharge circuit (Figure 3.42(a)). The main fuse is placed mid-battery, as is a safety disconnect plug that a technician removes before working on the high-voltage system. For safety, the battery may also include a mid-battery contactor and a post-discharge circuit (Figure 3.42(b)).

The 12V supply powers the safety contactor (K4) through a small emergency switch; the technician opens that switch before working on the high-voltage system. Using a contactor instead of a safety disconnect has two advantages:

TABLE 3.14
Options for Controlling and Driving the Contactors

The BMS includes drivers for contactors	The BMS includes intelligence to turn on contactors in the proper sequence	Options
		The VCU drives the contactors and relays directly (Figure 3.41(a))
✓		The VCU controls the drivers in the BMS (Figure 3.41(b)); the VCU doesn't use the BMS's drivers; it drives the contactors directly
✓	✓	The VCU requests that the contactors be on, and the BMS is in charge of this process (Figure 3.41(c)); the VCU overrules the BMS and takes charge of this process, controlling the drivers in the BMS; the VCU doesn't use the BMS's capabilities at all; it drives the contactors directly

FIGURE 3.41
Contactor control logic:
(a) by VCU directly,
(b) by VCU through
BMS, and (c) by BMS.

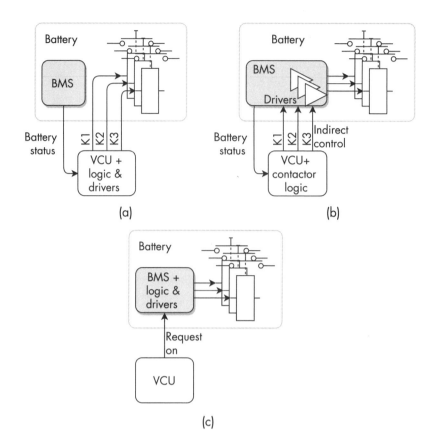

- In small quantities, the contactor is cheaper and more available than a safety disconnect switch of the same rating.
- When the low-voltage supply is off, the contactor automatically turns off, even if the technician has forgotten to turn off the safety switch.

Immediately after the battery is turned off, the capacitors across the input of the motor driver remain charged, which is dangerous. For safety, a post-discharge circuit quickly discharges those capacitors after the traction battery is turned off. The simplest solution is to use a resistor across the high-voltage bus. However, this would waste power whenever the battery is on. The solution is to add in series with the post-discharge resistor the normally closed contacts of a relay (K5) whose coil is powered only when the battery is turned on.

When turning on the traction battery, the contactor controller goes through this sequence (see Volume 1, Section 5.13.4):

- Check the battery isolation to detect any ground faults (Figure 3.43(a)).
- Check that no relay or contactor is welded shut (Figure 3.43(b)).
- Precharge the load capacitance, checking that the precharge relay and the negative contactor can close and that there is no short across the load (Figure 3.43(c)).
- Connect the battery directly to the load, checking that the positive contactor can close; end precharge (Figure 3.43(d)).

Should there be a ground fault, the contactor controller aborts the turn-on sequence for safety. Should the precharge fail (the precharge resistor is open, the

FIGURE 3.42
Power circuit in a traction
battery for an HEV: (a)
typical, and (b) advanced.

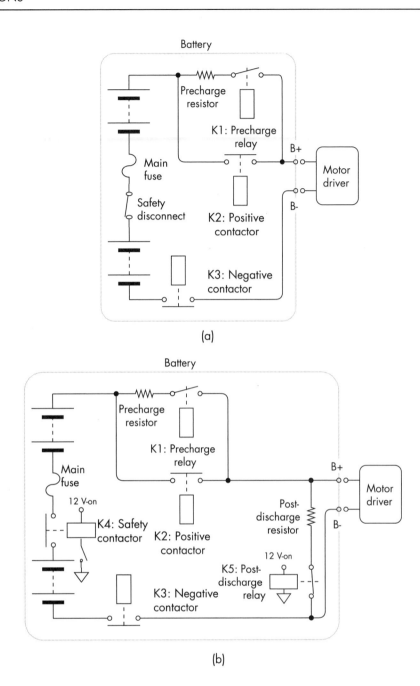

(a)

(b)

precharge relay did not close, the positive contactor is welded) the controller aborts
the turn-on sequence to avoid damage (see Volume 1, Section 5.13.2).

3.9.1.7 BEV Power Circuit

The power circuit for a plug-in EV is similar, except for the addition of a charger,
either onboard or off-board. A car with an onboard charger has an AC power inlet.
The output of the charger is directly connected to the string of cells (Figure 3.44(a))
(see Section 3.3.2.2). A car that uses an off-board charger has a DC inlet. There is
a second set of contactors and a precharge circuit between the DC inlet and the
battery to isolate the battery from the DC inlet and to precharge the capacitors on the
external charger (Figure 3.44(b)).

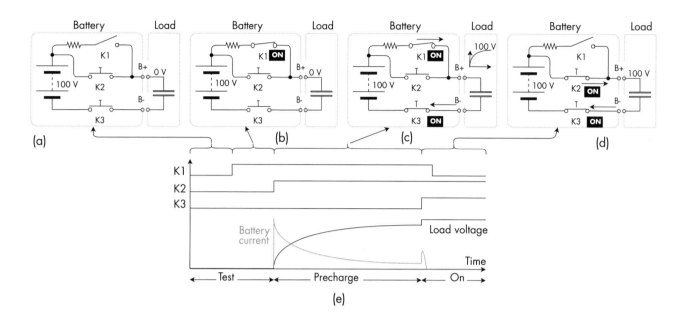

FIGURE 3.43   Simplified precharge sequence: (a) test ground fault, welded contacts, (b) test open or welded contacts, (c) precharge, test precharge, (d) fully on, and (e) plot.

3.9.1.8   V2G Inverger

Vehicle-to-grid (V2G) is a technology that allows the power company to buy energy back from a plug-in vehicle to meet a short-term shortage in the energy supply. For example, it's a hot summer afternoon, and a car is plugged in, at work, charging (Figure 3.45(a)). The power company has used up all its resources powering air conditioners; rather than bringing an old, stinky, and expensive power plant online, it can buy energy from EVs that are plugged into an outlet for charging (Figure 3.45(b)). This is unlikely to happen more than a few times a year. The power company needs power for only 15 minutes, during which time it can extract at most 6% of the battery energy[65].

Normally, pollution from the power company is increased when EVs are plugged in for charging; however, V2G has the counterintuitive effect of reducing pollution, since it replaces a dirty power plant.

V2G invergers have the same issues as those used in residential applications, including implementing anti-islanding (see Section 2.5.2.2). They also require a way for the power company to request energy, and a contract between the two parties to regulate the transactions.

3.9.2   BMU Selection

Full-size production EVs normally use a custom BMU (Figure 3.46) or a semi–custom BMS from companies that sell exclusively to auto manufacturers[66]. Other applications can use an off-the-shelf BMS.

A BMU explicitly designed for passenger automotive use should have the following characteristics:

---

65.   Assuming that the EV's charger can fully charge the battery in 4 hours.
66.   I + ME ACTIA (Germany).

FIGURE 3.44
Plug-in EV power circuit:
(a) onboard charger, and
(b) off-board charger.

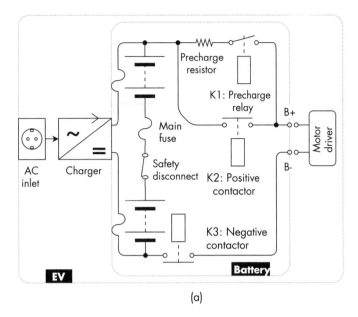

(a)

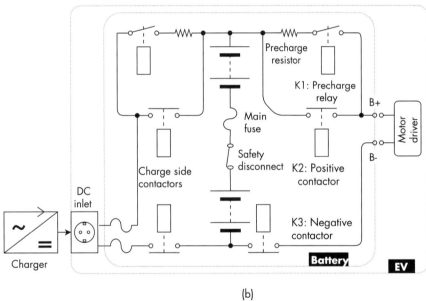

(b)

- Support a large number of cells in series: 30 cells for a 96V battery, 110 cells for a 350V battery.

- Support temperature sensors, up to one sensor per cell.

- Be configurable.

- Have two power supply inputs (see Volume 1, Section 4.10.1) able to withstand the rigors of the automotive 12V network without damage (see Section 3.15.1.1):

  - Charging power: 12V present whenever there is AC power; possibly generated by a power supply connected to the AC power inlet. Not required for HEVs.

  - Ignition power: 12V present whenever the car is on; this is the standard 12V bus for the car.

FIGURE 3.45
V2G: (a) charging
from the grid, and (b)
discharging to the grid.

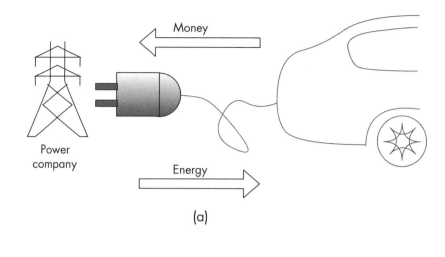

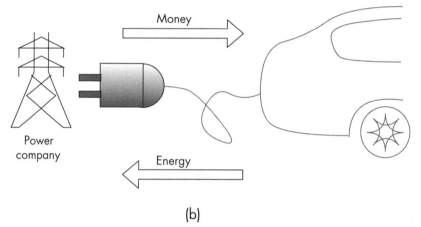

FIGURE 3.46
Custom BMUs for
traction batteries: (a)
Coda Automotive EV,
and (b) Enerdel BD3
traction battery.

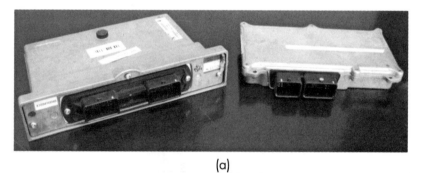

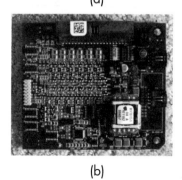

- Support for two current sensors: a low current one for charging and a high current, bidirectional (for regen braking) one for when operating.
- Evaluate SoC.
- CAN bus.

Optionally, it may also have

- Contactor drivers, precharge drive and sensing;
- Ground fault detection;
- Support of OBD II messages;
- Compatibility with a CAN charger.

The following BMUs meet most or all of the above criteria (Table 3.15). Unless otherwise noted, all these BMUs include a CAN bus and do not include protector switch or ground fault detection (see Volume 1, Section 5.14.3). Many are master/slave, which is fine, though not required for a traction battery for a car because it is not that big.

TABLE 3.15
Off-the-Shelf Digital
BMUs for Passenger EVs

Brand	Model	Cells	Topology	Notes
123 Electric	123SmartBMS	255	Distributed	CAN optional, large prismatic cells, ground fault detection
Batrium	WatchMon	249	Distributed	Large prismatic cells, ground fault detection
Emus	EMUS BMS	256	Distributed	Large prismatic cells
E-Pow (Huizhou EPower Electronics)	EV05	?	Master/slave	
	LEV05	?	Master/slave	
Ewert	Orion BMS	192	Centralized	Ground fault detection
Elithion	Lithiumate Pro	256	Distributed	Ground fault detection
Elithion	Vinci EV	256	Master/slave	Ground fault detection
EVPST	BMS-4	?	Master/slave	
I + ME ACTIA	IME	140	Master/slave	Only to large manufacturers
Ion Energy	FreeSafe LT	180	Master/slave	
JTT	X-Series	240	Master/slave	
John Elis	BMS ver 2	254	Distributed	Large prismatic cells
Ligoo	AUTO_EMS_2.0, EHUG BMS, LIGOO BMS, EK30	104	Master/slave	
Lithium Balance	LiBAL sBMS	256	Master/slave	Ground fault detection
O'cell	BMU LV-150	218	Master/slave	Ground fault detection
REC	BMS	255	Master/slave	
Shenzhen Klclear Technology	BMU06-I-12V-400V	100	Centralized	
Zeva	EVBMS	192	Master/slave	
Moviecom	—		Master/slave	Wired slaves for up to 12 or 18 cells CAN open, J1939, Modbus RTU, Ethernet

3.9.3   Passenger EV Projects

There are particular requirements for different types of passenger electric cars.

3.9.3.1   EV Conversions

For a long time, tinkerers have joined clubs of EV enthusiasts and converted standard cars from the original internal combustion engine to electric propulsion; today, conversions are no longer as popular, given that inexpensive and reliable EVs are commercially available. Conversions now use Li-ion batteries, with any quality BMS that can handle enough cells in series. The biggest challenge is not the BMS, but the tinkerer who does not understand that the BMS must be wired so that it is able to disconnect the battery directly when any cell is low. Worse is the hobbyist who understands this but refuses to implement it ("I never let the battery go too low") (see Section A.5.5).

Cost-conscious hobbyists can successfully use an analog BMS (Table 3.16). Most hobbyists are better served by a digital BMU.

3.9.3.2   Hybrid Conversions

In the mid-2000s[67], a cottage industry sprung up to convert the Toyota Prius and other HEVs to PHEVs, by adding a larger traction battery, a charger, and a BMS.

The best BMS for a conversion of a Toyota Prius is the Orion BMS because the Ewert brothers thoroughly reverse-engineered the protocol between the original battery and the main ECU, to the point of achieving the holy grail of Prius conversions: EV operation at highway speeds.

3.9.3.3   Production Passenger Vehicles

When developing a production vehicle, an off-the-shelf BMS allows testing to progress while waiting for the custom-designed BMS to be ready. An off-the-shelf BMS is also needed when building a proof-of-concept vehicle (a mule). A versatile and well configurable BMS is desirable in these applications.

3.9.4   Charging Station (EVSE)

A plug-in vehicle can be charged from a generic power cord from a simple AC outlet (Figure 3.47(a)). For safety, the AC power cord has a female plug that is mated to a male jack on the vehicle.

EVs may also be charged through a dedicated cable from an EVSE[68], a stationary charging station (Figure 3.48), which is the street's version of marine's shore power (see Section 2.6.1.2). The EVSE may feed AC to the onboard charger inside the vehicle (Figure 3.47(b)), or it may include a charger and feed DC to the vehicle (Figure 3.47(c)). SAE[69] defines six charging levels (Table 3.17).

TABLE 3.16
Off-the-Shelf Analog
BMUs for Passenger EVs

Brand	Model	Cells	Topology	Notes
EV Power	LFP BMS	Any	Distributed	Prismatic
Low Carbon Idea	BMS mini series	Any	Distributed	Prismatic

---

67. The decade between 2000 and 2009.
68. Electric vehicle supply equipment.
69. Society of Automotive Engineers.

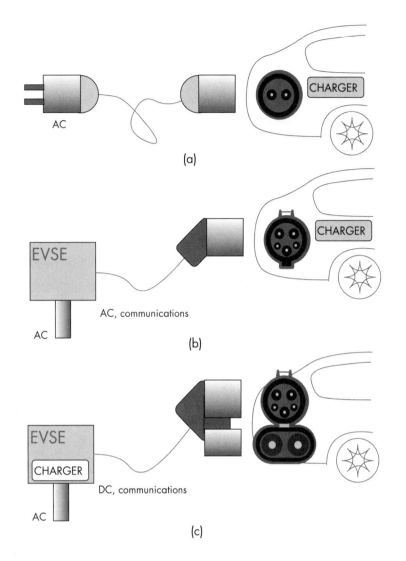

FIGURE 3.47
Charging: (a) direct,
(b) EVSE-AC, and
(c) EVSE-DC.

FIGURE 3.48
EVSE: (a) ready for use,
(b) Chinese plug for
DC charging, and
(c) Tesla car charging.

3.9.4.1 EVSE Standards and Connectors

Unfortunately, the industry defined multiple, mutually incompatible EVSE standards. Tables 3.18 and 3.19 list the main EV conductive charging connectors and the applicable standards.

TABLE 3.17
SAE Charging Levels

Level	Voltage	Maximum Current [A]	Maximum Power [kW]	Charger Location
AC Level 1	110 Vac, 1 φ	12, 16	1.4, 1.9	Vehicle
AC Level 2	240 Vac, 1 φ	80	19.2	Vehicle
AC Level 3	TBD, 1 or 3 φ	TBD	>20	Vehicle
DC Level 1	200~450 Vdc	80	36	Charging station
DC Level 2	200~450 Vdc	100	90	Charging station
DC Level 3	200~600 Vdc	400	240	Charging station

IEC 62196	Connector on Vehicle	Gender, Make	Standard/Country	SAE level	Voltage [V]	Current [A]	Pins/Signaling
Type 1		Male, Yazaki	SAE J1772, IEC 6, USA	AC 1	120 Vac	16	Five pins, two signal: analog
				AC 2	208~240 Vac	30~80	
		Male	SAE J1772, plus 2 DC power pins, IEC System C COMBO1, USA	AC 1 or 2	120~240 Vac	200	Up to seven pins, two signal: analog, PLC
				DC 2, 3	200~450 Vdc	200	
Type 2		Male, Mennekes	VDE-AR-E 2623-2-2, IEC 62196 type 2, Europe	AC	480 Vac, 3 φ	32	Seven pins, two signal: analog
			SAE J3068, USA				Seven pins, two signal: LIN
		Female	Guobiao, System B, AC GB/T, 20234.2-2015, China	AC	480 Vac, 1 or 3 φ	32	Seven pins, two signal: CAN
		Male	COMBO2, SAE combo CSS 2, IEC 62196 type 2, plus 2 DC power pins, Germany, IDC System C	DC	200~1,000 Vdc	200	Up to nine pins, two signal: analog, PLC

Notes: The European IEC 62196 standard specifies four types of connectors, plus two variants (combo) that add two large pins for DC power. The Chinese GB/T standard defines two connectors, one for AC and one for DC. The Tesla connectors are not defined under any standard

TABLE 3.18   EVSE Connectors, Types 1 and 2

It is worth noting that:

· The European IEC 62196 standard specifies four types of connectors, plus two variants (combo) that add two large pins for DC power;

· The Chinese GB/T standard defines two connectors, one for AC and one for DC;

· The Tesla connectors are not defined under any standard.

Given that the EVSE does not energize the contacts until the plug is securely mated on the vehicle, it doesn't matter whether the vehicle has a male or a female connector. Indeed, either gender is used in these standards.

IEC 62196	Connector on Vehicle	Gender	Standard/Country	SAE level	Voltage [V]	Current [A]	Pins/Signaling
Type 3		Female	SCAME Type 3A, Italy, France	AC	240 Vac 1 φ	16, 32	Seven pins, three signal: analog
			SCAME Type 3C, Italy, France	AC	480 Vac 3 φ	16, 32, 63	
Type 4		Female	CHAdeMO, JEVS G105-1993, IEC System A, UL2551, Japan	DC 3	500	125	Ten pins, eight signal: logic, CAN
NA		Female	DC GB/T, IEC System B, China	DC	400~570 Vdc	63~250	Nine pins, seven signal: CAN
NA		Female	Tesla High Power Wall Connector (HPWC) USA	AC 2		16, 32, 40, 62	Five pins, three signal: analog
			Tesla Super-charger USA	DC 3	480		Five pins, three signal: analog

TABLE 3.19   EVSE Connectors, Type 3, 4, and Tesla

3.9.4.2   Communications

The EVSE can deliver power to the EV and communicate with it in various ways (Table 3.20).

Before turning on the power, a charging station communicates with the vehicle to confirm its presence and possibly to establish the vehicle's charging requirements. This is of interest to the battery designer because, while charging, the VCU is likely to be turned off, so the only smart device that can interface with the EVSE is the BMS. A BMS that is directly compatible with the chosen type of EVSE is preferable but rarely available. Adapter modules are available to provide this function.

Communications may use dedicated logic, analog lines, CAN bus, power line communication (PLC), or wireless (Table 3.21). EVSE communications may be quite simple or rather complex. Some standards are proprietary.

3.9.4.3   SAE J1772

The J1772 [14, 15] standard is used in the United States and is quite straightforward. It uses analog signaling. The two signal lines, Pilot and Proximity, are used by both devices to send and receive the other device's status (Figure 3.49).

The cable has three low-power pins:

- Proximity Pilot (PP):
  - The connector handle notifies the vehicle of its presence and whether it's locked in, using two different values of resistance.

TABLE 3.20
EVSE Power Transmission and Communications

		Power to Vehicle	Medium	Device	Communication
	1	AC voltage	Conductive	Connector	None
	2				Wired
	3	DC voltage			
	4	High-frequency magnetic field	Inductive	Paddle inserted in car	Wireless
	5			Coil near car	

TABLE 3.21
Communications for Each
EVSE Standard

Type	Standard	Signal Lines	Signaling	Notes
1	J1772	2	Analog, bidirectional	
2	EU	2	Analog	Same as SCAME
2	China	2	CAN	
3	SCAME	2	Analog	Same as Type 2, EU
4	CHAdeMO	8	Analog, unidirectional CAN	EVSE controls the contactor inside the vehicle, between the connector and the battery; proprietary standard

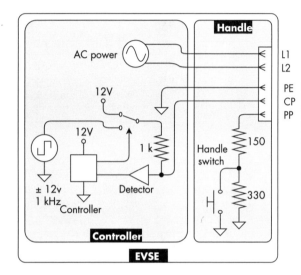 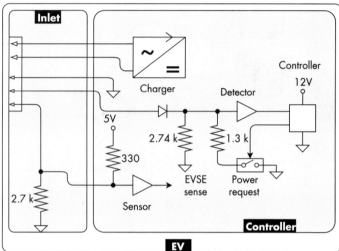

FIGURE 3.49   J1772 circuit.

- Control Pilot (CP): communications between vehicle and EVSE:
  - The EVSE notifies the vehicle:
    - Before connection: steady 12V;
    - After connection: 1 kHz, +/−12V, PWM wave; duty cycle (10% to 96%) indicates maximum available current (6A through 96A).
  - The vehicle notifies the EVSE by loading down the PWM wave with resistors to ground:
    - State A: not connected; 12Vdc (no wave);
    - State B: connected and ready; 2.74 kΩ, top of PWM wave is at 9V;
    - State C: Charging; 882Ω, top of PWM wave is at 6V;
    - State D: Charging; 246Ω, top of PWM wave is at 3V.
- Pilot Earth (PE): reference for the above lines.

If the BMS is not natively compatible with J1772, adapter boards are available[70].

3.9.4.4   EU Type 2, SCAME

This standard is used in Europe and may use either a Type 2 connector (male on the vehicle) or a Type 3 connector (SCAME). The cable has three low-power pins (same

---

70. Modular EV Power, modularevpower.com.

as J1772), though the protocol is somewhat different. The charging cable from the EVSE includes a resistor from PP to ground, to tell the vehicle the available current[71]. While the PP line should have full communications, it can be faked with simple diode and resistors from CP to ground: 2.74 kΩ when off, or 880Ω when on (Figure 3.50).

3.9.4.5 CHAdeMO

Many Japanese plug-in cars use CHAdeMO [16, 17] charging. This standard is proprietary and more complex than other standards, which can be challenging; yet it is now widely adopted, so you may need to work with it.

CHAdeMO[72] uses a combination of unidirectional analog signaling and verbose digital communications between the charging station and the vehicle (Figure 3.51):

- Six CAN messages must go back and forth before the charging station turns on.

- Two messages are required to turn off.

The cable has two high power pins, 4 signal lines, two CAN lines, ground plus a spare:

1. FE, IM: Ground;

2. DCS1, CP1, Charger start/stop 1: drives contactor coil in the vehicle, negative.

3. (spare);

4. DCP, CP2, Charging enable/disable: from the vehicle; grounded to enable the EVSE;

5. DC- power;

6. DC+ power;

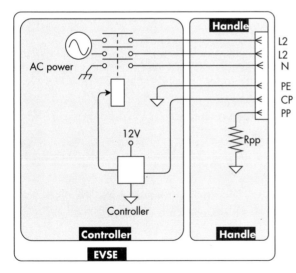
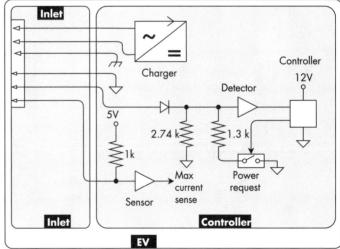

FIGURE 3.50   EVSE type 2 circuit.

---

71. 100Ω for 13A, 220Ω for 20A, 680Ω for 32A or 1.5 kΩ for 63A.

72. チャデモ is officially an abbreviation of "CHArge de Move." Actually, it is a pun in Japanese, because お茶でも, pronounced "ocha demo", literally means "how about tea," implying that charging can occur while you're having a cup of tea.

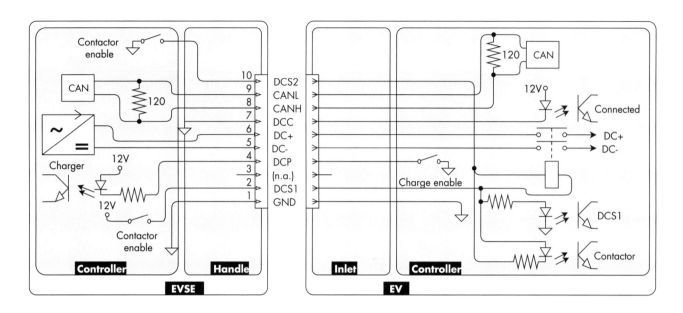

FIGURE 3.51   CHAdeMO circuit.

7. DCC, CS, Connection check: grounded in the EVSE; the vehicle detects the presence of the EVSE;

8. CANH: CAN bus high;

9. CANL: CAN bus low;

10. DCS2, CP3, Charger start/stop 2: drives contactor coil in the vehicle, positive.

The BMS is unlikely to be natively compatible with CHAdeMO; I do not know of any adapters.

3.10   RACING APPLICATIONS

A racing vehicle that is ideal for a given race is inappropriate for another race, due to the difference in the characteristics of various types of races. Table 3.22 lists a few types of EV races and their limitations.

This section explores a few races and the preferable characteristics for an EV that races in it.

3.10.1   Dragstrip Racing

The goal of a dragstrip race is acceleration, the shortest time and highest speed over a short distance (Figure 3.52). The race is run on a straight strip. The speed and time

TABLE 3.22
Comparison of EV races.

Race Type	EV Type	Goal	Limitation
Dragstrip	BEV	Fastest acceleration	Distance
Land speed	BEV	Top speed	Drag
Racetrack	BEV	First to finish	Energy storage
Racetrack	HEV	Endurance	Reliability
Solar	BEV	First to finish	Energy source
Solar	Batteryless EV	Top speed	Power source

FIGURE 3.52
Larry McBride's Lawless
Electric Rocket,
EV dragstrip record
holder. (Courtesy of
Chip Gribben.)

to reach a certain distance are clocked. The vehicle is designed for acceleration; it is barely able to steer. Electric motorcycles fare better than cars in this sport.

The record for an EV on a 0.4 km (quarter-mile) strip is 324 km/h and less than 7 seconds (Larry McBride, 2012). The battery charge is barely tapped; a run lasts less than 10 seconds. Therefore, high-power cells are used, operated close to the maximum power point (see Volume 1, Section 1.5.1). There is no need for cooling because there is not enough time for heat to propagate; yet there may be thermal management, not to cool the cells, but to heat the cells to 60°C to minimize their resistance; yes, hot cells degrade faster but, in this case, it's only for an hour or so.

The MPP may be determined based on cell voltage or total battery voltage. If all cells have the same resistance, then these two approaches are the same. If based on cell voltage, the battery load is adjusted so that the cell with the highest resistance (the limiting cell) operates at its MPP. All the other cells operate below the MPP, producing more power than the limiting cell but still less power than they could produce if each were operated at its MPP. If each cell were operated at its MPP, the efficiency would be 50%.

If based on battery voltage, then

- Cells with average resistance are operated at their MPP.
- Cells with lower resistance produce more power than the other cells, at a higher efficiency, but still less than they could produce if each were operated at its MPP.
- Cells with higher resistance produce far less power and suffer from poor efficiency.

It turns out that, if the battery doesn't have any weak cells, then it is most advantageous to operate it at the MPP based on battery voltage. Even though the highest resistance cells contribute little power, more cells operate at the MPP, and therefore the battery produces more power. However, if the cell resistances are mismatched, the power must be limited by the weakest cell, or else it risks having its voltage reversed, which is very dangerous.

During a run, the BMS may be disabled to keep it from issuing an undervoltage fault. Also, a distributed BMS, which is powered directly from the cells, won't work (see Volume 1, Section 4.13.1.2). Regardless of whether the BMS is used during a run, it must be used at all other times, when charging or moving the EV.

In the past, a successful approach has been to use two DC motors. The motors are connected in series at the beginning of the run for higher torque and then in parallel for higher speed. Today, AC motors and inverters provide even better performance (see Volume 1, Section 3.6.2).

3.10.2 Land Speed Record

The goal of open field racing is speed over a relatively long distance (compared to a drag strip), regardless of the run time. A run consists of two passes, one in each direction; the average speed is taken to cancel out the effect of wind. Bonneville Flats, a salt lake in Utah, is the principal location for these races. Its track is 16 km long (Figure 3.53(a)).

EVs designed to break land speed records look more like missiles than cars, are low to the ground, and can barely steer. The key for these vehicles is low drag and high power; given that each pass lasts only a couple of minutes, the vehicle uses a power battery (see Volume 1, Section 5.1.4).

In 2016, Venturi's VBB3 car, in cooperation with Ohio's State University's Center for Automotive Research, broke the record with a top speed of 545 km/h. It has a 2.25-MW AC motor and an 812V battery. The battery consists of 2,000 cells in series: 14Ah LFP cells from A123. The total energy is 92 kWh. The discharge rate is about 10 C. The battery energy is hardly used: I estimate that the SoC drops by only 20% in the two passes.

A distributed BMS would have worked, as the cell voltages do not drop excessively. However, I believe that this vehicle used a wired BMS.

I believe that the Venturi would gain by going away from the A123 energy cells and switching to power cells. That would make a battery that weighs only one-third as much that would be discharged at 10 C. I suspect that the A123 cells were chosen because of corporate sponsorship rather than for technical reasons.

3.10.3   Racetrack

EV race cars and motorcycles are designed to compete with other such vehicles on the same race track, with the goal of finishing first (Figure 3.53(b)). Unlike a drag race EV, the BMS in these vehicles must operate all the time. A distributed BMS won't work because, during acceleration, the traction battery may operate near the maximum power point (see Volume 1, Section 1.5.1). The standard BMS's voltage measurement rate of one reading per second is too slow. The BMS should read voltages two to ten times per second.

Unlike race vehicles described above, SoC measurement is important. Measuring the wide operating current (0.2A to 2,000A) accurately is a challenge. An open-loop Hall effect sensor won't cut it. A dual-range Hall effect sensor (see Volume 1, Section 5.9.2.2) or a shunt sensor (see Volume 1, Section 5.9.2.1) with a high-resolution analog-to-digital converter is required.

Race EVs tend to be very noisy (electrically) because of the high switching currents present. This noise can create havoc with the BMS, trying to read millivolt differences in the voltage of cells that are "shaking" 350V peak-to-peak (see Volume 1, Section 4.12.3). This noise is also a problem for any data acquisition and telemetry

FIGURE 3.53
(a) Eva Hakansson in the KillaJoule (courtesy of Eva Hakansson), and (b) Lightning Motors EV racing motorcycle (courtesy of Alex Tang).

(a)   (b)

system in the EV. A BMS with a high level of isolation (such as through fiber optic cables) may be required.

Cells reach excessive temperatures because the battery dissipates nearly as much power as the motor uses; temperature sensing is far more critical than for the previously described race vehicles. Measuring the temperature of each cell is ideal because the BMS needs to know if a cell is about to go into thermal runaway. A rate of one reading per second should be sufficient. Rugged contactors are required, in case they need to interrupt the full load current in an emergency.

Winning a race may be more important than the health of the cells: the professional racer may prefer a ruined battery to a vehicle that has been stopped by the BMS. For this reason, the BMS's ability to turn off the traction battery may be delayed. I do not condone this, but I do acknowledge that it is done.

3.10.4 Formula Races

Formula race vehicles[73] are little more than four wheels and a single seat; unlike the standard Formula vehicle, a Formula Electric vehicle uses a purely electric drive, and a Formula Hybrid vehicle uses an HEV drive[74] (Figure 3.54(a)).

Since 2006, university teams around the world have competed in races such as the Formula Hybrid race[75], Formula Student Electric[76], and the Formula SAE Electric[77].

Most teams design a battery, and many teams design a BMS. An off-the-shelf BMS saves the team great effort. Yet, as race rules [18] become more and more stringent, few off-the-shelf BMSs can meet the requirements. For example, no off-the-shelf BMS measures cell temperatures exactly as specified in the rules, inviting the team to consider questionable workarounds.

The driver, strapped in the vehicle, may have difficulties escaping in case of an accident. Therefore, it is essential that the BMS be allowed to shut off the battery if it detects any dangerous condition, to minimize the chance of an event (Figure 3.54(b)).

FIGURE 3.54
University of Vermont, Aero Formula Hybrid, 2010: (a) team, and (b) battery using 26650 cells and bank boards.

(a) (b)

---

73. Thanks to Tim Pattison, Robert Wilson Rowland (Waterloo Formula Electric), and Dustin Byrtus and Abhilash Arora (Mahindra Electric Mobility) for help with this section.

74. In 2009, the Brigham Young University team took the term "hybrid" literally and cleverly came up with a drive train that used an engine and a hydraulic drive. The judges were not amused and disqualified the vehicle. Then the rules were changed to be more specific that hybrid drive means engine plus electric. Yet the team took home the "Dean's Award for Innovation" and the "Best Hybrid System Engineering" award.

75. New Hampshire, United States, since 2006; formula-hybrid.org/.

76. Hockenheim in Baden-Württemberg, Germany; www.formulastudent.de.

77. Lincoln, Nebraska, United States, since 2013; www.sae.org/attend/student-events/formula-sae-electric/.

Every year, I remind students that they must grin and bear the race rules, finding consolation in the knowledge that, once they graduate and move to the industry, the rules in the real world will be more realistic[78]. However, I understand that some students perversely take some of those overscrupulous rules with them into their new workplace after they graduate.

Here is my advice for Formula-Hybrid designers:

- Study the performance of past years' vehicles carefully, talk to past team members from other institutions, read the forums (see Section C.1.3), and learn from their mistakes and performance data.

- While the maximum battery energy is 4.5 kWh, it does not mean that this is the ideal energy; a smaller and lighter battery, using power cells, may actually perform better in a hybrid than a full-sized battery; you do not have to operate cells between 30% and 70% SoC, as the Prius does; you can go further because you don't care about battery life the way Toyota does; this lets you use an even smaller battery.

- Use cells of the desired capacity; do not achieve the desired capacity by adding cells in parallel[79] because the rules require that a fuse in series with each parallel path (see Volume 1, Section 3.3.12); this eliminates small cylindrical cells and leaves pouch, large cylindrical, small and large prismatic cells.

  - The rules require you to place three temperature sensors for every 10 cells; this is one more reason to use a single large cell instead of a block of cells in parallel:

  - A distributed BMS has one sensor per cell, exceeding the requirement; however, the rules specify that the temperature sensor must be in direct contact with the cells, so you need to convince the judges that the cell boards are the sensors and that, since the cell boards are connected directly to the cells, the rule is met. Indeed, a team connected the cell boards to the bus bar and successfully argued that the measurements reflected the cell temperatures sufficiently.

  - A wired BMS does not have nearly enough sensors; a standard solution is to supplement it with a separate system to measure temperatures.

- Use power cells, not energy cells; this eliminates large cylindrical and large prismatic cells, leaving only pouch cells and, soon, small prismatic cells[80].

- The rules require a fuse for each tap wire, close to the cell; for this reason alone, a distributed BMS, which does not use tap wires, may be better.

- Use an isolated BMS, and make sure you do not defeat this isolation (see Volume 1, Section 5.9.1); only a limited number of high-voltage wires can leave the battery enclosure: the charging port and the load port to the motor driver.

- Make sure that there is no load that can drain the battery while you are all taking finals and not minding the car for 3 weeks; establish a schedule for someone to check that the an unintentional load does not unexpectedly discharge the cells, and check hourly for the first 6 hours after assembly, daily for the first week, and then weekly. Condensation may create an unexpected path for discharge[81].

---

78. As I write this, an overly meticulous judge asked a team to provide the part number of a surface mount component buried inside a commercial cell board!

79. For example, if you need 12 Ah, get 12-Ah cells, don't use five 18650 cells in parallel.

80. Small cylindrical cells also come in high-power versions, but I eliminated them earlier because the rules require one fuse per cell.

81. In 2017, at the Formula SAE Lincoln, bringing the vehicle from inside a cool trailer into the 40°C heat resulted in condensation, which reportedly damaged components in batteries.

- Keep thermal design connected to the real world:
  - If you assume continuous heat generation, you'll overdesign the thermal management; instead, use the real profile of heat generated during a race; again, talk to last year's teams.
  - Because each run in the race does not last long, rely on thermal mass rather than cooling; the race is over before the cells get a chance to get too hot; this 40°C limit is for continuous operation; 3 minutes at 60°C won't kill your cells and is still a long way from the self-ignition temperature.
  - If you need any cooling, you picked the wrong cells (i.e., cells with too high an MPT); switch to power cells instead; you certainly don't need liquid cooling; it may be sexy, yes, but too expensive, too heavy, and too many headaches for too little return.
- For safety, build your vehicle in an access-controlled room, to keep curious people's wandering fingers from poking where they shouldn't (see Volume 1, Sections 5.14.1 and 7.2.1); attend an HV safety training course before building a battery (see Volume 1, Section 7.2.4).
- Use social media as a resource: the casual environment can be much more conducive to sharing experiences than the official forum for the race (see Section C.13).

Because the race rules are incompatible with real-world BMSs, team members may decide that it's easier to meet those rules by designing their own BMS from scratch, than to augment a proven off-the-shelf BMS with workarounds (see Section A.5.4). If so, I suggest

- KISS (keep it simple, student).
- Don't implement balancing: prebalance the battery manually; the cells will remain in balance during the few weeks of actual use.
- Mount a thermistor on each cell; use this simple multiplexer circuit to select one thermistor at a time: use a decoder to ground one end of the thermistor at a time to measure its temperature, leaving the other ones open; connect the other end of all thermistors together and use a single pull-up resistor and send the signal to an A/D converter.
- Isolate the BMS at one place only: the data link that goes outside the battery box.

When begging for free products, cut the corporate lingo and be yourself; talk to a manufacturer's contact like the human beings you both are. Your promise of corporate visibility is not fooling anyone: no one sees the manufacturer's logo on your car but fellow students, and your website and all its links will disappear in a year. A corporation gives you free product if it has a budget for it, not because "it's a unique investment opportunity."

Don't neglect last year's team's vehicle; reuse their battery if you can. Don't assume that it's fine as is; understand it fully, improve on it. Use it as a time-saver, not as a way to skip on the learning experience.

3.10.4.1 Hybrid-in-Progress and Electric

New teams often start with a hybrid-in-progress, a simple BEV instead of an HEV. The assumption is that the first vehicle, a BEV, prepares the school for an HEV the following year. This is not entirely true, as the requirements for a BEV are quite different from an HEV: it uses a larger motor and an energy battery instead of a power battery.

Some teams participate with no intent to switch to HEV in the future. These are not hybrid-in-progress vehicles but true BEVs. Their battery is limited to 5.4 kWh.

My advice for a BEV is slightly different, in these respects:

- Use large prismatic cells; this reduces the design effort significantly.
- Allow the discharge current to be as high as 2 to 4 C; battery life is not your main concern.
- Limit regen to 0.5 C, and do not do fast-charge when parked: no sense degrading the battery when charging.

3.10.5   Solar Race Vehicles

Solar race cars (Figure 3.55)[82] compete in long-distance races, particularly the World Solar Challenge in Australia. In 2009, Tokai University's vehicle completed the 3,021-km race in the shortest time: 29 hours and 49 minutes, with an average speed of 100.5 km/h!

Solar panels power the vehicle through a solar charge controller. The battery starts full at the beginning of the race; extra solar energy and regenerative braking recharge the battery.

Energy management strategies are crucial in a solar race. These vehicles are incredibly light and efficient: somehow, they manage to squeeze about 1-kW peak worth of solar panels on the top of the car. The fuel efficiency is as good as 37 Wh/km. The typical traction battery is usually around 100V and can hold 5 kWh of energy. That can propel the vehicle for typically 400 km without sunshine. Regenerative braking efficiency is a respectable ~60%.

Lately, solar race vehicles have reached the legal speed limit of the roads on which they race. Therefore, solar races have switched from prizes based on speed to prized based on the advancement of technologies that could eventually lead to a practical solar-powered passenger car, a lofty although questionable goal (see Section 3.2.5.1).

FIGURE 3.55
Equipo Solar race car
in Chile. (Courtesy of
Matías Rivera Campos.)

---

82. Thanks to Matías Rivera of Corporación Equipo Solar, Chile; Nathan Coonrod, Iowa State University; and Quang Phuc N. Kieu of MIT for help with this section.

A curious anomaly is that the people who wrote the America Solar Challenge race rules use the term battery protection system (BPS) instead of BMS. Just call your BMS a BPS, and the judges will nod approvingly. You may want to use a redundant BMS that includes both a digital BMS and an independent analog fault protector.

Here are some unusual items in the race rules that affect the BMS (unusual in the sense that they do not apply to a BMS in a commercial vehicle):

- The tap wires are limited to 1 mA (which is insufficient for balancing the battery) or being fused; fuses that are rated for the battery voltage are bulky and may be hard to place, so you should consider a distributed BMS, with no tap wires, rather than a wired one, which requires fuses or forgoes balancing.
- The BMS must turn off the traction battery if it detects a fault in the BMS itself, but not a fault in the traction battery!
- Rule 8.3.B.1 says that the BMS must actively shut off the traction battery in case of fault; however, rule 15.4.S acknowledges that the car may keep on going with a BMS fault, by specifying penalties.
- In case of overcurrent, the traction battery must be shut down by the BMS and not by the main fuse; the fact is that the motor takes so little power, it never comes close to asking more current than the battery can provide; if the BMS is set for the maximum the cells can provide, it will never trip (the fuse blows in case of a short circuit); on the other hand, in a solar vehicle, the BMS is set for the maximum current required by the load (typically 40A): this allows for a smaller fuse, lighter power wires, and a smaller motor driver.

Here are some unusual items that affect the traction battery:

- Two emergency push buttons are required: one accessible to the pilot, and one external; activating either of them shuts off all power sources: the battery and the solar panels.
- While manual isolation testing is required, a permanent ground fault detector is not.
- The maximum weight of a Li-ion battery depends on the Li-ion technology: 40 kg for LFP; otherwise, it is 20 kg.
- The main fuse must be an actual fuse (a circuit breaker is not acceptable).
- The traction battery must include means for it to be impounded to make sure no one charges it surreptitiously.
- Curiously, the rules require an auxiliary lead-acid battery to power a flashing light if the traction battery is shut down. This weight is an unfortunate handicap.

Every small detail counts for a successful race, including the selection of BMS. I recommend working with a BMS company with solar racing experience; REAP and Tritium have been used in the past, but no longer sell BMSs; as far as I know, this leaves only the Orion BMS.

Many teams prefer designing a BMS from scratch. For example, the PrISUm team from Iowa State University uses a distributed architecture, with wireless communications from each cell to the BMS master. The team from MIT designed a master/slave BMS around the LTC6804 IC that uses a simple twisted pair of wires between adjacent slaves. In many cases, instead of designing a BMS, the team designs an energy management system, controlling all systems (battery, PV, solar chargers, and motors) in a holistic way.

3.10.6   Solar Speed Vehicles

A solar car designed to break the speed record does not have a traction battery: it only uses the power from the sun available at the time. The race occurs on a 500-m stretch in both directions. The current record is 88.8 km/h, set by the University of South Wales in 2011, and with only 1.2 kW to the motor!

In 2007, the South Whidbey High School[83] team set a record of 29.5 seconds in a quarter-mile drag race, reaching a speed of 40 km/h.

3.11   PUBLIC TRANSPORTATION

Public transportation offers a partial solution to environmental concerns and electric public transportation, even more so. Vehicles range in size from tuk-tuks (see Section 3.7.5) to passenger trains. Several BMUs are explicitly offered for vehicles used in public transportation (Table 3.23).

3.11.1   Personal Rapid Transit (PRT)

Pods in a personal rapid transit carry just a few passengers and may run on demand. They are automated (there is no driver) and run along guided paths, such as lanes, tracks, or monorails or are suspended from a guideway.

PRT is a type of automated transit network (ATN) and automated guideway transit (AGT). The path contains power rails to provide propulsion power to the pods. DC power is preferred because it requires only two rails; three-phase AC power requires three or four rails. High voltage is preferred because a given power requires less current so that the contact resistance between a brush and a rail is less of a concern. Standard voltages are 600 Vdc, 750 Vdc, and 1.5 kVdc.

Pods use power from the grid. When braking, power is returned to the grid. Each station has an inverter (see Volume 1, Section 1.8.4.1) with high-voltage DC output to feed the power rails along the path and back-feed regen power back to the grid. Each pod has an AC motor and a bidirectional AC motor driver (Figure 3.56(a)).

If power fails, the pod stops, which is a problem if the pod is suspended or on a monorail, as the passengers have no way to walk away. A small power battery (see Volume 1, Section 5.1.4) could power the pod to let it crawl to the nearest station (Figure 3.56(b)). As this battery absorbs regen energy, the station can use a simple power supply rather than an inverter. A bidirectional DC-DC converter between the battery and the high-voltage line converts between their two voltages and limits the charging current into the battery. The battery could be high-voltage (1.2 kV, for

TABLE 3.23
Digital BMUs for Public
Transportation

Application	Brand	Model	Cells	Communication	Additional Features
Train	E-Pow (Huizhou EPower Electronics)	RT02	?	CAN	Master/slave
		RT03	?	CAN	Centralized
Bus	Shenzhen Klclear Technology	BMU06-I-12V-400V	100		
	Huizhou EPower Electronics	E-Pow	?		

---

83. Dang. My high school didn't even have any laboratories of any kind.

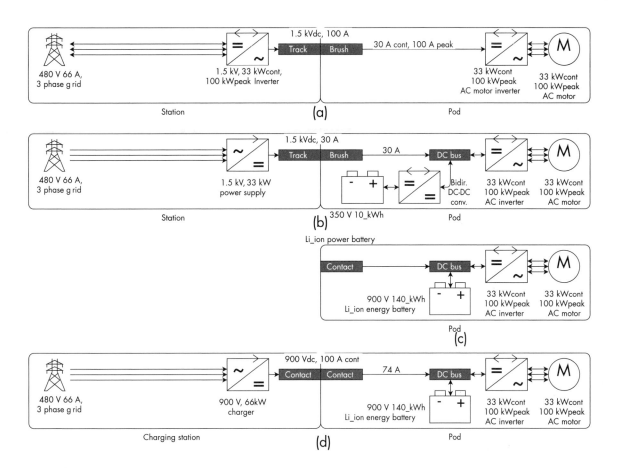

FIGURE 3.56 PRT power: (a) batteryless, (b) with small battery, (c) with traction battery, running, and (d) with traction battery, charging.

maximum efficiency of the DC-DC converter) or medium-voltage (350V, to use a standard traction battery for a passenger car).

Because power rails are expensive[84], a PRT may use instead a large traction battery that can power it for a few hours (Figure 3.56(c)). The pod is then taken out of service and directed to a charging station to recharge for 2 to 4 hours (Figure 3.56(d)).

3.11.1.1 Case Study: Spartan Superway

A team at the San Jose State University in California has been developing a PRT called the Spartan Superway (Figure 3.57).

Instead of one of the standard solutions, the Spartan Superway includes two extraordinary goals: to eliminate both the power rails and the grid, all without stopping long enough to charge a Li-ion battery (Figure 3.58(a)):

· Solar panels at the station charge a super-capacitor.

· A DC-DC converter powers a short power rail just within the station with low voltage.

· Another super-capacitor in the pod is charged while at the station.

· This super-capacitor powers moving the pod between stations.

---

84. About $100/meter.

FIGURE 3.57
Spartan Superway PRT
pod, half-size model.
(Courtesy of Professor
Burford Furman, San
Jose State University.)

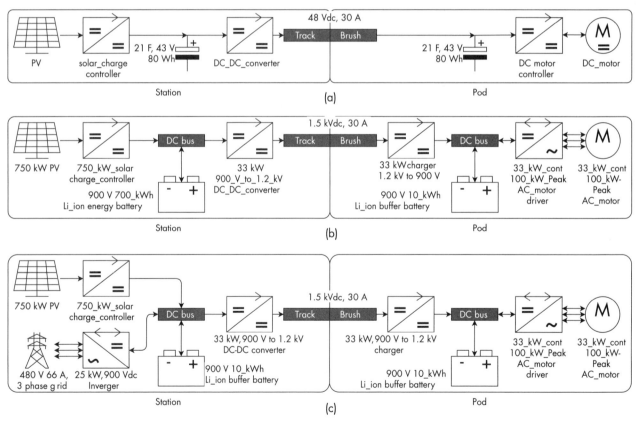

FIGURE 3.58   Spartan Superway solution: (a) proposed, (b) replacing the super–capacitor with a battery, and (c) adding a grid connection.

At first sight, these choices make sense:

· No power rails: saves money.
· Super-capacitors: the stop at a station is too short to charge a battery.
· Solar energy: wonderful.

However, once I plugged in the numbers—33 kW to move the pod[85] at 45 km/h in the 1.6 km between stations, 2-minute stop at each station—a very different picture emerged:

· The 80 Wh of storage in the pod is way too small to power the pod to the next station; 1.2 kWh would be needed, and a super-capacitor that size would cost about $250,000.
· Solar panels limited to the roof of the station would not provide enough power for the system; instead, the panels would have to cover the entire path, with a height of 8m.
· Without a connection to the grid, the 80 Wh of storage in the station would be utterly inadequate to power the system at night (no sun); instead, 700 kWh would be required to handle 4 days without sunshine; a super-capacitor this size would cost about $146 million.
· The 48V track at the station is limited to 30A and can transfer only 43 Wh during the 2-minute stop, much less than the 1.2 kWh required.

It appears that the team did not do these calculations.

Yes, it is possible to implement something similar to this design, though with significant changes and at a high cost (Figure 3.58(b)):

· In the station:
  · 750 kW solar panels per station, which would extend the entire way to the next station;
  · 750 kW solar charge controller;
  · 700 kWh Li-ion battery to hold enough energy to power the station for 4 days without sun;
  · 33 kW DC-DC converter to power the station's track at 1.2 kV (since the current is limited to 30A, the voltage must be increased to transfer 33 kW);
  · Estimated cost for just the power equipment to be $1.2 million per station.
· In the pod:
  · A 33-kW charger in the pod, 900V output for highest efficiency of the power electronics.
  · A 10 kW Li-ion buffer battery (see Volume 1, Section 5.1.4) that charges by 10% during a stop at a station, and discharges that same 10% while traveling to the next station; 90% of the capacity is not used, but a large capacity is required to reduce the specific current to below 3 C; this has the added benefit of increasing the lifetime of the battery; if LTO cells are used, a 5 kW battery could be used, but the cost is the same (LTO being more expensive).

---

85. This power level seems like an order of magnitude too high. Yet the team confirmed that this is the correct value. If this value turns out to be too high by a certain factor, then my analysis is off by that same factor.

· A 33-kW AC motor driver to drive the high-voltage AC motor and convert regenerative braking to recharge the battery.

· Estimated cost for just the power equipment is about $350,000 per pod.

If we assume an average of one pod per station, the total cost is $1.55 million for the power systems for one station and one pod.

If the off-grid requirement is relaxed, then the grid can be used as a battery, and the station only needs a smaller buffer battery, 10 kWh, of which only 10% to 20% is used, just to even out the power from and to the grid (Figure 3.58(c)). If so, the cost of the station drops to about $800,000, and the total drops to $750,000 per station and pod.

Yet, if the Spartan Superway would use the standard solution of using power rails (Figure 3.56(a)), the cost would be $25,000 for the station for the inverger, $5,000 for the pod for the motor inverter, and $160,000 for the power rail between stations. The total would be $190,000 for one station and one pod. A small, 1-kWh battery would allow the pod to crawl to the nearest station in case of failure of power from the grid (Figure 3.56(b)).

Finally, if the Spartan Superway would use the standard solution of using an onboard traction battery (Figure 3.56(c)) and rotating pods so that some are in use and some are charging (Figure 3.56(d)), the estimated cost would be $12,000 for the charging station (for a charger) and $75,000 for each pod (one running, one charging: $70,000 for a battery and $5,000 for the inverter) for a total of $162,000 per station and pod.

Either one of the standard solutions would be significantly more affordable, which is a requirement for the success of this project.

Some 100 bright students, under the guidance of forward-thinking faculty advisors, have spent the last seven years working on this project; I applaud those who devised this concept for thinking outside of the box; however, I would have expected that, at some point, someone would have calculated the actual costs of the chosen approach[86].

3.11.2 Electric Trains

Electric trains include Automated People Movers[87], subway trains, trams or light electric rail (LER), and many long-distance trains. An overhead line powers these trains. The train lifts a pantograph to connect to the high voltage line above. The return path for the current is the grounded track on which the train runs. You may assume that the wheels' ball bearings are also used to provide a low-drag electrical connection to the wheels. In reality, running current through a bearing rapidly damages it due to thermal expansion: and the lubricant offers high resistance; the bearing dissipates high heat and sparking pits the bearing's components. An axle brush or an earth return unit is used to connect the return current directly to the wheels, bypassing the bearings.

---

86. In the industry, a project manager would have considered real-world data and performed this analysis within the first few hours of the project. It would appear that academia works differently.

87. An Automated People Mover is an automated electric train to transport people within an airport or small area. It is also a type of automated transit network (ATN) and automated guideway transit (AGT), just like a PRT. However, unlike a PRT, it is not on demand.

Other trains are powered by a high-voltage third rail along the track[88] (Figure 3.59(a)). The train uses a sliding shoe to connect to the electric rail. Again, the grounded running rails provide a return path for the current.

People mover trains for short runs may receive power from two or more electric rails and not the running rails. These trains do not require a traction battery. In case of power failure, people can disembark and walk away. If a light rail train stops in the middle of an intersection, it can be pushed out of the way. 15. However, batteries have been added to these trains to meet special requirements, as in the following cases.

In one case in which a tram did require a traction battery, the train had to cross a large park, in which unsightly and dangerous overhead lines were not allowed. The train required enough energy onboard to cross the park, stop and restart at a station in the park, and possibly restart from an emergency stop. Working at 1.5 kV (the line voltage) presented multiple challenges:

- The battery voltage was 1.5 kVdc so that it could power the traction system just as it would be from the overhead line.
- For political reasons, the customer specified Gaia large cylindrical LFP cells; yet these cells had too high an MPT and were therefore ill-suited for the job, a full discharge over 15 minutes.
- An off-the-shelf BMS capable of such high voltage was not available at the time.
- Designing a DC-DC converter/charger to charge the battery from the line voltage (both at about 1.5 kV) was not straightforward.
- The battery had to be placed on top of the train, which exposed it to the heat of the day, so air conditioning had to be added.

It appears that the project was never completed.

The county of West Midlands, England, is currently deploying a fleet of trams fitted with batteries that can be propelled by either overhead lines, if present, or by its battery when running on recently laid tracks in areas where overhead lines are considered a blight.

Siemens designed a hybrid storage system for trains[89] with a Li-ion battery and ultra-capacitors (see Section B.2). It has several advantages:

- It supplements power from the overhead lines when accelerating, avoiding voltage sag the overhead grid.
- It stores regen energy locally, avoiding surges on the overhead grid.
- It propels the tram for up to 2.5 km when the pantograph is lowered.
- It reportedly results in energy savings of up to 30%.

A few BMUs are explicitly designed for electric trains (Table 3.23).

3.11.3 Diesel Trains

Most train locomotives in the Americas (Figure 3.59(b)) use batteryless, series hybrid electric traction (see Section 3.1.3). A diesel genset powers an electric motor. This two-step traction takes advantage of the strengths of each technology:

---

88. These voltages are lethal, which is why the infamous third rail has become a colloquial expression for a subject that must not be touched.
89. Sitras HES, Hybrid energy storage system for rail vehicles.

 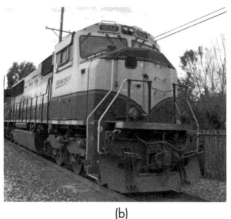 

(a)   (b)   (c)

FIGURE 3.59   Electric trains: (a) third rail between two tracks for an airport transporter, (b) diesel–electric locomotive going by my office, and (c) light train in Denver.

- A diesel genset is used because of the high specific energy of diesel fuel.
- Electric motors have humongous torque at stall, unlike diesel motors.

The locomotive uses regenerative braking. Large heaters and fans at the top of the locomotive dissipate the regen power.

While most light trains are electric, in some areas, diesel-powered light trains are used instead. They also use series hybrid electric traction. Yet some diesel hybrid light trains are now incorporating a traction battery to recover energy from regenerative braking and to help during acceleration. This allows the use of a slightly smaller engine, reduces pollution, and results in quieter operation. The first such train was deployed near Frankfurt, Germany.

Over the last decade, pilot programs have tested the feasibility of adding batteries to the locomotive [19]. In a car, a battery provides the peak power and permits the use of a smaller power source to provide the average power. On the contrary, in a train, a battery does not help because a train runs at full power for long periods. Therefore, the gains are not sufficient to justify the costs and risks of transitioning locomotives from the existing, tried-and-true, batteryless design. So far, battery locomotives are rare.

The traction battery in a train is a buffer battery, which is charged and discharged over just a few minutes. For this reason, LTO cells may be most appropriate, due to their ability to charge in 10 minutes. Toshiba is a major manufacturer of LTO cells, yet, surprisingly, Toshiba locomotives use GS Yuasa batteries.

3.11.3.1   Light Rail

Small electric trains for urban mass transit are called light rail[90], trams, or light electric rail (LER). A high-voltage overhead line (on the order of 2 kV) powers them; the rails provide the ground return. Regen braking feeds electric power back to the overhead line. There is no battery; in case of power failure, passengers get off and walk away along the tracks.

---

90.   "Light" as in "not heavy," not "light" as in "not dark" or "light" as in electric lights.

3.11.3.2 Monorails and Suspended Trams

Monorails pods are powered by the rail and do not normally require traction batteries. However, unlike a light train, in case of power failure, a monorail car is likely to be stuck between stations, in an area where passengers cannot disembark and walk away. Therefore, a traction battery is required to bring the car to the nearest station. The traction battery can be relatively small, just enough to let the car crawl slowly to the station, in about 10 minutes at 10 km/h.

A power battery is used. The power is on the order of 10 kW, and the energy is about 1.6 kWh. LFP power cells are considered safer than other chemistries and would provide the high discharge current required, about 6 C.

3.11.4 Vacuum Trains

A vacuum train rides in an evacuated tube to eliminate air resistance and achieve speeds that are faster than the cruising speed of an airplane. For the comfort of the passengers, acceleration and deceleration may be limited to 0.5 G.

A vactrain that accelerates during the first half the time and decelerates during the second half would travel from Los Angeles up to San Francisco in 44 minutes (Figure 3.60(a)). The train would use about 100 kW to accelerate during the first half of the trip, reach a top speed of 1,800 km/h at the halfway point, and then regenerate about 80% of the energy while braking on the second half of the trip. Of course, on the return trip from San Francisco down to Los Angeles, it would use much less energy.

The vactrain doesn't use wheels (too much friction, too high a speed). Instead, it is suspended above the track, either on an air bearing, a puff of air is ejected beneath the train's skis to lift them off the track, or using electromagnetic suspension, magnetic levitation (maglev).

For forward propulsion, the train uses a linear motor. Half of the motor is the track, which includes permanent magnets or is magnetizable, and half is the train car, which has electromagnets whose polarity is reversed at a high frequency.

In some implementations, the tube propels the train. The linear motor is inverted: the train has permanent magnets, and the tube has electromagnets. The train itself needs just a little bit of house power, which is generated from the forward motion of the train, backed up by a small battery.

In some implementations, a sizable onboard battery powers the train. In other implementations, the tube powers the train; power rails along the tube are impractical because maintaining contact with power rails at such speeds is problematic, and doing so adds drag. Instead, coils in the tube transmit power to the train, wirelessly. At the very end of a trip, the speed is too low for regen braking, so mechanical braking is required, which would probably require wheels.

In case of power loss, the train must crawl to the next station, powered by an internal battery:

· At the moment of power loss, the battery must immediately start powering the suspension; otherwise, the train crashes on the track and stops suddenly.

· The train coats without any forward traction, decelerating until it reaches the crawl speed.

· At this point, it may use wheels in a retractable landing gear to eliminate the power consumption of the suspension.

· The battery powers the train as it crawls to the next station.

This backup battery would be an energy battery, as it needs to power the train over a long period. However, unlike an energy battery, it is not kept at 100% SoC because it needs to be able to accept regenerative energy in case the train needs to brake hard (because there's a stopped train ahead, or it is already close to a station). An SoC of 80% may be appropriate to both accommodate regenerative braking and to power the train as it crawls to the next station.

3.11.4.1 Hyperloop Concept

Hyperloop is a variation of a vactrain. A pod travels in a partially evacuated tube, at 10% of the atmospheric pressure, which is easier to maintain. There is still some air drag, and therefore the top speed is limited.

Initially, SpaceX presented the following concept in 2013. Since then, further development revealed that other techniques and working parameters should be considered. The pod accelerates during launch, coasts at a nearly constant speed (1,200 km/h max), and then decelerates before arrival (Figure 3.60(b)). The launch consumes the majority of the energy. Afterward, the pod only requires enough power to overcome the residual air resistance. Finally, the pod coasts to a stop, using regenerative braking to slow down.

The travel time for the same trip between Los Angeles and San Francisco is 46 minutes[91], just 2 minutes longer than for a vactrain trip, even though the top speed is significantly lower. The launch is at full power, the coast uses a bit of power so that the speed doesn't drop too much, and regen recoups about 80% of the launch power and about 50% of the energy. The deceleration doesn't last as long as the acceleration and starts at a lower speed. An internal traction battery powers the pod. An energy battery of about 500 kWh would propel the pod for this 46-minute trip.

Over a longer distance (e.g., Paris to Moscow), a vactrain would be significantly faster than Hyperloop because a vactrain's top speed is almost unlimited. A Hyperloop pod uses air bearings for suspension. A large fan at the front of the train sucks in the little air left in the tube to reduce the drag further, add some forward motion, and power the air bearings.

Recharging the battery is a challenge. Here are some approaches:

1. The battery is recharged while the pod is stopped at a station: however, there is not enough time to do so. By comparison, the Shinkansen bullet train stops only 1 minute at a station; such a short stop would be not enough to recharge this train's battery, even if using LTO cells that can charge to 80% in 10 minutes.

2. After each trip, a pod is taken out of service for recharging: this gives the pod plenty of time to charge at a reasonable rate (which increases battery life), but it means that passengers continuing onto the next leg of a trip must change pods.

3. The spent battery is replaced with a freshly charged one at the station: this is probably the best solution, as it allows for slow charging and lets continuing passengers remain in the same pod.

Solar panels on the outside of the tube could be sufficient to power the trains. A small, stationary, buffer battery evens out the load on the grid as a train starts and

---

91. The commonly stated time is 35 minutes, but it's physically impossible to go between those two cities in 35 minutes if the acceleration is kept below 0.5 G.

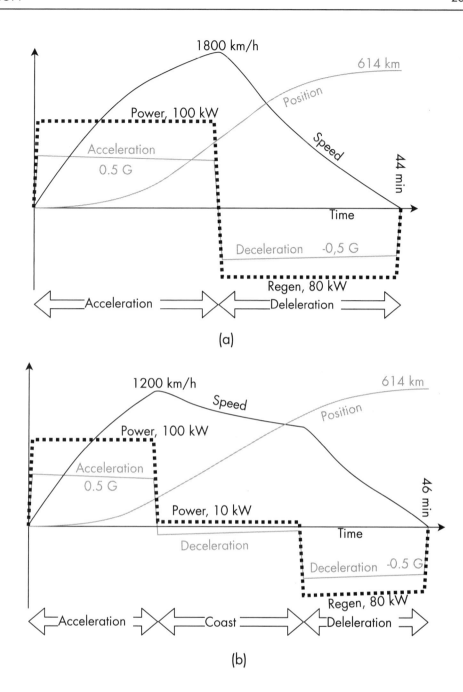

FIGURE 3.60
Trip between Los
Angeles and San
Francisco: (a) Vactrain,
and (b) Hyperloop.

stops charging. This battery is not used to store solar energy for use at night: the grid acts as the battery.

This is all theoretical at this point. The actual pod, if and when it is built, will probably look very different.

3.11.4.2 SpaceX Hyperloop Contest

Today's Hyperloop is very different from the original Hyperloop concept. Several teams, mostly associated with universities, are competing in a XPrize contest [20] to reach top speeds within the short distance of a 1.6-km linear tube built by SpaceX.

This well-intentioned contest has had the unintended consequence of diverting the energy of design teams towards the short-term goal of maximum speed in a short tunnel, rather than the long-term goals of developing technology that may ultimately lead to a practical Hyperloop (Table 3.24). These pods are abusing Li-ion batteries and have to replace them after a few runs. They use batteries designed for UAVs, stacks of pouch cells with a balance connector, and undersized battery terminals. They are discharged much faster, damaging them even more than UAVs do.

As the batteries are operated at a terrible efficiency, they dissipate much heat; yet they cannot be cooled using standard convection in the near-vacuum in the tube. Even liquid cooling requires a radiator, and radiators work mostly through convection. This application is perfect for heat storage, especially phase change material (see Volume 1, Section 5.17.7), to absorb the significant heat generated for a short time.

I wish that, instead of a straight tube, the SpaceX team had developed a toroidal test track, which would allow testing an actual Hyperloop pod (see Section B.10).

3.11.5 Buses

Electrifying city buses may be one of the most effective ways of reducing pollution. Many European cities have used electric trolleybuses (trams) powered by overhead lines since the 1930s. A trolley on top of the bus is raised to connect to two parallel lines are powered by 750 Vdc (Figure 3.61(a)).

Since the turn of the century, Denver, Colorado, has deployed hybrid buses offering free transportation downtown. These buses use an ICE generator powered by natural gas, lead-acid batteries, and electric motors (Figure 3.61(b)).

Rome, Italy, deployed 60 full-electric buses in 2010; today, only two remain; the rest caught fire or are broken down beyond repair [21]. Double-decker battery EV buses provide service in London (Figure 3.61(c)). Newer designs use Li-ion batteries; the voltage is typically 700V, and the energy ranges from 5 kWh for a hybrid bus and 75 kWh to 300 kWh for a pure EV bus. While certain BMSs are offered explicitly for buses (Table 3.23), they do not support a sufficient number of cells in series. Therefore, a BMS for industrial vehicles should be used instead.

3.12 HEAVY-DUTY APPLICATIONS

Today, industrial EVs are used where diesel engines are not desirable:

- In mining, where combustible fuels and exhausts are prohibited (Figure 3.62(a)).

TABLE 3.24
Comparison of Original
Hyperloop Goals and the
Contestants' Work

	Original Hyperloop Goals	Contestants' Work
Passengers	20	0~6
Acceleration	Comfortable: 0.5 G max	Insane: 6 G
Distance	>500 km	1.6 km
Speed	1,200 km/h	500 km/h
Propulsion	Train and track for a linear motor	Often, standard rotary motor
Suspension	Maglev or air bearing	Often, wheels
Battery energy	500 kWh	1 kWh
Battery type	Power battery	Ave Imperator, morituri te salutant
Discharge time	15 minutes	30 seconds
Specific current	4 C	120 C
Battery life	10,000 cycles	10 cycles

FIGURE 3.61
Electric buses: (a) old trams
at the Seashore Trolley
Museum, Kennebunkport,
Maine; (b) hybrid bus
on the 16th Street mall,
Denver, Colorado,
and (c) battery EV bus
in London. (Courtesy
of Vantage Power.)

(a)   (b)

(c)

FIGURE 3.62
Heavy-duty vehicles:
(a) mining excavator,
recharging (courtesy of
RDH–Sharf, underground
mining leaders in
battery technology),
and (b) tractor for the
Port of Los Angeles.

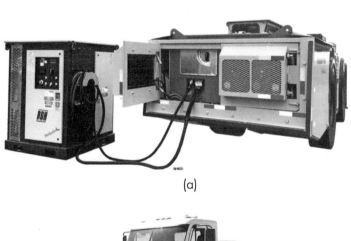

(a)

(b)

- Maritime port operations, to move containers (Figure 3.62(b)); the ports of Los
  Angeles stipulated that all tractors be emission-free to reduce the high level of
  pollution.

- Electric agricultural tractors will appear soon, offering the same torque as their hydraulic counterparts but promising higher reliability due to the simplicity of electric traction; reliability is essential to the farmer in the middle of a tight harvest window.

3.12.1 Technical Considerations

The requirements for a heavy-duty vehicle are more stringent than for a passenger car.

3.12.1.1 Battery

The traction battery in a heavy-duty electric vehicle typically has a voltage of 700V. A hybrid vehicle uses a buffer battery (see Volume 1, Section 5.1.4) of about ~5 kWh. A BEV uses a large energy battery of 10 kWh to 50 kWh.

3.12.1.2 BMUs

These applications require a sophisticated digital BMU that can handle up to 200 or so cells in series and has CAN bus communications supporting the J1393 standard (see Volume 1, Section 5.10.3.9) (Table 3.25).

If the traction battery is split into multiple modules (hopefully in series[92]), a master/slave BMS or a distributed BMS are best because they only use low-voltage communications cables between battery modules. A centralized BMS is inappropriate because it would route high-voltage sense wires throughout the vehicle.

3.12.1.3 Chargers

Unfortunately, 700V chargers are not as widely available as 350V chargers. Some 350V chargers can be wired in series to provide 700V (check with the manufacturer)

TABLE 3.25
BMUs for Large Industrial
and Agricultural Vehicles

Brand	Model	Topology	Modular Battery	Strings in Parallel
Batrium	WatchMon	Distributed	✓	✓
Emus	EMUS BMS	Distributed	✓	
Elithion	Vinci EV	Master/slave	✓	
Elithion	Lithiumate Pro	Distributed	✓	✓
EVlithium	Mini BMS	Distributed	✓	
Jon Elis	ver.2	Distributed	✓	
JTT	X-Series	Master/slave	✓	
Lithium Balance	sBMS	Master/slave	✓	
Lithium Balance	nBMS	Master/slave	✓	
O'Cell	BMU-LV	Master/slave	✓	
Orion BMS	Standard size	Centralized		
REC	BMS 9M	Master/slave	✓	
Tritium	IQ	Master/slave	✓	
Zeva	EVMS	Master/slave	✓	
STW	Power MELA-mBMS	Master/slave	✓	

---

92. A traction battery that uses strings in parallel is less than ideal. It requires a BMS that can handle that.

(Figure 3.63(a)). This is the best solution because there is only one battery, one charging current, one BMS.

If the chargers cannot be wired in series, then you can add a center tap to the battery, which splits it into two batteries (see Volume 1, Section 6.5). Then you can use one charger for the top half and one for the bottom half (Figure 3.63(b)). The problem is that, while charging, you have two separate batteries, each with its current and each with each SoC. By the time you stop charging, the two halves are likely to have differing SoC, and therefore the battery could be severely unbalanced. Therefore, you should have two BMSs when charging (each keeping track of the current and SoC of its half), yet only one BMS when operating.

If you wish to use a single 350V charger, you can split the battery in two, connecting both halves in parallel while charging at 350V (Figure 3.64(a)) and in series during operation at 700V (Figure 3.64(b)). However, this requires eight perfectly synchronized contactors[93] (Figure 3.64(c, d)); plus, this takes considerable care as the half-batteries cannot be safely connected directly in parallel (see Volume 1, Section 3.3.6).

A clever solution uses two diodes and only one contactor with normally closed contacts:

- Plugged in for charging (Figure 3.65(a)):
  - The contactor is powered and turns on, separating the two halves of the traction battery.
  - The rectifier diode for the half with the lowest voltage is forward-biased and the charger chargers it.
  - Once the voltage of this half-battery increases and reaches the voltage of the other half, the other rectifier diode is forward-biased, effectively connecting the two battery halves in parallel.
  - The charger charges both halves as if they were connected in parallel.
  - Yet they aren't connected directly in parallel: the diodes prevent current from flowing between the two halves.
- Unplugged (Figure 3.65(b)):

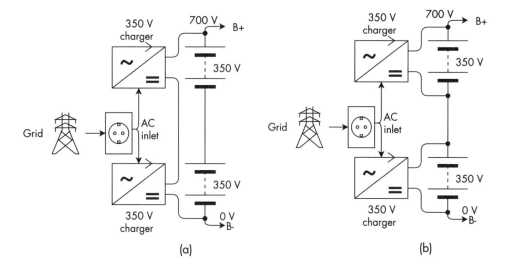

FIGURE 3.63
Charging a 700V battery with two 350V chargers:
(a) in series, and
(b) center-tap split battery.

(a)   (b)

---

93. If not synchronized, they may short across a battery.

FIGURE 3.64
Split battery, parallel
charge, series discharge,
complex circuit:
(a) charging in parallel,
(b) discharging in series,
(c) circuit to charge
in parallel, and
(d) same circuit,
discharging in series.

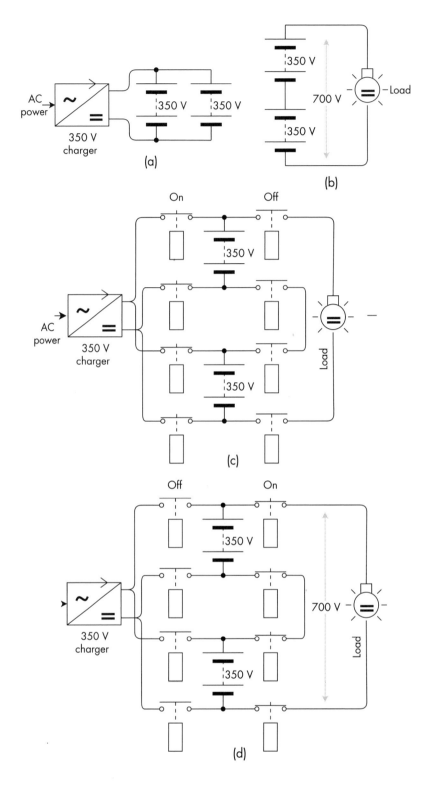

- The contactor is not powered and turns off, connecting the two battery halves
  in series.
- The total battery voltage of 700V is available to the vehicle.
- The diodes are reverse-biased (off) isolating the traction battery from the output
  of the charger.

FIGURE 3.65
Single charger, split
battery, one–contactor
circuit: (a) plugged
in for charging, and
(b) vehicle in use.

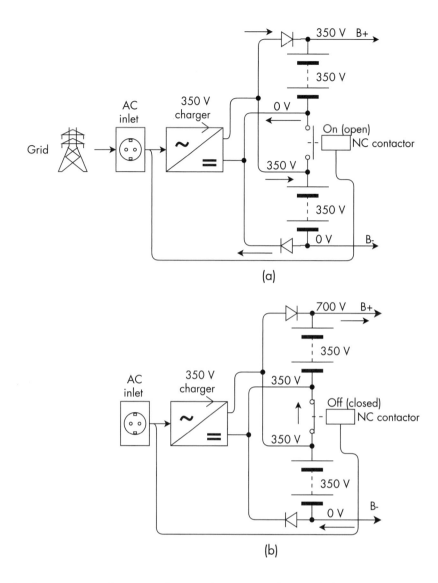

As there are two batteries during changing and a single battery when running, this solution requires two BMSs when charging and one BMS when operating. If a single BMS is used

- While charging, the BMS reports twice the actual voltage (e.g., 700V instead of 350V); if the BMS configures the chargers, it requests a constant voltage setting of 700V, not 350V; even if the charger only goes up to 400V, it is still dangerous.

- The current sensor is mounted on only one of the two half-batteries, so the BMS only knows the current and the SoC of that half-battery; because the two chargers are not identical, the SoC of the other half-battery is different, forcing the BMS to do a lot of balancing; even if the chargers were absolutely identical, they would behave differently once they go into the constant voltage mode because of differences in the cells in the two halves of the battery.

- When any cell voltage is high, the BMS shuts off both chargers, even the one charging the half-battery that can still accept current; this prevents fully charging both half batteries and forces the BMS to do a lot of balancing.

A BMS that is compatible with this solution[94] needs to be configured for it and to be notified whether the battery halves are in series or parallel so that it can report the correct battery voltage. However, such a BMS only solves the first of the problems listed above. If two BMSs are used, one for each half-battery, the same problems arise when the two halves are connected in series as a single battery (see Volume 1, Section 6.3).

3.13 OFF-LAND APPLICATIONS

All the EVs that we discussed up to this point operate on land. The ones in this section operate on water, under water, and in the air.

3.13.1 Marine Traction

The shipping industry is responsible for around 3% of global CO_2 and greenhouse gas emissions. It electrification is a key component of addressing climate change. Vessels (e.g., boats, yachts) may use Li-ion batteries for house power (see Volume 1, Section 2.6) or for traction. This section[95] discusses the latter. For example, sailboats may use a Li-ion battery for house power (so the engine can be left off most of the time) or a traction battery to travel without using wind power. Li-ion has a bad rap in the yachting world, due to some nasty fires (no one has been hurt to date) (see Volume 1, Section 5.2.2.2).

3.13.1.1 Low-Voltage Battery

Low-voltage traction batteries provide a quiet form of traction in small boats, such as sailboats and fishing boats, while trolling. In practice, there are three options for low-voltage marine traction batteries:

- Off-the-shelf batteries: with protector BMS;
- Off-the-shelf modules: several modules form a battery, require external protector switch;
- Custom made.

A low voltage traction battery is likely to use one of the many off-the-shelf, 12V to 48V batteries from companies that specialize in marine products. Some of these batteries are wholly self-contained and look like a lead-acid battery[96]. They include a high-power protector BMS. They may be connected in series to increase the voltage, up to 48V or so; check the specifications. Some manufacturers claim that batteries can be connected in parallel to increase the capacity; yet, with rare exceptions, this is dangerous (see Volume 1, Section 6.3).

Battery modules (see Volume 1, Section 6.2) include the BMS but not the protector switch[97], which must be implemented by the boat builder separately. These batteries can be safely connected in parallel, as long as their voltages match at the time of connection, and in series. Each battery has a communication connector, to report the state and to control the shared protector switch.

---

94. Elithion Lithiumate Pro.
95. Thank you to Kristian Sylvester-Hvid of Clean-e-marine for help with this section.
96. NexGen, SmartBattery.
97. MasterVolt, Lithionics.

Finally, custom batteries are designed for a given application, usually by a company that specializes in marine batteries[98]. They use large prismatic cells and a waterproof BMU.

3.13.1.2 High-Voltage Battery

For large, high-voltage batteries, the traction battery may consist of battery boxes placed in multiple locations in the vessel, using large prismatic cells. The BMS and the power components may be mounted on a bulkhead or in an equipment rack. If so, distributed or master/slave BMUs are ideal, because they use fewer cables. In contrast, a wired BMS would have many high-voltage wires from the BMS to the cells.

3.13.1.3 Battery Design

The marine environment is notoriously a challenge for electronic products. Salty water permeates everything, corrodes connections, and leaves conductive deposits on the surface of PCBs. Ideally, the BMS is completely sealed and uses sealed connectors[99].

Today, no off-the-shelf BMS is explicitly designed for marine use[100]. Instead, a standard BMS for traction batteries for passenger or industrial vehicles must be adapted to use in a marine environment (see Section 3.9.2):

- Enclose the BMS in a sealed enclosure.
- Install sealed glands in the enclosure, and run cables to the batteries through them.
- Place in a dry equipment area.
- Make gas-tight connections: properly crimped terminals are best, and snug ring terminals are good.
- Coat all connections generously with anticorrosion spray or paste.

The constant rocking of a vessel puts a different kind of mechanical stress on batteries than vibration does for land vehicles. Hard bus bars may be acceptable in land vehicles but are inappropriate in a vessel because they put stress on the terminals of large prismatic cells as they sway with the waves. Instead, use resilient spacers between cells and braided jumpers (see Volume 1, Section 5.5.3).

3.13.2 Submarines

Electric traction in submarines offers considerable advantages compared to diesel traction: no need to carry air for the internal combustion engine and no fumes to force out against the water pressure (Figure 3.66(a))[101].

Submarines use two hulls:

1. *Light hull:* exterior, not water-tight, used to provide a hydrodynamically efficient shape. The pressure is the same inside it and outside of it; inside it, the pressure increases as the submarine dives deeper.
2. *Pressure hull:* interior; strong; the pressure inside it is at sea level.

---

98. Genasun.
99. The typical automotive-grade connectors are not sealed.
100. Genasun makes one, but no longer sells it by itself.
101. Thanks to Carme Parareda of Ictineu, Catalonia, for help in this section.

FIGURE 3.66
Submarines: (a) Deepflight
Dragon submarine EV
(courtesy of Deepflight),
and (b) oil-filled, sealed
submarine battery.

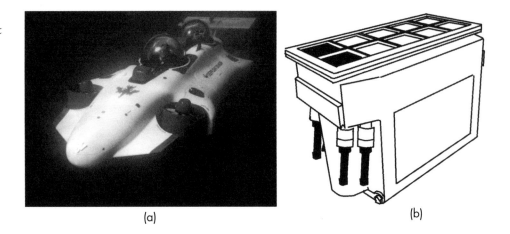

(a)   (b)

If the traction battery is inside the pressure hull (Figure 3.67(a)), then there are no additional requirements compared to a boat. If the battery is inside the light hull (Figure 3.67(b)), it is exposed to the full pressure of the water at depth.

If you thought that a Li–ion battery on fire is bad in a sailboat, think how much worse it is in a crewed submarine! That by itself is a good reason to place the battery outside the pressure hull. Regardless, for safety reasons, proper battery design is crucial in a submarine. Surfacing must not rely on the Li–ion battery alone, as it may not be available at the time.

FIGURE 3.67
Submarine battery:
(a) battery inside the
pressure hull, and
(b) battery outside
the pressure hull.

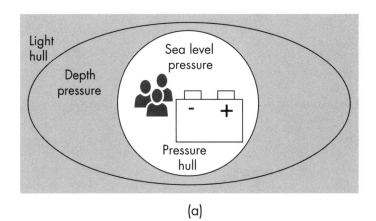

(a)

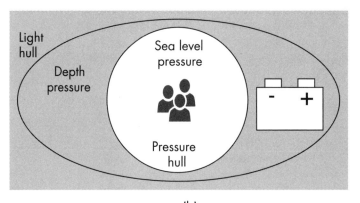

(b)

There are two methods to let devices in the light hull handle this pressure: submersion in a noncompressible oil or encasing is a hard material [22]:

- *Oil*: to handle the pressure, every enclosure, including the traction battery case, is filled with noncompressible oil (Figure 3.66(b)); the oil distributes the pressure and keeps the enclosure from crushing; there cannot be any open spaces nor any gasses in the oil bath because they would be crushed under the pressure; typically, mineral oil (transformer oil) is used because:

  - It is inert and dielectric: it won't affect the components encased in it.

  - It is clear: allows visual inspection.

  - It doesn't cling: leaving the components clean when removed from the oil.

- *Hard encasing*: when an empty space is unavoidable, entomb the component into a thick, hard epoxy; for PCB components, use first a layer of a soft compound to absorb dimensional changes, without which the component may be ripped off the PCB, then cover with a hard layer.

3.13.2.1 Cells

If the battery is in the outer hull, the cells must be such that they won't crush under pressure and must be sealed; not all cell formats meet this requirement:

- *Pouch*: while they are vacuum packed and therefore should contain no empty spaces, they may generate some trapped gas, which would be compressible; regardless, constant changes in pressure are likely to degrade the cells.

- *Small prismatic*: may be OK; studies are required.

- *Large prismatic*: they include empty spaces as well, plus they include a safety vent, which is not sealed.

- *Small cylindrical*: they are sealed and solid, but the metal wall is as thin as possible to maximize energy density; the cylindrical shape inherently resists pressure. However, any pinpoint force may cause them to buckle like a soda can; regardless, the caps are flat and not designed to withstand an external force of that magnitude.

- *Large cylindrical*: these are more solid than any other cell, thanks to their shape and their hard case, so they appear to be the most likely to survive the pressure in a submarine's outer hull[102].

Therefore, large cylindrical cells may be the best cells to handle depth pressure. Pouch cells have been used successfully in outer hulls even though they contain an air space.

For safety, hard cells include safety vents that open up in case of high internal pressure and before the cell bursts. Those vents assume that the external pressure is no higher than sea-level pressure and may not work when exposed to higher external pressure. On the contrary, they may allow foreign material to seep into the cell. Pouch cells are soft and do not use safety vents, which may be advantageous; however, the seal between the two outer layers in a pouch cell is not considered to be very reliable.

---

102. Yet, while the wall is sturdy, the ability to handle hoop stress may be low due to the large diameter.

3.13.2.2 Electronics

Most modern electronic components that are likely to be found in a traction battery are pressure tolerant. Exceptions include

- Aluminum electrolytic capacitors;
- Crystals;
- Sealed mechanical relays.

Electronic assemblies can be converted to pressure tolerant by

- Replacing aluminum electrolytic capacitors with solid tantalum capacitors.
- Entombing crystals first into a soft compound (to absorb dimensional changes without ripping the component off the PCB), and then into a thick, hard epoxy; alternatively, replace the crystal with a ceramic resonator, although their frequency tolerance is not as good.
- Replacing mechanical relays with solid-state switches: transistors, solid state relays.

Open relays are pressure tolerant, though they operate more slowly in oil, depending on the viscosity. The operating times increase as the pressure increases.

Sealed contactors for EV use (see Volume 1, Section 5.12.5.1) do include an open space (they are filled with an inert gas) but are very tough and may be able to withstand the pressure. I have not found any contactors that are specifically rated to be pressure-tolerant.

3.13.3 Aircraft

Aircraft are the last vehicles to be successfully converted to BEV[103]. Challenges include weight, mass, and reliability; a failure in the electrical propulsion system can be fatal. Long-range flights are not yet possible because increasing the size of the traction battery becomes counterproductive, as a heavier battery requires disproportionately more energy to lift. A fuel-powered plane becomes lighter the longer it flies; a battery-powered plane does not.

EV conversion of a small plane does not work; the aircraft is too heavy. A few have tried to no avail. An electric airplane must be designed from scratch to be lighter than a standard airplane.

Battery designers want the BMS to be able to shut down the battery to protect it. Designers of electric aircraft fight back, not realizing that an unprotected battery is more dangerous than a battery that shuts down. We must acknowledge that an electric aircraft may include a bypass switch that allows the pilot to overrule the BMS and keep the battery turned on until the plane lands. The result is a dead battery, but at least the plane landed. The design focus should be on handling BMS warnings to avoid a battery shut down in the first place.

The Solar Impulse 2 plane, which circumnavigated the Earth in 2015 and 2016, offers an example of what happens when the BMS is not allowed to shut off the battery. During the longest leg of its journey, from Japan to Hawaii, the cells overheated due to an inadequate thermal design; normally, the BMS would have protected the cells by shutting down the battery. It wasn't allowed to do so. Consequently, the plane was stuck in Hawaii for months, as the traction battery had to be rebuilt.

---

103. A few were presented at the 2019 Paris Air Show, starting with air taxis and announcing longer-range airplanes.

Redundancy may be used to increase reliability. Having two independent traction systems is beneficial. For safety, there can't be a switch that connects two batteries directly in parallel (see Volume 1, Section 6.3). In any case, connecting two batteries at different SoC in parallel wastes a lot of energy in heat. The traction battery in an aircraft must be designed meticulously, and it must include a very reliable BMS.

3.14 BATTERY TECHNOLOGY

This section discusses the technology inside the traction battery.

3.14.1 Sensing

The BMS senses the battery current and temperature.

3.14.1.1 Current Sensing

Hall effect current sensors are commonly used in traction batteries (see Volume 1, Section 5.9.2). Accuracy can be improved by using two current sensors, a low current one for charging, one a high current one for ignition. Dual-range Hall effect sensors in a single package are available[104].

Shunt current sensors are used as well. Some sensors include an electronic assembly, that reports the current on the CAN bus[105].

For 12V and 24V batteries, there are intelligent battery sensor (IBS) modules[106] using the LIN bus. However, they are not isolated, and a BMS won't support LIN; I mention them just because they are useful to monitor the 12V low-voltage battery.

3.14.1.2 Temperature Sensing

Production-level traction batteries use only a few, strategically placed temperature sensors, such as one against the cooling system and two to four on the cells. An extensive thermal analysis is performed to ensure that those few sensors can detect a thermal event before it is too late. Without a rigorous thermal analysis, it is unsafe to use such a low number of sensors. Instead, the battery should use as many sensors as the BMS allows (for a wired BMS, it is typically a sensor for every six cells or block of cells in parallel).

Yet, ironically, thermistors may actually cause a fire, should their wires create an accidental short circuit. This danger increases as the number of thermistors increases, and more wires snake through the battery. A safer solution is to use a distributed BMS, which includes a sensor in every cell board.

3.14.2 Safety

A traction battery can be lethal due to its high voltage. As it is capable of delivering currents up to 2 kA, a traction battery can cause some severe damage. The following measures may minimize accidents (Figure 3.68):

1. Ground isolation, coupled with ground fault detection;
2. Galvanic isolation between the battery and the vehicle, when the battery is off;

---

104. LEM DHAB series.
105. Sendyne SFP200MOD (isolated power and CAN, also senses voltage).
106. Hella IBS 6KP series; Vishay WBP.

FIGURE 3.68
Safety in a traction battery
for a passenger vehicle.

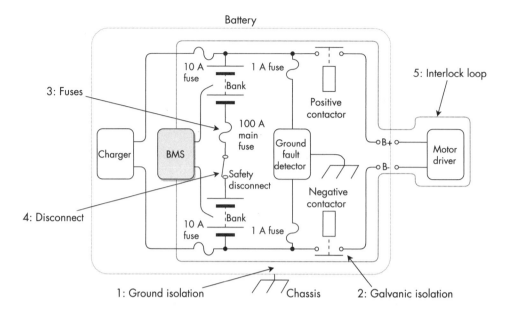

3. Fuses;

4. Safety disconnect;

5. Interlock loop.

These are discussed in the following sections.

3.14.2.1   Ground Isolation

For safety, a traction battery should be isolated from the vehicle's chassis (ground) (see Volume 1, Section 5.1.4). That means that all the following devices must be isolated between the string of cells and the low-voltage power supply and signals:

- In the battery:
  - The cells.
  - The BMS, between cell voltage sensing and low voltage.
  - The current sensor, between the high current conductor and the low voltage.
  - The contactors, between the contacts and the coil; MOSFETs are not isolated; however, an isolated gate drive may be used, or the entire solid-state protector could be isolated.
  - A ground fault detector.
- The motor driver;
- The charger, if any;
- A DC-DC converter powered by the traction battery, if any.

3.14.2.2   Galvanic Isolation

A traction battery of 60V or more should be galvanically isolated 106.5 when it is not powering the vehicle (see Volume 1, Section 5.13.3). This is achieved by opening mechanical switches (contactors) that disconnect all battery terminals, not just the positive side. While solid-state devices (MOSFETs and solid state relays) can interrupt current, they do not provide galvanic isolation. If the vehicle has a charging inlet for fast DC charging, when not connected to an EVSE, the charging inlet must be galvanically isolated from the traction battery.

3.14.2.3   Safety Disconnect

Except for small vehicles, the traction battery must include a safety disconnect that is turned off during manufacture, transportation, service, or after an accident. It must be accessible to emergency personnel without the need for tools.

This safety disconnect may be implemented as

- A removable component, which may contain the main fuse (Figure 3.69); for passenger vehicles, a safety disconnect is an orange component on the traction battery. It is inside the car, yet it can be easily accessed without tools; first responders are trained to find it in various vehicles.
- A connector (also known as a loop key).
- A circuit breaker.

Just like for a fuse, the best placement for a safety disconnect is mid-battery, but not in the middle of a BMS bank (see Volume 1, Section 8.3.2.4).

3.14.3   Fuses

Fuses protect in case of excessive current. Contrary to some misunderstandings, they do not offer any other protection.

3.14.3.1   Main Fuse

A properly selected and placed main fuse offers protection in case of excessive battery current and, specifically, a short circuit across the battery terminals (see Volume 1, Section 5.12.6).

For a fuse, consider

- 48V battery:
  - Littelfuse BF2: 20A to 300A, 58 Vdc, 1-kA breaking.
- 350V battery:
  - Eaton Bussmann EV20, EV25, EV30: 50A to 400A, 500 Vdc, 20-kA breaking;

FIGURE 3.69
Nissan Leaf safety disconnect: (a) plug, and (b) opened to access main fuse.

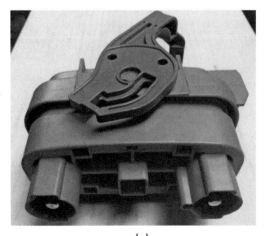

(a)   (b)

- PEC EVFP series: 60A to 250A, 450 Vac, 2-kA breaking;
- Littelfuse LC HEV: 10A to 50A, 425 Vdc, 10-kA breaking.

For safety and reliability, the fuse must be installed in a fuse holder designed for it.

3.14.3.2 Charger Fuse

With a high-power charger, no additional fuse may be required: if the charger current is on the same order as the rating of the main fuse, and the wires to the charger are rated for a current that exceeds the rating of the main fuse, then the main fuse is sufficient. In particular, if the vehicle has a DC fast charging inlet and the main fuse is already sized for the wire gauge to the charging inlet, no additional fuse may be required.

Normally, however, a traction battery is charger slowly and discharged fast. If so, the charger current is significantly lower than the rating of the main fuse and the charger uses small gauge wires. If so, each wire to the charger requires a fuse placed near the cells rather than the charger.

The current rating of the fuses must be less than the current handling of the wires. The fuse must be rated for DC and the total battery voltage. The breaking capacity requirement is not as critical as it is for the main fuse because the wire resistance limits the short-circuit current. Use a chassis-mounted fuse holder, designed for a particular fuse (Figure 3.70(a)).

Some EVs experience this hard-to-troubleshoot problem: the charger fuse blows when the user accelerates hard. This may happen when an onboard charger is connected directly to a traction battery that is too weak for the application; its internal resistance is too high. The fuse blows because the voltage of the traction battery drops suddenly and current rushes from the charger's output capacitor to the battery.

There is no easy solution:

Attempting to slow down the current with inductors in series with the charger line won't help because the time frame is too long for a reasonably sized inductor to make a difference. This may actually cause more problems because the inductor and the capacitor in the charger form an LC tank, which resonates and generates terrible spikes in voltage, approaching twice the battery voltage.

A diode won't help because, if placed in one direction, it would prevent charging; if placed in the other direction, it won't prevent the fuse from blowing.

If the charger's output capacitor can handle those high current spikes (which occur continuously throughout the life of the vehicle), consider using larger gauge wires and slow blow fuses with a higher current rating. Otherwise, the only solution is to add a set of contactors and precharge between the battery and the charger.

3.14.3.3 Auxiliary Load Fuses

The lines feeding low power devices (e.g., DC-DC converter, battery voltage sensing, ground fault detector) should use low gauge wire (e.g., 18 AWG) and low-value fuses (e.g., 1A), placed on the cell side of the wires. They also must be rated for DC and the battery voltage.

For high-voltage batteries, consider the cartridge fuses for digital multimeters, which have great specifications for the price: 440 mA or 11A, 1 kVdc.

A sudden drop in battery voltage may blow the fuse powering a DC-DC converter, just like the charger fuse, as explained above. Unlike with a charger, a diode placed in series with the line powering the DC-DC converter does keep the fuse from blowing

FIGURE 3.70
Unexpected fuse
blowing: (a) sudden
acceleration blows the
charger fuse, and (b)
diode to prevent blowing
DC-DC converter fuse.

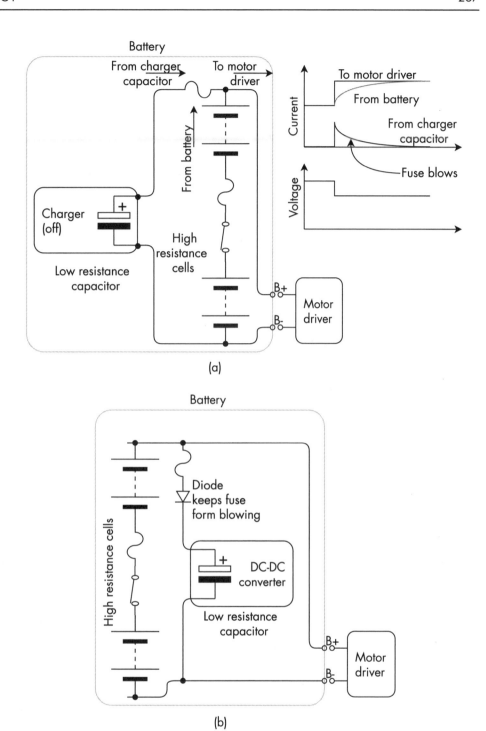

(a)

(b)

when the battery voltage drops (Figure 3.70(b)): it prevents the large capacitor on the input of the DC-DC converter from feeding current back to the battery. However, when the user brakes hard, regen may increase the battery voltage suddenly, and then the fuse may blow. If so, a diode won't help. Using a slow blow fuse may help.

3.14.4 Protector Switch

A small traction battery may use a PCM with a solid-state protector switch, while a large one uses contactors.

3.14.4.1   Single-Port Versus Dual-Port Batteries

The traction battery may be single-port or dual-port (see Volume 1, Section 4.3.1). A dual-port battery (one port controls charging and one controls discharge) is the most energy-efficient and most straightforward to implement. It is acceptable if (see Volume 1, Section 4.3.3.4)

- The charger is permanently connected to the battery.
- The motor driver doesn't do regenerative braking.

Otherwise, a single-port battery is required.

Yet most traction batteries are dual-port, even when regen is possible. The problem is that, generally speaking, the BMS cannot prevent charging through the discharge port in a dual-port battery because

- With a solid-state protector switch, it's physically impossible for the BMS to prevent current from flowing into the battery through the discharge port.
- With contactor protector switch, the BMS may not be smart enough to turn off the discharge contactor to prevent overcharging.

3.14.4.2   Precharge

The large capacitance of loads (motor driver, charger, and DC-DC converters) connected to the traction battery requires precharge (see Volume 1, Section 5.13). EV battery designers tend to consider only the first one and to neglect the last two; yet precharge may be required for chargers (see Section 3.3.2.2) and is required for large DC-DC converters.

Traction batteries with a solid-state protector rarely include precharge. They rely on the nonideal characteristic of the MOSFETs to provide a certain amount of current limiting. Also, a vehicle with a small traction battery is likely to use a relatively low-power charger and motor driver, which presumably have relatively small capacitors. Still, the current inrush stresses the MOSFETs.

3.14.5   Power Supply

In a small traction battery, the PCM is powered directly from the cells. The vehicle powers the BMU in a larger traction battery with a low-voltage supply (e.g., 12V) (see Volume 1, Section 5.8.1). In a plug-in vehicle, the power comes from various sources:

- *Plugged in for charging*: by a power supply powered by the AC power outlet;
- *Ignition on*: from the vehicle's low-voltage supply;
- *At all times*: optionally, by a 12V supply that is always on.

3.14.6   Mechanical Design

The points in this section are specific to traction batteries and are in addition to the general points discussed earlier (see Volume 1, Section 5.18).

3.14.6.1   Design for High Vibration

Lighter vehicles are subjected to more intense vibrations, requiring special attention. Here are some suggestions:

- Consider using small cylindrical cells because of the inherent strength of their enclosure.
- Create parallel blocks of cells that are mechanically compact, sturdy, and stiff, using cell interconnections with small hoops to absorb any small, relative motion between cells.
- Install blocks in resilient holders to allow each block to move freely relative to other blocks.
- Connect these blocks in series with flexible cables to absorb the significant relative motion between blocks. Hard bus bars may be fine for a homemade EV, but may not last for 1 million km required of a production EV. Any connecting method other than flexible connections needs rigorous vibration testing.
- Mount the entire traction battery on rubber mounts to reduce the vibration transferred to the battery.
- Use automotive connectors, which are designed to withstand fretting; gold plated contacts are not acceptable.

Things to watch out for are bolts loosening, thin bus bars shearing, mechanical stress on cell terminals that then stress the interior of a cell, and, in extreme cases, displacement of the jelly roll inside small cylindrical cells.

3.14.6.2 Design for Accidents

Consider what happens if the vehicle is turned on its side or upside down: the battery must remain in place, and cells must not slide out of place and contact a metal enclosure. Do not rely on gravity to keep battery components in place.

Design the battery enclosure so that it is not crushed in case of an accident. For example:

- Place the traction battery in the middle of the vehicle (Figure 3.71(a)); in case of an accident, the battery is protected in the same way that the passengers are; for example, the Toyota Prius.

FIGURE 3.71
Battery crushing avoidance: (a) interior placement, (b) armored enclosure, and (c) displacement.

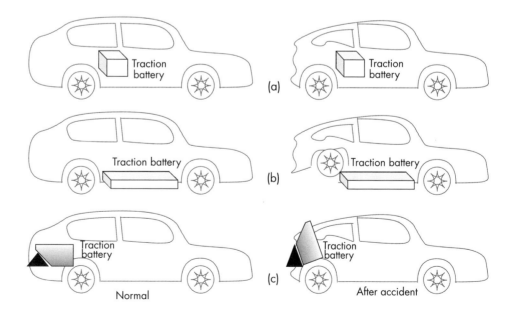

- Use a robust enclosure (Figure 3.71(b)); in case of an accident, the battery resists crushing; for example, the Nissan Leaf.
- Create a wedge shape in the battery and the chassis (Figure 3.71(c)); in case of an accident, the wedge shape pushes the battery up and away, for example, the Ford Escape hybrid.

Include a low-voltage safety interlock loop that is opened in case of an accident, and wired such that it opens the battery's contactors. Include also an orange safety disconnect, which the emergency personnel can quickly locate, reach, and remove.

3.14.6.3 Environmental

In particular, in automotive applications, the battery must be protected from the road grime and the indirect splashes of water. A hobbyist's EV conversion may get away with having an unsealed battery with large prismatic cells. A distributed BMS places the cell boards in the same environment as the cells, while a wired BMS doesn't place any electronics by the cell; in this respect, a wired BMS is better. In production vehicles, the battery should be sealed.

3.14.7   Second Use into a 48V Battery

Some may try to convert a used traction battery to a 48V battery for residential solar applications only to discover that it is impractical; it's either too messy or too expensive (see Section 2.12.6).

3.15   SYSTEM INTEGRATION

This section discusses the interface between the battery and the vehicle; it covers certain technologies in the vehicle that relate to the battery.

3.15.1   Low-Voltage Systems

Other than in a small EV, low-voltage systems provide the house power in a vehicle. Although not part of the traction battery system, they are discussed here because they form an integral function in the vehicle.

3.15.1.1   12V System

In a standard passenger car, a 12V lead–acid battery powers the electrical system. This voltage has been standard since the 1950s and is used in other vehicles as well. An alternator driven by the engine charges the battery.

The 12V voltage is poorly regulated and can peak to more than 100 V or go negative under exceptional conditions:

- *Load dump*: This occurs when a high power load is suddenly turned off. It takes some time for the regulator to react and reduce the field current in the alternator. During this time, the voltage on the 12V network can exceed 100V!
- *Disconnected SLI battery*: The 12V system relies on the SLI battery as a short-term storage device like a capacitor to even out the current. Running the engine without a battery results in nasty spikes on the 12V line that may damage sensitive electronics and reportedly even blow up the rectifier diodes in the alternator (see Section 2.2.1.1).

- *Reverse jump-start*: If someone jump-starts a car by connecting the jumper cables in the wrong direction, −14V appears on the 12V network.
- 24V jump-start: Some people like to jump start a car at high power from a 24V truck battery; this places about 28V on the 12V system.

Devices connected to the 12V network must withstand these conditions.

In a BEV, there is no engine and, therefore, there is no alternator. The 12V battery is charged in one of these ways:

- With an AC charger in a plug-in vehicle: whenever the vehicle is plugged in (Figure 3.72(a)).
- With a small DC-DC converter powered by the traction battery (Figure 3.72(b)): this works at any time and is required in an HEV because it is never plugged into AC power.
- Both (Figure 3.72(c)).

The DC-DC converter is low power because it only needs to provide the average power used by the 12V system. The 12V battery provides the peak power required

FIGURE 3.72
Charging the 12V battery: (a) from AC power, (b) from traction battery, and (c) both.

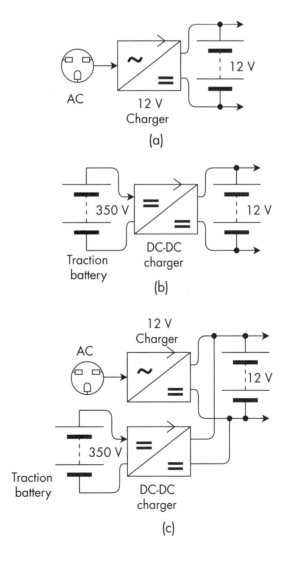

only occasionally, for example, power-steering and opening all windows while the high beam headlights are on.

Some vehicles do not have a 12V battery and power the 12V system directly with a high-power DC-DC converter. This presents some challenges:

· The converter needs to be larger because it must support the peak power of the loads, not the average power.

· If the string of cells powers the converter directly, before the protector switch, the converter's constant drain will kill the battery sooner or later (Figure 3.73(a)); don't do this!

· If the converter is powered after the protector switch (Figure 3.73(b)), when the traction battery shuts off it disables the DC-DC converter; this kills the 12V supply, removing power from the ignition key and the BMS; the vehicle is bricked[107].

The above demonstrates the importance of a 12V battery.

If you absolutely cannot use a 12V battery, then minimize the chance that the vehicle is bricked and include ways to recover if it is bricked:

· Use a BMS with two protector switches with two different low-voltage cutoff thresholds:

  · A high power switch for the main loads with a moderate low-voltage threshold (say, 3.0 V/cell).

  · A low power switch for the 12V DC-DC converter with a low-voltage threshold; even after the BMS shuts off the main loads, it still has more energy left to power

FIGURE 3.73
Vehicle without a 12V
battery: (a) DC-DC
converter powered
before BMS (DON'T!),
and (b) after BMS.

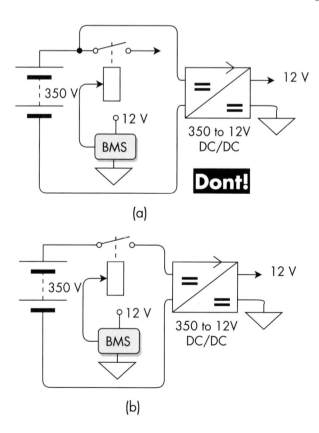

---

107. Bricked means nonfunctional, no more useful than a brick.

the converter (and itself) for a long time, until the cell voltages drop really low (say, 2.5V/cell).

- In a BEV, use a 12V power supply (Figure 3.74(a)) to recover by plugging the vehicle into AC power.
- In an HEV, add a Bootstrap switch[108] that temporarily powers the DC-DC converter from the traction battery (Figure 3.74(b)).
- In a PHEV, add both of the above.

The EV driver has access to the Bootstrap push-button switch. The timer prevents abuse; if the driver keeps the switch pressed, the timer limits how long the battery feeds the converter. The diodes prevent the output from the timer from back-feeding to the other high-voltage loads.

FIGURE 3.74
Bootstrap for vehicle without a 12V battery: (a) BEV, and (b) HEV.

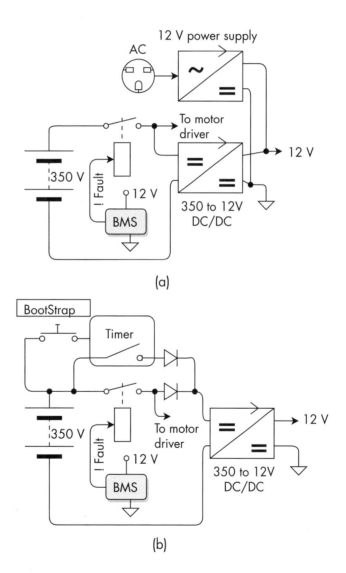

(a)

(b)

---

108. A bootstrap is a loop mounted to the back of a boot; you pull on it when putting on the boot. In this context, the term "bootstrap" comes from the English expression "pull oneself up by one's bootstraps," which is the impossible task of elevating yourself by pulling your legs up with your hands. In technology, it means "to start from the beginning without any outside help." Hence, in a computer, a bootstrap program is its integral BIOS that allows the computer to start when first powered and prepare to accept software from the hard drive.

3.15.1.2   4V, 36V, and 48V Systems

Most passenger cars today use 12V, while most large vehicles use 24V. Currently, the automotive industry is introducing a 48V network for high power loads: heaters, air conditioning, and motors/generators. This is not displacing the 12V system but complementing it. Passenger cars will have a 48V Li-ion battery in addition to the 12V battery. A 48V to 12V DC-DC converter charges the 12V battery from the 24V bus.

This trend is particularly useful for EVs, as it develops electrical versions of devices that in an ICE vehicle would be powered by coupling to the engine, such as pumps and compressors. This 48V system can also power some electrical functions that are only found in EVs such as the starter motor in a mild hybrid.

Of course, a 24V, 36V, or 48V Li-ion battery requires a BMS as well. To prevent bricking, the BMS in these batteries should be powered by either the charging source or by the battery itself (through the BMS), whichever is available. When the battery is empty, and the BMS shuts it down, the BMS removes its own power as well; later, when the charging source returns, it powers the BMS, allowing the battery to recharge.

3.15.2   High-Voltage System

The high-voltage electrical systems are connected to the traction battery. They include a motor driver, a charger (in a plug-in EV), and possibly one or more DC-DC converters.

In traction batteries higher than 60V, the externally visible insulation of every high-voltage component (cables, connectors, safety disconnect, fuse holder) is colored safety orange as a warning to technicians and emergency response personnel. The metal case of a high-voltage assembly is grounded to the vehicle chassis and is not painted orange.

The high-voltage cable connected to the traction battery may be shielded to reduce the EMI[109] generated by the motor driver. The mating high-voltage connector on the battery connects the shield of the cable to the vehicle chassis, to complete a Faraday cage.

A high-voltage battery may include an interlock loop that, if opened, it causes the traction battery to shut down immediately. The interlock loop consists of wires and switches connected in series, such that, if any of them is opened, the entire loop is opened. For example:

- A small-gauge wire is piggy-backed to the high voltage cable from the traction battery to the motor driver (Figure 3.68). If the cable is disconnected, or cut in an accident, the small wire is also disconnected or cut.

- A switch detects if the safety disconnect is removed.

- A switch detects if the cover of the battery is opened.

3.15.3   Communications

Other than small vehicles, EVs communicate through the CAN bus; production cars also use the local interconnect network (LIN) bus (see Volume 1, Section 5.10.3.5).

---

109. Electromagnetic interference. See Volume 1, Section 5.16.

3.15.3.1 CAN Bus

Several devices communicate through a CAN bus (see Volume 1, Section 5.10.3.9) (Figure 3.75). In a typical passenger vehicle, there may be multiple, separate CAN buses:

- A high-level CAN bus for communication between the VCU and the safety-critical devices, such as the brake, throttle, BMS, and motor driver.
- A low-level CAN bus for communication between the VCU and the noncritical devices, such as the dashboard and entertainment.
- The BMS may have a separate CAN bus to a charger.

In passenger vehicles, the speed of the bus is typically 500 kbps. It may be slower in industrial vehicles: 125 kbps or 250 kbps; 11-bit IDs and 29-bit IDs can coexist on the same CAN bus. An 11-bit ID is sufficient for full functionality in a vehicle. Each device on the bus may broadcast messages to whoever needs to hear them. In many cases, it may send a request message to another device. That device will reply with the requested data.

A lean-running CAN bus is best for reliability and hence for safety. The chance for errors increases with the complexity of the messaging scheme.

Therefore, I recommend

- Fewer, terse messages;
- Use 11-bit IDs if possible;
- Broadcast essential data only;
- For nonessential data, request it when needed (e.g., if the BMS sets a Fault flag in the State flags, the VCU requests what the fault is, and the BMS replies);
- Do not use multiframe messages (such as in the NMEA 2000 standard);
- Do not require multiple handshakes to transfer information.

3.15.3.2 Broadcast Messages

The BMS should broadcast the battery's data regularly:

- Essential data:
  - State flags: charge OK, discharge OK, warning, fault, contactors, and others;
  - Charge current limit (CCL);
  - Discharge current limit (DCL).
- Typical, but not essential auxiliary data:
  - Voltage, current, temperature;

FIGURE 3.75
CAN bus communications in a typical vehicle.

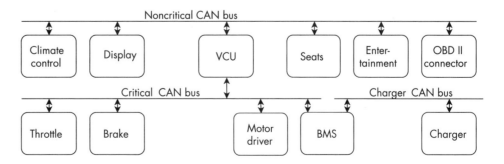

- State of charge;
- Warning and fault flags.
- Minimum, average, and maximum cell voltage;
- Minimum, average, and maximum temperature;
- Evaluated parameters such as the state of health, capacity, resistance, and how long at the present power.

The BMS can broadcast various messages at various rates, every 5 ms to every 1 second, depending on how quickly the particular data may change.

A single CAN message is sufficient for the BMS to broadcast all of the essential data plus some auxiliary data (see Volume 1, Section 5.10.3.9).

3.15.3.3   Request/Response Messages

Other electronic control units (ECUs) may request any other, nonessential data from the BMS, as needed (for example, any of the optional data listed above). The data in the request messages may include an ID for the requesting unit or an ID for the unit from which a response is requested (this is not the message ID, which, again, is not an address).

3.15.3.4   J1939

J1939 is a closed standard developed for trucks, although, it is being adopted by more industries. It requires a 29-bit ID to convey additional information:

- 3 bits of priority;
- 18 bits of Parameter Group Number (PNG) to identify the message's function and data;
- 8 bits if Source Address (SA) to identify the requesting device.

J1939 promised message standardization by defining PNGs for many standard functions in an engine-powered vehicle. The reality for electric vehicles is different because few of the predefined PNGs apply to the functions in an EV. Still, J1939 is useful to define a general format for messages, even though each company defines its own sets of PNGs with little coordination within the industry.

3.15.3.5   OBD II, PIDs

Onboard diagnostics (OBD) is a technology implemented by all production vehicles that allows a technician to

- Diagnose and query the status of a vehicle.
- Scan for past status (freeze frames), and see stored, permanent, and pending diagnostic trouble codes (DTCs).
- Clear diagnostic trouble codes and stored values.
- Run tests.
- Get vehicle information.

The technician connects a scan tool to the OBD port (under the steering wheel in passenger vehicles) to communicate with the vehicle.

The scan tool sends an 8-byte parameter identification (PID) request using a specific CAN ID (e.g., 0x7E0), and the vehicle responds with an 8-byte message whose ID is 8 higher than the request (e.g., 0x7E8). The response contains the requested data. The standard defines some PIDs; other PIDs are proprietary, defined by each vehicle manufacturer.

ODB is a closed, proprietary standard. Access to it is effectively reserved to large automotive manufacturers. However, volunteers have reverse-engineered and published a large number of PIDs [23, 24].

3.15.3.6 Messages to Other Devices

Motor drivers (see Section 3.3.1.1), chargers (see Section 3.3.2), and charging stations (see Section 3.9.4) communicate specific CAN messages.

REFERENCES

[1] Wikipedia, "Wind-Powered Vehicle," en.wikipedia.org/wiki/Wind-powered_vehicle.

[2] "Downwind Faster Than the Wind," Wired, www.wired.com/2010/06/downwind-faster-than-the-wind/.

[3] "Under the Ruler, Faster Than the Ruler," www.youtube.com/watch?v=k-trDF8Yldc.

[4] Tuvie, "Mercedes Benz Formula Zero Concept for the World of Motor Sports Racing," www.tuvie.com/mercedes-benz-formula-zero-concept-for-the-world-of-motor-sports-racing/.

[5] "Electric Vehicle Battery Being Charged Explodes and Starts a Fire," November 25, 2019, https://redd.it/e3slp7.

[6] "It can take approximately 3,000 gallons (11,356 liters) of water, applied directly to the battery, to fully extinguish and cool down a battery fire..." *Tesla Model S Emergency Response Guide*.

[7] New Eagle, neweagle.net/.

[8] enviromotors.com/wiki/index.php/Sparrow/ElithionThundersky.

[9] "Electric Vehicle in Kathmandu Nepal," energyhimalaya.com/energy-efficiency/transportation.html.

[10] "Forklift Batteries: Lead-Acid versus Lithium-Ion" na.bhs1.com/forklift-batteries-lead-acid-vs-lithium-ion/.

[11] Samsung SDI, www.samsungsdi.com/automotive-battery/products/prismatic-lithium-ion-battery-cell.html.

[12] Green car reports, "Nissan Buys Back Leaf Electric Cars Under Arizona Lemon Law," https://www.greencarreports.com/news/1079475_nissan-buys-back-leaf-electric-cars-under-arizona-lemon-law.

[13] Cleantechnica, "2019 Nissan LEAF — Still No Liquid-Cooled Battery?" https://cleantechnica.com/2018/12/05/60-kwh-nissan-leaf-still-no-liquid-cooled-battery/.

[14] SAE, "SAE Electric Vehicle and Plug in Hybrid Electric Vehicle Conductive Charge Coupler J1772_201710," standards.sae.org/j1772_201710/.

[15] Wikipedia, "J1772," en.wikipedia.org/wiki/SAE_J1772.

[16] CHAdeMO, "CHAdeMO," www.chademo.com/.

[17] Wikipedia, "CHAdeMO," en.wikipedia.org/wiki/CHAdeMO.

[18] "SAE Rules," www.fsaeonline.com/cdsweb/gen/DocumentResources.aspx.

[19] Toshiba, "Hybrid Locomotives," www.toshiba.co.jp/sis/railwaysystem/en/products/locomotive/hybrid.htm.

[20] SpaceX, "Hyperloop," www.spacex.com/hyperloop.

[21] Wanted in Rome, "Another Electric Minibus Catches Fire in Rome," www.wantedin-rome.com/news/another-electric-minibus-catches-fire-in-rome.html.

[22] Barnes, H., "A Review of Pressure-Tolerant Electronics" U.S. Department of Commerce, 1976.

[23] Wikipedia, "OBD-II PIDs," en.wikipedia.org/wiki/OBD-II_PIDs.

[24] The OBD-II Home Page, www.obdii.com/.

CHAPTER 4

HIGH-VOLTAGE STATIONARY BATTERIES

4.1 INTRODUCTION

The traditional power grid, which was developed more than a century ago, suffers from the limitation that, at a given moment, it must generate exactly as much power as it being consumed. Adding an energy storage system (ESS) overcomes this limitation. An ESS stores excess generated energy and releases it when needed to meet demand. Additionally, an ESS improves service and reduces costs (see Section 4.5.1). ESSs are slowly being integrated into grid and off-grid applications[1], using a variety of storage technologies[2].

An ESS that uses a battery is a battery energy storage system (BESS) (Figure 4.1). BESSs are preferable for their energy efficiency, durability, quick response, and high specific power; unlike pumped hydro[3] or compressed air storage, they can be placed anywhere. However, they are relatively expensive and don't scale up as well as some mechanical solutions. These batteries are stationary[4] and operate at high voltages. A high voltage is preferable because high-voltage inverters are more efficient than those that use a 48V battery and a transformer to generate the high AC voltage[5].

This book defines high-voltage stationary batteries as ranging from 450V to 1,800V, and from about 3 kWh to 100 MWh. Note the difference between BESS and battery: the battery is but one component inside the BESS, which also includes high power converters.

High power electronics convert the voltage between a high-voltage stationary batteries and an AC[6] line, either in one direction or both (see Section 4.2.3). They are not isolated, meaning that, when connected to a floating battery, they reference the battery to the AC line voltage: the battery becomes more hazardous, and a battery isolation tester would report a ground fault (see Volume 1, Section 5.14).

Applications for high-voltage stationary BESSs range from a wall-mounted unit for an off-grid home to many cabinets of backup power for a power company. They may be classified as on-grid or off-grid:

---

1. Thanks to Guilherme Bonan of WEG Group and Sébastien Maes of AllCell Technology for contributing to this chapter.
2. Including pumped hydro, standard batteries, flow batteries, flywheels, pressure chambers, and ice storage. See Section B.9.
3. Two lakes at different altitudes joined by a pipeline. When energy is available, water is pumped up from the lower lake. When energy is needed, water flows down from the top lake, running generators.
4. Traction batteries in plug-in vehicles may also be used for electrical storage; vehicle-to-grid (V2G) technology is covered in Chapter 3.
5. The efficiency of a transformer-less DC-DC converter is maximum when the input and output voltages are closest to each other. The efficiency of any DC-DC converter of a given power is highest at low currents, which, for a given power, is achieved by operating at high voltage.
6. Alternating current (not air conditioning).

FIGURE 4.1
High-voltage stationary
battery. (Courtesy of
HNU Energy.)

- *On-grid*: A system connected to the power utility grid; the high voltage stationary battery exchanges power with the electrical grid; if the grid fails, the system may or may not operate off-grid.
- *Off-grid*: A stand-alone system.

This chapter discusses BESSs that use Li-ion, high-voltage batteries, which can have a much lower MPT than other battery chemistries (see Volume 1, Section 1.7).

4.1.1.1 Tidbits

Some interesting items in this chapter include:

- The power company is in the clairvoyance business, predicting if you're about to turn on a switch (Section 4.3.1.1).
- The power company uses big batteries to procrastinate (in upgrading transmission lines) (Section 4.3.2).
- Large energy users pay more for the one time when they use a lot of power than for the total energy they use (Section 4.5.4).
- Some companies make money by storing and trading electricity, the way Wall Street makes money by trading stocks (Section 4.3.3.1).
- It's possible to detect a dead squirrel across high-voltage lines; indeed, it's essential to do so before it starts a fire (Section 4.9.1).

4.1.1.2 Orientation

This chapter starts with a discussion of some of the devices used in high-voltage stationary batteries. It then covers a range of applications (utility, microgrid, energy users, residential), discussing practical issues that arise in each when using Li-ion and gives possible solutions. It discusses technical details about the art of high-voltage design, strategies to avoid battery shutdown, the high-voltage battery, and its integration into a system. Finally, it looks at high-voltage battery arrays.

The focus of this chapter is the battery (and specifically the BMS). It is not a complete guide on designing high-voltage applications.

4.1.1.3 Applications Classification

This chapter divides applications for high-voltage stationary batteries into six groups (Figure 4.2).

Table 4.1 lists these applications and their typical characteristics.

4.2 TECHNICAL CONSIDERATIONS

This section discusses technical issues common to high voltage stationary batteries. Issues specific to a particular application will be discussed in the sections for that application.

4.2.1 Battery Voltage for a Given AC Voltage

Most inverters for high-voltage stationary batteries do not use a transformer for simplicity and improved efficiency. Without a transformer, there is no isolation between the battery and the AC line, and the voltage of the AC line determines the battery voltage.

Single-state inverters generate the AC by dropping down the battery voltage. Therefore, the minimum battery voltage must exceed the peak of the line voltage (Figure 4.3(a)). For example, with a 240-Vac line, assuming a ±6% tolerance in the line voltage:

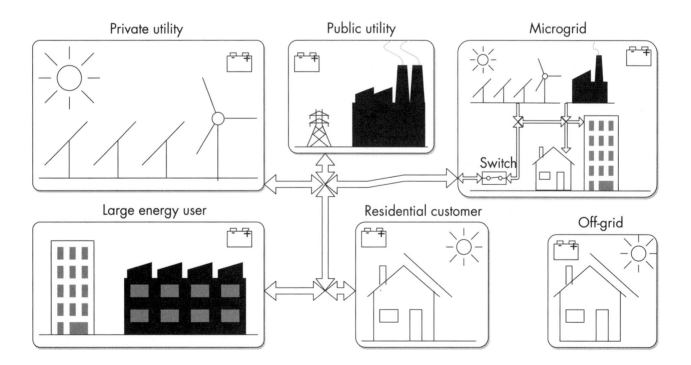

FIGURE 4.2 Application groups; each group includes a high-voltage stationary battery.

Group	Description	Grid-Tied	Battery Type	Voltage	Power [MW]	Energy [MWh]
Public utility	The electric power company (or someone on its behalf); generates electrical power and sells it to customers	Yes	Energy, power, buffer	900~1,800	0.1~100	0.5~100
Private utility	Investor-owned and privately-owned companies that store and/or generate electric power and trade it with the power company	Yes	Energy, power, buffer	450~1,800	0.1~10	0.1~10
Microgrid	A local, smaller version of the full grid	Either	Energy	450~900	NA	10~100
Large energy user	Major customers of the power company: industrial, commercial, government	Yes	Energy, buffer	450~900	0.05~0.2	0.02~1
Residential customer	Minor customers of the power company, in a single house	Yes	Energy	450	NA	5~20
Off-grid	A stand-alone site with no access to the grid: a house, a telecom site	No	Energy	450	NA	10 ~50

TABLE 4.1   Application Groups Using High–Voltage Stationary Batteries

$$240\,\text{Vac} + 6\% = 254\,\text{Vac} = 360\,\text{Vpeak}^7 = 720\text{V}_{\text{peak-to-peak}} \tag{4.1}$$

For lead-acid batteries, a higher nominal voltage of 960 Vdc is required to accommodate voltage sag.

Rather than a single battery, a split battery may be used (see Volume 1, Section 6.5), each half having one-half of the total voltage (Figure 4.3(b)).

Table 4.2 lists the battery voltage for a given AC line voltage and type. It depends on the inverter's topology (see Section B.6.1) and whether it uses a single battery or a split battery. Table 4.2 assumes that the maximum AC voltage is 6% higher than the nominal and that the minimum cell voltage is 3.0V. The notation 2 × means a split battery.

The values in Table 4.2 apply only to single-stage inverters. A dual-stage inverter allows a given battery voltage to generate higher AC voltages thanks to a DC-DC converter stage that raises the battery voltage to a voltage that exceeds the peak-to-peak voltage of the line (Figure 4.4).

Typically, batteries and inverters don't go any higher than these voltages[8]. For higher voltages, a line-frequency transformer is required. An auto-transformer may also be used to step up the AC voltage a bit and allow the use of a battery with a slightly lower voltage[9]. These points also apply to chargers and invergers.

4.2.2 Power Quality

Power quality is the measure of how ideal the grid voltage and current are (Figure 4.5):

---

7.   The peak voltage of a sine wave is $\sqrt{2}$ times the RMS AC voltage value: for example, for 110 Vac, the peak voltage is 156V. The peak-to-peak voltage is twice this: 311V.

8.   The limitation is the maximum voltage of the power switching components: IGBT modules. They can handle 1.2 kVdc comfortably, perfect for 240 Vac. Above this voltage, it's really hairy: either multiple IGBTs must be connected in series (tricky) or you need one of a few IGBTs for 3.3 kV, which are available but are very expensive.

9.   An auto-transformer is smaller, more efficient than a standard transformer of a given power. It is not isolated.

FIGURE 4.3
Minimum battery voltage
must exceed maximum
AC line peak voltage:
(a) single battery, and
(b) split battery.

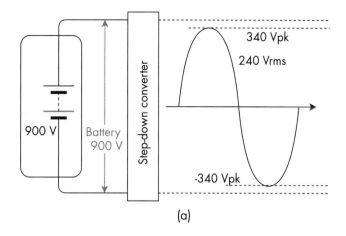

(a)

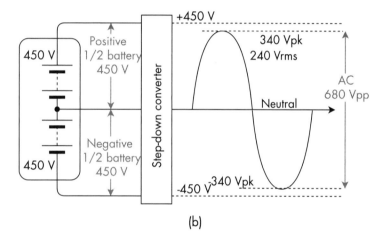

(b)

- The grid voltage, supplied by the power company: the user wants it to be a perfect sine wave at the correct voltage.
- The grid current, drawn by the user: the power company wants it to be a perfect sine wave synchronized with the voltage.

In reality, neither the voltage nor the current are ideal. A BESS can help.

4.2.2.1 Voltage Quality Issues

Power quality issues on the grid voltage include (Figure 4.6) [1–3]

- Transient disturbances:
  - Impulsive: <200 μs, unidirectional;
  - Oscillatory transients: low frequency (<500 Hz), medium frequency (500 Hz~ 2 kHz), high frequency (> 2 kHz);
- Voltage disturbances:
  - Short duration (0.5~30 cycles): voltage sags (10%~90%) and swells (105%~173%);
  - Long duration (>30 cycles): undervoltages and overvoltages;
  - Interruptions (complete loss of voltage): momentary (<2 seconds), temporary (2 seconds~2 minutes) and long-term (>2 minutes).
- Fundamental frequency disturbances:

Phase	Nominal Phase Voltage [Vrms]	Nominal Line Voltage [Vrms]	Inverter Topology	Maximum Line Voltage [Vrms]	Peak Voltage [V]	Peak-to-Peak Voltage [V]	Minimum Battery Voltage [Vdc]	Nominal Battery Voltage [Vdc]	Cells in Series (3.65V Cells)
1 Φ	120	—	Half bridge	127	180	360	2 × 180	2 × 225	62
	120		Full bridge	127	180	360	180	225	62
	240	—	Half bridge	254	360	720	2 × 360	2 × 450	123
	240		Full bridge	254	360	720	360	450	123
Split	120	240	Half bridge	132	180	360	2 × 180	2 × 225	2 × 123
3 Φ Y	120	207	Half bridge	127	180	360	2 180	2 225	2 123
	240	415		254	360	720	2 × 360	2 × 450	2 × 123
	480	831		509	720	1,439	2 × 720	2 × 900	2 × 246
3 Φ Δ	—	120	Full bridge	127	180	360	180	225	62
	—	240		254	360	720	360	450	123
	—	480		509	720	1,439	720	900	246

TABLE 4.2   Battery Voltage for a Given AC Line Voltage and Phases

FIGURE 4.4
Minimum battery voltage
must exceed maximum
AC line peak voltage:
(a) single battery, and
(b) split battery.

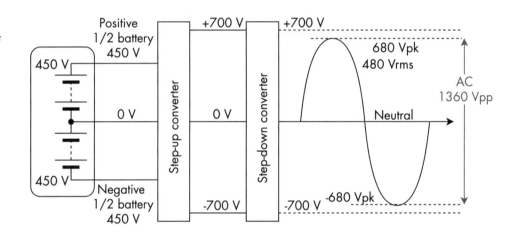

FIGURE 4.5
Grid voltage and
load current.

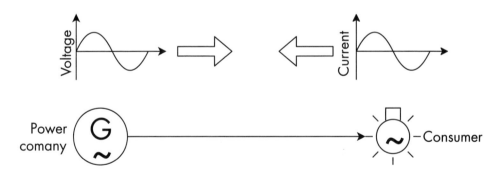

- Within ±20 mHz: acceptable time correction for synchronous clocks;
- Between ±20 and ±50 mHz: corrected by a governor in the generator;
- More than ±50 mHz: to be corrected by BESS;
- Beyond −1.25% ~ +2.5%: grid shuts down.
- Variations in steady-state:
  - Grid voltage harmonic distortion: misshapen waveform, ≤3 kHz;

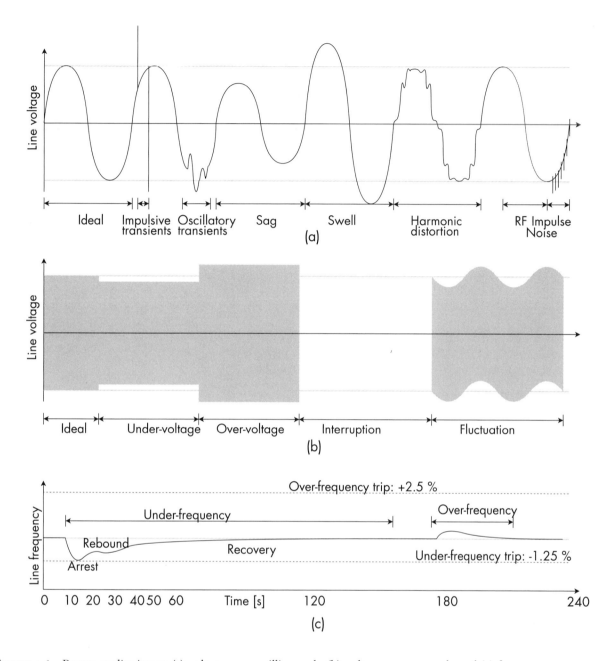

FIGURE 4.6   Power quality issues: (a) voltage, over milliseconds, (b) voltage, over seconds, and (c) frequency.

- Voltage fluctuation: intermittent variations in voltage, <25 Hz, this result in a visible flicker from lamps[10];
- Noise: uncorrelated to line frequency, >3 kHz.

---

10. Flicker and fluctuation are often conflated, but are different: fluctuation refers to the measurement of the line voltage, while flicker refers to the human eye perception.

4.2.2.2 Current Quality Issues

Power quality issues on the grid current include

- Load current harmonic distortion: misshapen waveform, ≤3 kHz;
- Power factor: load current out of phase with the grid voltage.

Ideally, a load draws power from the AC power line as if it were a resistor; the shape of the current that it draws is identical to the sinusoidal shape of the voltage applied to it (Figure 4.7(a)). In reality, the typical charger draws current in pulses that look nothing like a sine wave (Figure 4.7(b)). This misshapen current reduces the efficiency of the power transmission grid. Current that is not in phase with the voltage results in power that is transferred back and forth in the power company's transmission lines without doing any work; the power company cannot charge the user for it, yet it pays for power that only heats the transmission lines needlessly.

4.2.2.3 Power Company's Perspective

Ideally, the grid provides a perfectly shaped sine wave exactly at the nominal voltage and frequency. In reality, the grid deviates from this ideal:

- The frequency and voltage drop when heavily loaded; the generators truly slow down and may overshoot upon recovery.
- The voltage drops out suddenly in case of a dysfunction in the grid or another customer's load; it takes a little bit for breakers to trip, for the fault to be isolated, and for service to be reestablished.

FIGURE 4.7
Current drawn by a
load: (a) ideal, and
(b) typical charger.

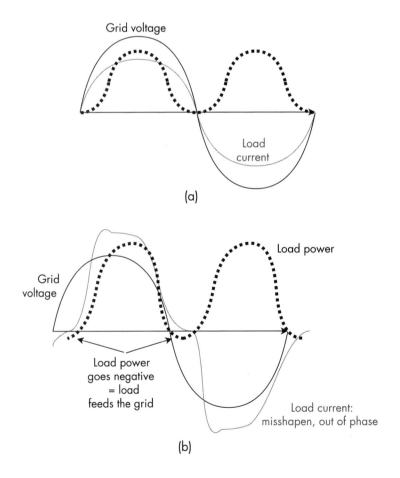

- Transformers and other nonlinear loads at other customers' sites distort the sine wave.
- Other customers add large transients when they suddenly open a large load, due to the drop in current and the inductance of the power company's transformer.
- Solar flares and lightning bolts add extremely high-voltage surges.

Power quality issues cost the power company in terms of inefficient power transfer and damaged equipment.

4.2.2.4 Customer's Perspective

Ideally, the customer draws current from the grid with a sine wave that perfectly matches the grid voltage's shape and phase, the way a simple resistor would. In reality, it deviates from this ideal:

- Inductive loads (e.g., motors, transformers) draw a current that is delayed relative to the voltage waveform from the grid.
- Nonlinear loads draw current unevenly and distort the waveform: harmonic distortion.
- Large power supplies temporarily short-circuit the grid when first turned on.
- Standard power supplies draw current in short pulses at the top and bottom of the sine wave (PCF[11] power supplies do not); other loads draw current unevenly over the cycle, distorting the waveform of the current.
- Electric arc furnaces, static frequency converters, and cycloconverters cause fluctuations.
- Current goes up and down as loads are turned on and off; the current surges every time a high-power load (anything from an incandescent bulb to a motor) is turned on.
- A large transient is added to the grid when suddenly opening a large load.
- Switching power electronics place evenly spaced pulses at the PWM frequency onto the AC line.
- High-power radio sources (e.g., induction heaters) place radio frequency onto the AC line.

Power quality issues cost the customer in terms of disruptions and downtime, equipment damage and service, and energy to restart machinery.

4.2.2.5 Equipment Perspective

The data-processing industry specifies the level of disturbances that equipment such as computers must be able to withstand without damage or loss of functionality (Figure 4.8) [4].

The CBEMA curve[12] shows that the range in overvoltage and undervoltage is reduced as the duration of disturbances becomes longer.

4.2.3 Unidirectional Versus Bidirectional BESS Topology

A BESS can be classified based on its topology as

---

11. Power factor corrected.
12. Computer and Business Equipment Manufacturers Association.

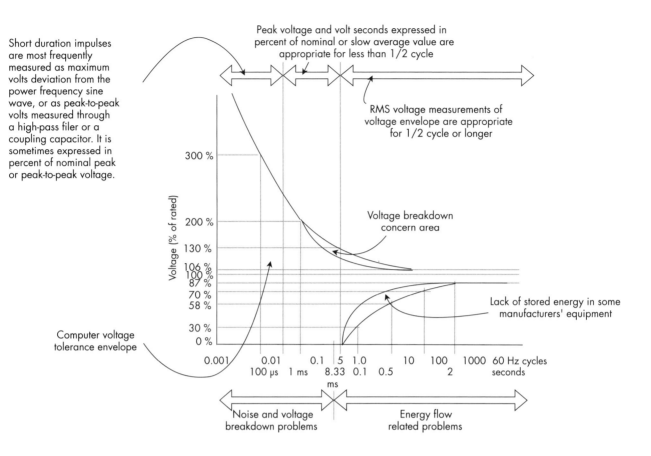

FIGURE 4.8   CBEMA curve.

- *Unidirectional*: from the grid to the battery, then to the load (Figure 4.9(a));
- *Bidirectional*: from and to the grid (Figure 4.9(b)).

Generally, each class of applications tends to use one topology or the other one:

- Public utility, private utility, microgrid, residential customer: bidirectional;
- Large energy user, off-grid: unidirectional.

Some in this industry have redefined the terms "UPS" and "ESS" to mean unidirectional and bidirectional, respectively. I won't. In this book, I use these terms according to their original definitions:

- *UPS*: uninterruptible power supply, regardless of the battery voltage;
- *ESS*: energy storage system, regardless of the direction of the flow of energy and regardless of storage technology used.

4.2.3.1   DC-Coupled Sources

Regardless of the topology, additional sources may charge the battery directly, such as DC-coupled solar through a solar charge controller, or a wind generator through a charger or a DC-DC converter. In most cases, though, any additional resources would be AC-coupled, externally to the BESS.

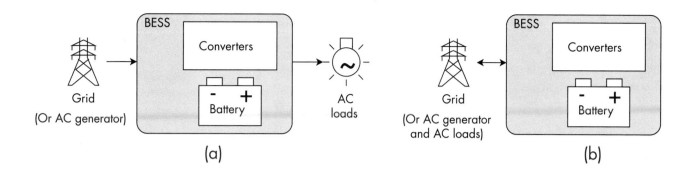

FIGURE 4.9   BESS topology classification: (a) unidirectional, and (b) bidirectional.

4.2.4   Unidirectional BESS

The unidirectional topology receives power from the grid[13], stores it, and powers the loads (Figure 4.9(a)). Mainly, unidirectional BESSs are used at the customers' sites (see Section 4.5).

4.2.4.1   Unidirectional BESS Devices

The unidirectional topology uses these devices:

- Unidirectional devices: charger (see Volume 1, Section 1.8.3) and inverter (see Volume 1, Section 1.8.4), off-grid (see Section 2.2.4);

- Bidirectional devices[14]: inverger (see Volume 1, Section 1.8.4.2), grid-interactive.

The inverter is off-grid, meaning that it generates its own 50/60-Hz voltage reference. For the sake of efficiency, some inverters produce a squarish approximation to a sine wave, which some loads cannot handle.

The inverger is grid-interactive, meaning that it switches automatically between grid-tied mode (when an AC reference such as the grid is present) and off-grid mode (when absent).

Optionally, an automatic voltage regulator (AVR) uses a ferro-resonant auto-transformer to regulate the line voltage coarsely. It usually includes a filter to attenuate RF noise and a transient voltage absorber to clamp surges.

4.2.4.2   Unidirectional BESS Topologies

Unidirectional BESSs may use one of a variety of fundamental topologies, including the following five and variations thereof. Different companies use different terms for a given topology. At times, different companies may use the same term for two different topologies; it's a mess. In this book, I use the following terms:

- Standby, off-line, voltage and frequency dependent (VFD) (Figure 4.10(a))[15];

---

13. Alternatively, this could be an off-grid AC line with an AC generator and AC loads.
14. Even though the overall topology of the BESS product is unidirectional, internally, it may use a bidirectional device.
15. Some companies (including AEG, Uninterruptible Power Supplies Ltd., Elprocus) call the off-line topology line-interactive or voltage independent (VI) if it includes an AVR.

FIGURE 4.10
Unidirectional BESS
topologies: (a) standby,
(b) online, (c) hybrid,
(d) delta, and (e)
grid–interactive.

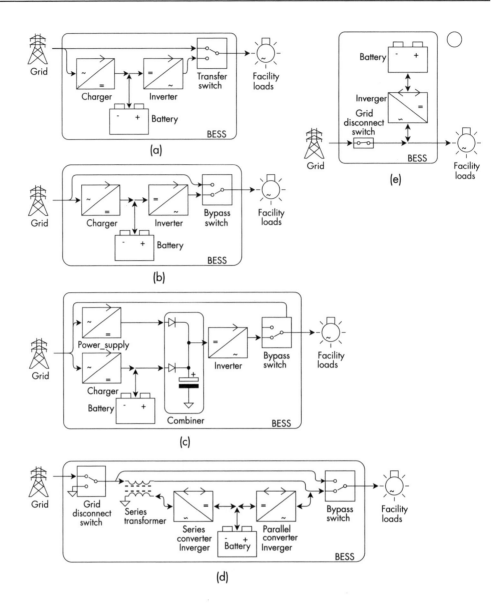

- Online[16], double conversion, voltage and frequency independent (VFI) (Figure 4.10(b));
- Hybrid, standby online hybrid (Figure 4.10(c));
- Delta conversion (Figure 4.10(d));
- Grid-interactive (Figure 4.10(e))[17].

The difference between standby and online topologies is not immediately apparent from their block diagrams; to distinguish them, note the position of the switches in the two diagrams:
- *Standby topology* (Figure 4.10(a)): a transfer switch is normally "up": from the grid;

16. When discussing online and off-line, I can never remember which is which because, with an off-line BESS, the load is normally on the line (the grid), and online UPS it is not. I guess the idea is that in an online BESS, the inverter is normally online (as in, it is running), and in an off-line BESS, the inverter is normally off.
17. Some companies (including American Power Conversion (APC)) call the grid-interactive topology line-interactive.

- *Online topology* (Figure 4.10(b)): a bypass switch is normally "down": from the inverter.

Table 4.3 compares the characteristics of each unidirectional BESS topology [5]. Three main issues arise in a unidirectional BSS:

1. *Grid voltage*: The imperfect grid voltage is applied to the load, which may affect it or even damage it.
2. *Load current*: Some of the components in the misshapen load current drawn from the grid do not do any work and waste power heating the transmission lines.
3. *Transitions*: Ideally, when the AC power goes away, and a UPS takes over, the load sees no interruption in voltage and phase; in reality, there may be a short transition without power to the load, and there may be a step in the phase of the voltage; similar disturbances occur when AC power returns (Figure 4.11).

The various topologies may handle these issues differently or not at all:
- Load voltage:
  - The online and hybrid topologies power the load with an ideal voltage source from their converters.
  - The delta topology compensates for voltage level and distortion.
  - The standby and grid-interactive topologies may include a ferro-resonant transformer to regulate the grid voltage, although not the waveform.
  - The standby, delta, and grid-interactive topologies may include an AVR[18] to reduce noise and surges.
- Grid current:

Type	Operation	Power [kW]	Efficiency	Load AC Voltage	Grid Current	Grid Off, Return	Cost
Standby	Remains idle until needed; then, it transfers the load from the grid to the inverter	0.5–2	Very high: the converters are normally idle	Dirty: the load sees the grid (1)(2)	Dirty: the grid sees the load	Abrupt (3)	Low
Online	The load is always powered by the inverter, regardless of what's happening on the grid	5~5000	Low: the converters are always in use	Clean: generated by the inverter	Clean: uses a PFC charger	Seamless	Medium
Hybrid		0.5–5					High
Delta		5~5000	High: the converters do little work	Medium: the inverger cleans waveform, not noise (1)	Clean: the inverger corrects the power factor	Seamless (3)	Medium
Grid-interactive	The load is powered by either the grid, the inverter, or both; if needed, the grid is disconnected	0.5–5	High: the converter is normally idle	Dirty: the load sees the grid			Low

Notes: (1) Optional AVR attenuates noise and clamps surges; (2) Optional ferro-resonant auto-transformer regulates the voltage; (3) Phase step may be eliminated with special algorithms.

TABLE 4.3   Comparison of Unidirectional BESS Topologies

---

18. Automatic Voltage Regulator. See Section 4.2.4.1.

FIGURE 4.11
Transition at loss and
return of AC power in
a unidirectional BESS.

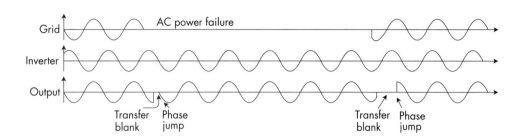

- The online and hybrid topologies draw ideal current from the grid with power factor corrected converters.
- The delta and grid-interactive topologies compensate for distortion and phase in the load current.
- Transitions:
  - The online and hybrid topologies isolate the load from the grid, so there are no transitions.
  - The delta and grid-interactive topologies respond immediately.
  - The standby, delta, and grid-interactive topologies may include algorithms to synchronize their converters to the grid to avoid steps in the phase.

4.2.4.3   Standby Topology

The standby topology is also known as off-line topology or voltage and frequency-dependent (VFD). If an AVR is included, it is also known as line-interactive or voltage independent (VI) [6]. If a ferro-resonant transformer is included on its output, it is known as standby-Ferro.

The operation is straightforward:

- Normally, the grid powers the AC load (Figure 4.12(a)).
- If required (e.g., a power failure, a disturbance, or the price of electricity is too high), the inverter is turned on, and a transfer switch connects the AC loads to the inverter rather than the grid (Figure 4.12(b)).
- When no longer required, the transfer switch reverts to powering the AC loads from the grid, and the inverter is turned off.

This topology has two advantages:

1. *High efficiency:* Most of the time, the inverter is off, and the grid powers the load directly; this results in nearly 100% efficiency and low stress on the inverter.
2. *Smaller charger:* The charger doesn't have to power the loads, and it can take a long time to recharge the battery.

However, there are also several disadvantages:

- A smaller charger cannot charge the battery fast enough if the grid is available for less than 50% of the time, which can be a problem in developing countries with unreliable service.
- The power factor of the load affects the grid.
- The AC to the load is not clean (Figure 4.13(a)) (see Section 4.2.2):

FIGURE 4.12
Operation of the standby
topology: (a) standby,
and (b) active.

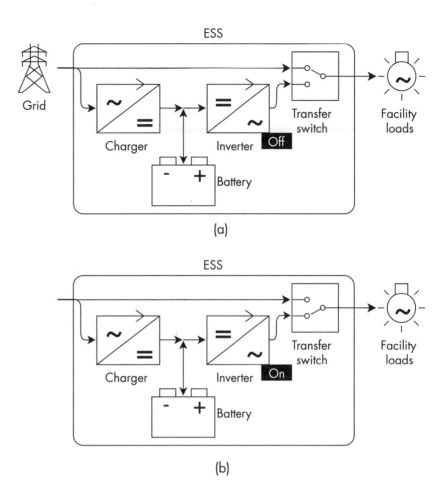

FIGURE 4.12
Operation of the standby
topology: (a) standby,
and (b) active.

- When the grid powers the load, it exposes the load to line disturbances.

- When the AC power drops out suddenly, it takes some time for the UPS to react (on the order of 1~10 ms) during which time the loads are not powered.

- If the transfer switch uses a contactor, there is a short blank time with no power when transferring between sources, on the order of 5~50 ms.

- There is a jump in the phase of the AC output when the transfer switch changes from the grid to the inverter; similarly, there is a jump when switching back to the grid.

Various techniques may be used to minimize these disturbances (Figure 4.13(b)):

1. *High-frequency noise and surges*: An AVR clamps overvoltage surges and attenuates noise.

2. *Transfer blank time*: Sophisticated algorithms may be able to predict a loss of AC power through careful analysis of the disturbances on the AC sine wave that typically occur before a loss; this may allow detection faster than a half-cycle of the line frequency; a solid-state switch reduces the switch time.

3. *Switch-on phase jump*: The jump when the AC goes away can be eliminated by synchronizing the inverter to the input AC voltage so that after the transition the phase of the AC power to the loads is unchanged.

4. *Switch-off phase jump*: Eliminating the phase jump when power returns is more complicated because, by the time the AC power returns, the phase has drifted,

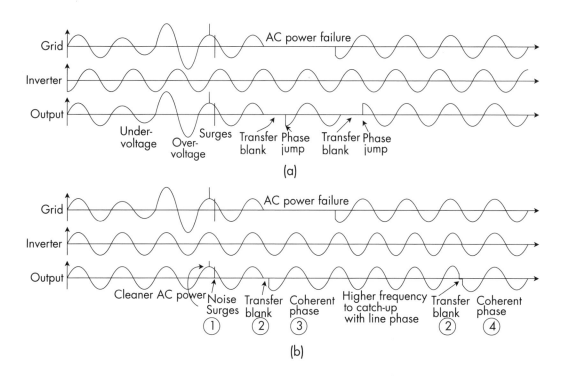

FIGURE 4.13   Standby topology handling of disturbances and glitches: (a) standard, and (b) with improvements.

and the inverter output is out of phase with the grid; after power returns, the inverter increases the frequency ever so slightly to let its phase slowly drift back in sync with the phase of the grid; at this point, the transfer switch is flipped back to the grid, without a phase jump.

If the transfer switch is electromechanical, no matter how quickly the loss of AC power is detected, there is always a disruption because the contactor requires tens of milliseconds to switch over. During this time, between two and four cycles of the AC power are lost; loads must be able to operate with about ~200-ms loss of power (~10 cycles). A solid-state transfer switch is capable of switching instantaneously[19]; however, it is slightly inefficient and doesn't handle high current as well as a contactor does.

High-power applications may use an AC motor with a flywheel as a storage device: during the disruption, the motor keeps on spinning, turning into a generator, powering the load.

4.2.4.4   Online Topology

This topology is also known as double conversion or voltage and frequency independent (VFI). Its function is quite simple: the charger keeps the battery full while the inverter powers the load (Figure 4.14(a)). In case of a fault in the BESS, a bypass switch switches automatically to power the load directly from the grid (Figure 4.14(b)).

The advantages of this topology are:

---

19. Although it must wait until the end of a half-cycle for the AC voltage, because it uses thyristors: TRIACs or SCRs.

FIGURE 4.14
Operation of the
online topology: (a)
normal, and (b) fault.

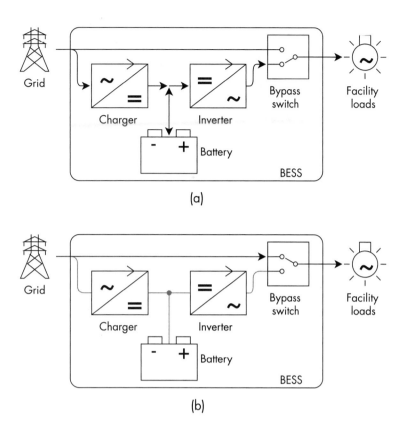

FIGURE 4.14
Operation of the
online topology: (a)
normal, and (b) fault.

- There is no transition; regardless of what may happen on the grid, the load continues to be powered as if nothing happened.
- Clean AC powers the load.

This topology is ideal for critical loads, that is, loads that may not turn off.

The disadvantage is that the power electronics operate constantly at full power, making this topology less efficient than the standby topology. The major difference between the standby topology and this topology is that in the standby topology inverter powers the load only when there is no power, while in this topology it does so all the time.

4.2.4.5 Hybrid Topology

Instead of switching the AC source, as the standby topology does, the hybrid topology switches DC sources onto a capacitor to maintain the voltage during the transition. This method would not work with AC. One DC source is the battery, through a DC-DC converter, and the other source is the grid, through a power supply.

The combiner selects whichever source is available and feeds the inverter, which, in turn, powers the load:

- *AC power present* (Figure 4.15(a)): Power flows from the grid to the power supply, through the top diode, to the inverter, and the load; the voltage of the power supply is higher than the battery voltage, which is why, when the top diode is on, the bottom diode is off; the charger also charges the battery.
- *AC power absent* (Figure 4.15(b)): Power flows from the battery, through the bottom diode, to the inverter, and the load.

FIGURE 4.15
Operation of the hybrid
topology: (a) normal, (b) no
AC power, and (c) fault.

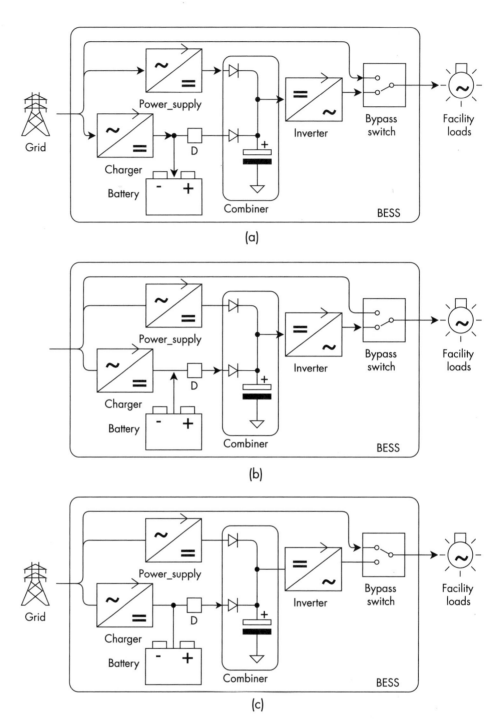

(a)

(b)

(c)

- *BESS fault* (Figure 4.15(c)): A bypass switch switches automatically to power the
  load directly from the grid.

In some implementations, there is a DC–DC converter in the location marked D
(Figure 4.15(b)). It is not necessary.

This topology combines some advantages of the online topology (the load is
driven by a clean signal at all times) with some advantages of the standby topology (a
smaller charger only works part of the time). However, this topology is more complex
than either of those and, therefore, is more expensive.

4.2.4.6 Delta Conversion Topology

The delta topology is the most effective unidirectional topology as it offers the advantages of both the standby and the online topologies, without any of their limitations. Its efficiency is somewhere between those two topologies. It lends itself to very high power implementations. It is currently patented by American Power Conversion.

This topology's hardware is quite complex, as are its control algorithms. It uses a current transformer in series, between the grid and the load, and two invergers:

- Series converter: grid-tied inverger (see Section 2.2.4) between the current transformer and the battery;
- Parallel converter: grid-interactive inverger between the AC output and the battery.

Normally, the grid powers the load, plus or minus a bit of power from the battery through the converters. In case of a grid failure, a grid disconnect switch opens, and the battery powers the load through both converters. The beauty of this topology is that it can supplement both the load voltage and the grid current:

- The series transformer and the series inverter allow it to add voltage to the grid voltage or subtract it from the grid voltage.
- The parallel inverter allows it to add current to the load or subtract current from it.

Therefore, this topology can correct waveform distortion in both the grid voltage and the load current. The online topology can do this as well, but it does so by using much more power.

This topology's many operating modes demonstrate its versatility:

- *Normal* (Figure 4.16(a)): The grid voltage and waveform are nominal; power flows from the grid to the load, unchanged.
- *No grid* (Figure 4.16(b)): The battery powers both invergers; the parallel inverger switches to the off-grid mode, leading the grid-tied series inverger; together, they produce the total power required to power the load; there is a slight transition as the grid disconnect switched from the grid to ground, which is required to switch the transformer from a series configuration to a parallel configuration.
- *BESS fault* (Figure 4.16(c)): The bypass switch selects the grid to power the load directly

The capabilities of this topology shine when correcting the grid voltage or the load current, moment by moment, to restore not only their average values but also their respective waveforms:

- *Low line voltage* (Figure 4.17(a)): If, at a given instant, the grid voltage is low, the series inverger generates the missing voltage; the series transformer adds this voltage to the grid voltage so that the voltage on the right side of the transformer is the expected value.
- *High line voltage* (Figure 4.17(b)): Conversely, if the grid voltage is high, the series inverger absorbs the extra voltage and charges the battery.
- *Misshapen current* (Figure 4.17(c)): Similarly, if, at a given instant, the load current is high, the parallel inverger sources the missing current so that the grid current

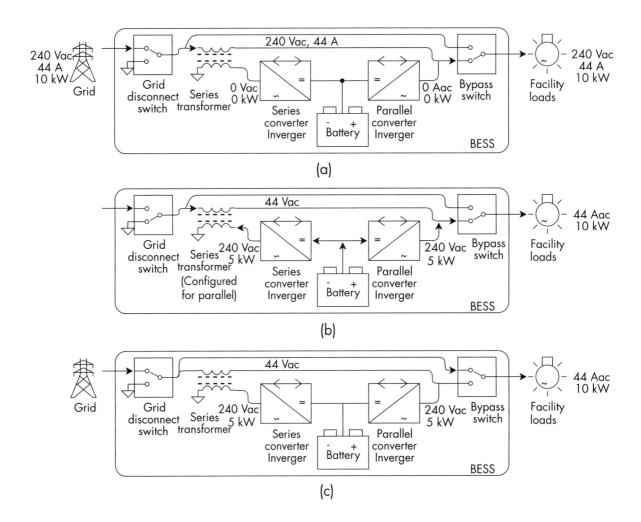

FIGURE 4.16   Operation of the delta topology: (a) no correction required, (b) no AC power, and (c) BESS fault.

is the expected value; conversely, if the load current is low, the parallel inverger draws the extra current and charges the battery.

The result is that the load receives a perfect sine wave at the correct level and that the grid supplies a current shaped like a perfect sine wave in sync with the voltage.

This topology's efficiency comes from the fact that, when the grid is present, the converters need to provide only the difference (the "delta") that is required to achieve the desired performance:

- From the load's point of view: It corrects only the difference between what the grid provides and what it should provide (e.g., if the voltage is 10% low, the converters only provide the missing 10%).

- From the grid's point of view: It corrects only the difference between what the load takes and what it should take (e.g., if the current has harmonic distortion, the converters only provide the harmonics).

However, when the grid is absent, the converters must power the load entirely. Therefore, the total size of the two converters is no less than the size of the inverter in the standby or online topologies.

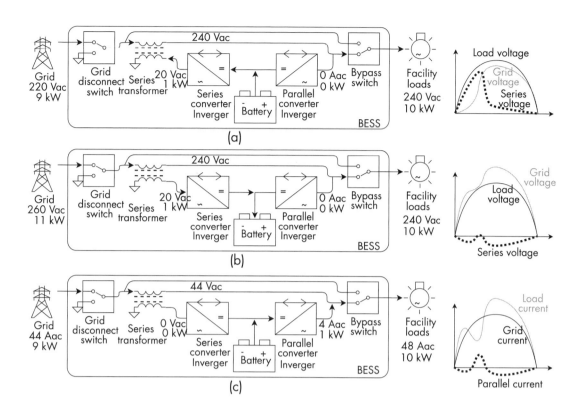

FIGURE 4.17 Operation of the delta topology: (a) low grid voltage, (b) high grid voltage, and (c) misshapen load current.

4.2.4.7 Grid-Interactive Topology

The grid-interactive topology is a subset of the delta topology. It performs nearly as well as the delta topology but with a single inverger. Its control algorithms are just as complex, and its efficiency is the same. It also handles all four classes of service, except that it cannot clean up the line voltage. Adding an AVR on the grid input brings the line voltage closer to ideal.

The operation is simpler than the Delta topology:

- *Normal or BESS fault* (Figure 4.18(a)): The grid powers the load; the inverger is off.
- *No grid* (Figure 4.18(b)): The moment the grid goes away, it opens a switch to disconnect from the grid and the inverger switches to the off-grid mode to power the load; therefore, there is no disruption; power comes from the battery; when the grid returns, the inverger takes some time to slowly drift the frequency until it is synchronized with the grid[20]; then the grid switch is turned back on and the inverger switches to the on–grid mode.
- *Misshapen load current* (Figure 4.18(c)): The inverger operates in the grid-tied mode and either generates the extra current to power the load (power comes from the battery) or the inverger absorbs extra current (power goes into the battery) so that the current that the grid sees is ideal.

---

20. At the transition, the voltage of the inverter matches the voltage of the grid, and there's no current between the two. Also, this ensures that there is no jump in the phase of the AC voltage. When powering an inductive load, a step in the phase of the AC power produces a current spike that settles after a few cycles of the line frequency.

FIGURE 4.18
Operation of the grid-interactive topology:
(a) no correction required or BESS fault,
(b) no AC power, and
(c) misshapen load current.

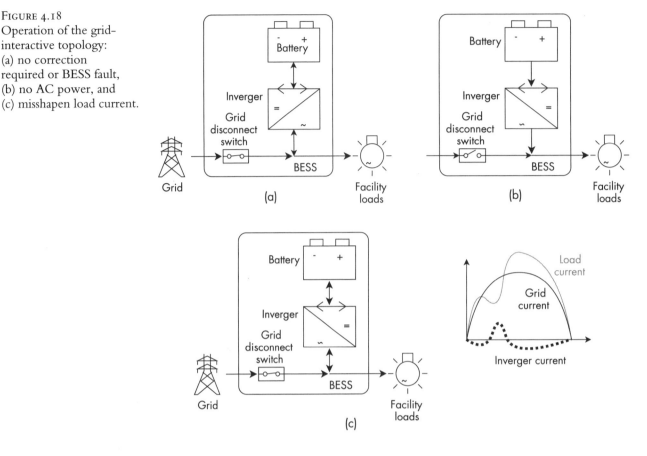

FIGURE 4.19
Switches in a unidirectional BESS; all switches are shown in the default position.

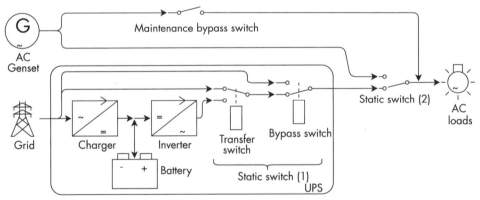

4.2.4.8 Transfer, Bypass, and Static Switches

These UPSs may use a switch to select the source for the output power (Figure 4.19). The switch takes different names depending on its function:

- Transfer switch: used in a standby UPS to select either the grid or the inverter; activated by the UPS automatically as the grid goes away and returns.

- Bypass switch: switches to grid power in case of UPS fault (e.g., output voltage or frequency is out of bounds).

- Static switch, meaning 1: a combination of the two functions above into a single switch.

- Static switch, meaning 2: selects either the UPS output or an alternate AC power source called a bypass supply (e.g., a genset).
- Maintenance bypass switch: used by the service technician to connect an AC input directly to the AC output.

Table 4.4 compares their functions.

4.2.5 Bidirectional BESS

The bidirectional topology transfers power to and from the grid[21] (Figure 4.9(b)); when back-feeding to the grid, it is a current source synchronized to the grid's voltage, frequency, and phase.

4.2.5.1 Bidirectional BESS Devices

The bidirectional topology uses a grid-tied inverger (see Section 2.2.4) that converts power between the high-voltage battery and the AC line in either direction. The inverger sets the direction of over flow by setting the phase angle of the current relative to the phase angle of the line voltage:

- To receive power from the grid and charge the battery, the inverger accepts a current that is in phase with the voltage on the grid (Figure 4.20(a)); because they are in phase, the power from the grid is positive, meaning it's leaving the grid; a simple charger functions this way as well.
- To power the grid from the battery, the inverger generates a current that is 180° out of phase with the voltage on the grid (Figure 4.20(b)); because they are out of phase, the power from the grid is negative[22], meaning it's going into the grid; a simple inverter functions this way as well.
- To act as a capacitor across the grid[23], the inverger sets up a current that leads the voltage on the grid by 90°; power changes direction four times every cycle of the line frequency (every 4.1 or 5 ms), twice each cycle power flows out of the battery, and twice each cycle it flows into the battery; therefore, the average power is 0 (Figure 4.20(c)).

TABLE 4.4
Comparison of UPS
Switches

	Default Selection	Alternate Selection	Controlled by	UPS Topology
Transfer	Grid	Inverter	UPS, loss of grid power	Standby
Bypass	UPS	Grid	UPS, fault in UPS	Any
Static, 1	UPS	Grid	UPS, loss of grid power or fault in UPS	Standby
Static, 2	UPS	Bypass AC	The system, fault in UPS, or grid	Any
Maintenance bypass	None	Bypass AC	Technician, during maintenance	Any

---

21. Alternatively, this could be an off-grid AC line with an AC generator and AC loads.
22. Power = voltage × current. At any given time, one is positive and the other is negative, so the result (power) is negative, meaning that it's flowing in the opposite direction: into the grid.
23. Capacitors are added across a transmission line to compensate for large inductive loads (motor, transformers) on the same line.

FIGURE 4.20
Power flow direction in a
bidirectional BESS: (a) to
the battery, (b) to the grid,
and (c) back and forth.

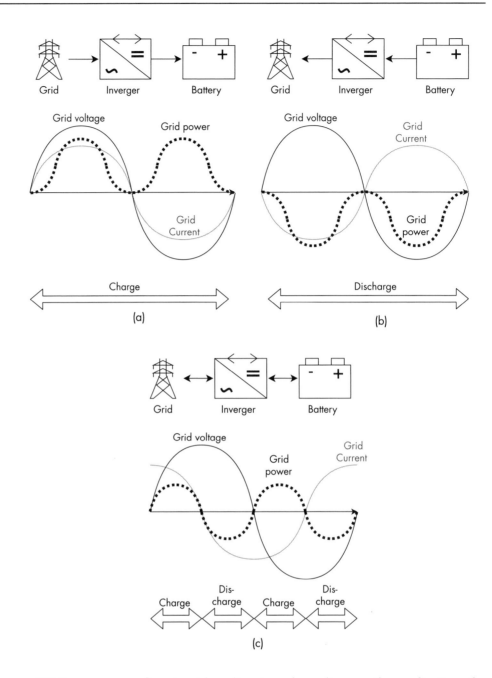

While power can flow in either direction, depending on the application, the
direction changes for different reasons, depending on the service provided (see
Section 4.3.2). For example:

- *Energy service* (Figure 4.21(a)): An external controller tells the inverger whether
  to charge or discharge; at specific times it may instruct the inverger to discharge
  the battery to generate AC power; sometime later, it may instruct the inverger
  to receive power from the AC source (e.g., grid) to charge the battery; the
  controller may tell the inverger to change direction every few minutes (on the
  order of 0.1 to 15 minutes).

- *Regulation service* (Figure 4.21(b)): Direction can change instantaneously, depending on the voltage, phase, and frequency of the AC line; this occurs automatically as a natural consequence of the physics of the coupling between

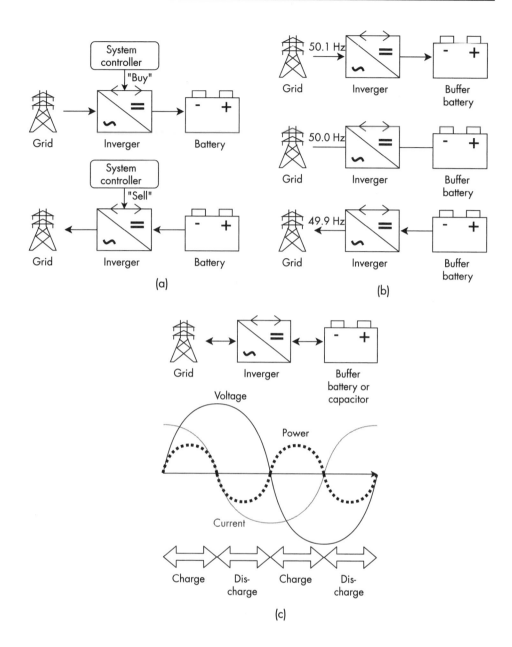

FIGURE 4.21
Power flow direction in applications for a bidirectional BESS: (a) energy service, (b) regulation service, and (c) reactive control.

the inverger AC output, which is reliably at the correct voltage, frequency, and phase, and the line, whose voltage, frequency, and phase fluctuate; power may change direction every few cycles (on the order of 0.1 to 10 seconds).

· *Reactive control* (Figure 4.21(c)): Used to change the phase angle of the current in an AC line; power changes direction within a single AC cycle; for example, to add capacitive load to the AC line, current flows to the inverger as the line voltage rises, and out of the inverger as the line voltage drops, just as it would in a capacitor.

Each of these services uses a different type of inverger.

4.2.6 Strategies for Battery Shutdown

The issues related to the shutdown of a high-voltage stationary battery are different from those for a low-voltage battery (see Section 2.11) and a traction battery (see

Section 3.2.4). We must distinguish the shutdown of a battery within a BESS from the shutdown of the entire BESS; we must also distinguish the shutdown of a unidirectional or a bidirectional BESS.

4.2.6.1 Unidirectional BESS

In a unidirectional BESS, such as a UPS, the shutdown of its battery may have little immediate consequence because the AC output continues to be powered by the charger. An automatic bypass switch overcomes the shutdown of the entire UPS by switching to the input AC power. Obviously, if an AC power failure were to occur before the causes of the shutdown have been corrected, the loads would not be powered.

4.2.6.2 Bidirectional BESS

The shutdown of a bidirectional BESS, whether due to the battery or the converters, may be acceptable. Any on-demand resource (e.g., arbitrage (see Section 4.3.3.1)) that is idle can simply declare itself as unavailable for the next demand period.

The sudden disappearance of resources that are in use may be a concern, particularly if they represent a significant portion of the generation or the load. An accurate estimate of the energy remaining in the BESS helps prevent an unexpected shutdown due to low battery voltage. That is also true for a high battery voltage or high temperature. Despite these precautions, a fault may shut down a BESS unexpectedly. Dividing the BESS into multiple smaller units minimizes the consequences of the shutdown of a single unit.

4.2.7 Levelized Cost of Storage (LCOS)

Regardless of technical considerations, financial considerations affect the decision an ESS should be installed and, if so, the choice technology and size. The estimated cost of building and operating an ESS within a given number of years is compared to the estimated value derived from it. This analysis entails two estimates:

1. *The levelized cost of storage (LCOS):* The ratio of the total cost of an ESS over the total energy discharged over a given period [7].
2. *The storage value:* The ratio of the savings and benefits from operating the ESS over that same period over the total energy discharged over that same period.

Both are measured in $/MWh. The investment is considered worthwhile if the latter exceeds the former. LCOS is a convenient measure when comparing multiple solutions.

These analyses consider various parameters, including

- Financing costs and discount rate (bank rate);
- Capital costs, initial construction costs;
- Operating and maintenance costs (O&M);
- Government subsidies and grants;
- Capacity factor.

The analyses are performed over various time periods.

Studies have shown that Li-ion BESSs are the most cost-effective form of ESS for most applications and will be until at least 2030. Pumped hydro, compressed air, and hydrogen are best for long discharge applications [8].

4.3 UTILITY APPLICATIONS

The power utility uses a bidirectional BESS (Figure 4.9(b)) to improve its service.

4.3.1 Technology

These items apply specifically to utilities.

4.3.1.1 Demand Period

A power utility manages power production and demand through integrated resources planning (IRP). It does so in blocks of time. Depending on the service, it may use blocks of 6 seconds (Figure 4.22(a)), 1 minute (Figure 4.22(b)), 5 minutes (Figure 4.22(c)), or 15 minutes (Figure 4.22(d)). For every block of time, the utility predicts how much power will be needed, decides where to produce it or buy it, and sets a price. If an ESS is available, before each of those blocks of time, the utility decides whether to charge it, discharge it, or neither.

A demand period of 6 seconds does not mean that the battery is discharged from full to empty in 6 seconds. It just means that, between 1:00 AM and 1:06 AM, the battery may be discharged a bit; between 1:06 and 1:12, it may be idle; between 1:12 and 1:24 in may be charged a bit; and so on. (Figure 4.22(a)). Indeed, the buffer batteries used in these applications are never fully charged or fully discharged.

4.3.1.2 Battery Characteristics

Grid-tied batteries operate at high voltage: typically 900V to 1.5 kVdc, although at times as high as 4 kVdc. The limitation is the availability of high-voltage transistors used in the invergers.

FIGURE 4.22
Demand periods: (a) 1 minute, (b) 5 minutes, and (c) 15 minutes.

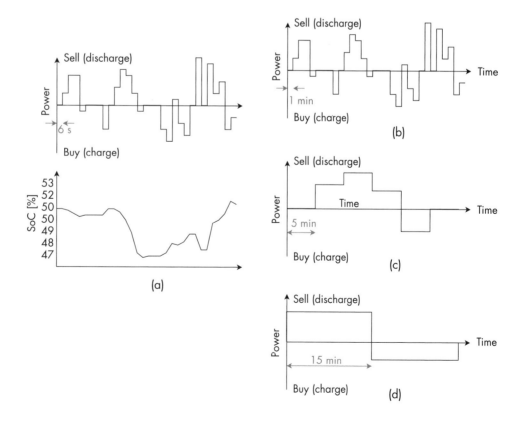

The voltage of transmission lines is much higher. Transformers convert the AC side of the inverger (typically 480 Vac) up to the transmission line voltage. The power ranges from 100 kW to 100 MW (such as a Tesla Mega-battery) because

Operating at lower power than that doesn't make economic sense. If more power is required, it makes more sense to set up multiple sites at various locations, each close to the point of generation or use.

Many of those services require an energy battery (see Volume 1, Section 5.1.4); it is normally full, it discharges as needed over 1 or more hours, and it is recharged later when extra energy is available. Other services require a buffer battery; it may charge or discharge, for a few minutes, as needed. The voltage support service requires storage that responds within milliseconds; within a single cycle of the line frequency, it charges for 2.5 ms and discharges for 2.5 ms[24]. Capacitors can perform this function far more efficiently than batteries. Li-ion capacitors (see Section A.3.4) are well suited for the job.

4.3.2 Public Utility Services

A power utility (or some other company, on its behalf) may own and operate an ESS to augment its services (Table 4.5)[25] [9]. The benefits include the following:

- Breaks the requirement that, at a given moment, generated power must equal consumed power.
- Reduces wear and tear of traditional (thermal) generation, from constant spinning up and down.
- Postpones the expenses of upgrading a transmission line to a given location with growing electrical needs by storing energy locally; the ESS supplies the varying demand for the local loads while presenting a constant demand to the old transmission line (see Section 4.3.2).
- Increases transmission line efficiency by maximizing the power transfer[26].
- Increases service reliability and power quality (see Section 4.2.2) to the customers.

4.3.2.1 Stacked Services

Table 4.5 lists each service separately as if a battery would only be used for a single purpose; rarely does this make economic sense. In reality, a BESS is likely to provide two or more of those services, paying for itself more rapidly by capturing multiple revenue sources. Doing so is called stacked services.

Large installations may combine buffer and energy batteries. For example, the Hornsdale Power Reserve in South Australia, which uses Tesla's Mega-battery, consists of two sections: a 70-MW buffer battery, and a 100-MWh energy battery. The first one is relatively low energy (~10 MWh usable, for 10 minutes), while the second one is relatively low power (20 MW for ~3 hours) [10].

---

24. Depending on the need, it acts like a capacitor or an inductor to change the phase angle of the current in the line, relative to the phase angle of the voltage.
25. See Section 4.3.2 for the bottom portion of this table. .
26. All of the power transferred is fully used when the current waveform matches the voltage waveform. If so, the power factor is equal to 1.0.

Group	Class	Description	Time Frame	Battery Type	Power or Energy
Bulk energy services	**Electric energy time shift (arbitrage)**	Buy when electricity is cheap, sell when it's expensive, every 15 minutes or so	15 minutes	Buffer	0.1~100 MW
	Electric supply capacity	Same as above, but over a longer time frame	Hours	Energy	2~3000 MWh
Ancillary services	**Regulation**	Fast response to even out generation by supplying momentary energy needs and absorbing momentary excess energy; in constant use	15 minutes	Buffer	10~40 MW
	Load following/ ramping support for renewables	Same as above, but also supporting variations in renewable energy generation	15 minutes	Buffer	1~100 MW
	Frequency response	Same as above, but specifically to handle the sudden loss of a generator and while the generator returns to the proper frequency and phase	Seconds	Buffer	10~100 MW
	Voltage support	Strategically placed along high-voltage transmission to add the appropriate reactance; dynamically compensates for variations in reactive loads, resulting in more efficient transmission	Milliseconds	LIC	1~10 MVAR
	Black start	Emergency power to restart the grid in case of complete shutdown while generating a frequency reference	Minutes	Buffer	5~50 MW
	Supplemental reserves	Rarely used, emergency only	Hours	Buffer	10~100 MW
Transmission infrastructure services	**Transmission stability damping, subsynchronous resonance damping**	Similar to voltage support but used only when there is a strong disturbance on the grid to help the grid's stability	Seconds	Buffer	10~100 MW
	Transmission upgrade deferral	For an area where growth has increased the demand to more than the transmission lines can handle; local storage supplies peak loads while leveling off power from the grid, to postpone the need to upgrade the transmission lines; may be needed few times a month	Hours	Energy	20~800 MWh
	Transmission congestion relief	Same as above, but to relieve a transmission line (not an area); may be needed once or twice a week	Hours	Energy	1~400 MWh
Distribution infrastructure services	**Distribution upgrade deferral and voltage support**	Same as above, but to relieve a single location	Hours	Energy	0.5~40 MWh
Customer energy management		See Section 4.5			

TABLE 4.5   Public Utility Services

4.3.2.2 Capacity Firming

Capacity firming refers to technologies that supplement renewable resources as their output varies. The solutions for short-term and long-term variations differ.

For short-term variations (over minutes), a BESS using a buffer battery provides power during drops in the generated output (e.g., a cloud passes over). When excess power is available, it used to recharge the battery. This service is listed under the Ramping Support for Renewables category in Table 4.5.

For long-term variations (over hours), certainly, a BESS using a long-term buffer battery (see Volume 1, Section 5.1.4) could provide this service.

In practice, however, a firming generator (typically burning natural gas) provides power during drops in the generated output. When excess power is available, proper management of all the available resources minimizes unused power [11]. Alternatively, a BESS using an energy battery may be used instead of a firming generator; excess power recharges the BESS. This service is listed under the Electric Supply Capacity category in Table 4.5.

If capacity firming is not a practical application for long-term buffer batteries, that doesn't leave many other uses for such batteries; therefore, this book does not discuss them.

4.3.2.3 Adoption Year

According to one study [8], some of these technologies are already economically viable and have already been widely adopted. Others will take more time:

- *2015*: Bill management, power quality, power reliability;
- *2020*: Black start, congestion management;
- *2025*: Energy arbitrage, response (primary, secondary, tertiary), peaker replacement;
- *2030*: T&D investment deferral;
- *Never*: Seasonal storage.

4.3.3 Private Utility Services

A private utility provides services to the public utility by

- Storing electric energy and trading it with the utility;
- Generating electric power and selling it to the utility.

A private utility may also provide services to the large energy user, in exchange for 50% of the savings in the electrical bill:

- Installing and paying for a solar array on their land;
- Installing and paying for a BESS in their building.

The power company uses an ESS to provide a long list of services. A private utility only implements the first two in this list, the Bulk Energy Services (Table 4.6). These are the profitable ones. The power and energy levels of these batteries are lower than for a public utility.

4.3.3.1 Arbitrage

An investor may connect a BESS to the grid to profit from energy arbitrage: buying energy when it's cheap, and selling it when it's expensive. The BESS site must use a buffer battery because energy must be transferred very quickly. At a given time, it can either buy or sell energy; typical power is 100 kW. Nonbattery ESS technologies react

TABLE 4.6
Private Utility Services

Group	Class	Description	Time Frame	Battery Type	Power or Energy
Bulk energy services	**Electric energy time shift (arbitrage)**	Buy when electricity is cheap, sell when it's expensive, every 15 minutes or so	15 minutes	Buffer	0.1~10 MW
	Electric supply capacity	Same as above, but longer time frame	Hours	Energy	0.5~20 MWh

too slowly to provide this service. An arbitrage site uses a buffer battery, an inverger[27], and a system controller that negotiates with the power company.

4.3.3.2 Energy Service

The private utility may

- Generate power from renewable sources, wind, solar, geothermal, and hydro, or nonrenewable sources, gas turbines[28]; it sells that energy to the public utility, as requested.

- Store extra energy when available, and sell it to the public utility at a later time, at a premium, when needed.

Even if the private utility does not generate power, it may maintain a fully charged battery, ready to sell energy to the public utility at a premium, when needed. It would use a power battery, for the 1–15-minute market, or an energy battery, for longer time frames.

4.3.3.3 Coordination with the Utility

The contractual and communication aspects of this service may be more complicated than the technical ones. Companies have emerged that specialize in setting up a private utility (or even a single user who produces electricity) to be incorporated into the power company's energy management system, and possibly to act as an intermediary between the two parties.

A computer at the power company communicates with computers at all the energy providers. The computers exchange offers to buy or sell a given energy block for a given duration at a given price, negotiate, and come to an agreement. This is akin to what occurs in the stock market. Once the two parties agree on a deal, the energy provider exchanges energy with the grid at the allotted time and for the allotted duration. At the end of the month, the private utility sends a bill to the power company for services rendered.

4.4 MICROGRID APPLICATIONS

A microgrid has all the attributes of the regular grid, generation, transmission, loads, and storage, except on a smaller, localized scale (Figure 4.23): its size ranges from a single building to a whole city[29] [12].

---

27. Such as a Princeton Power GTIB-100 Inverter.
28. Distributed generation (DG).
29. For example, Fort Collins, Colorado, United States.

FIGURE 4.23
Typical microgrid.

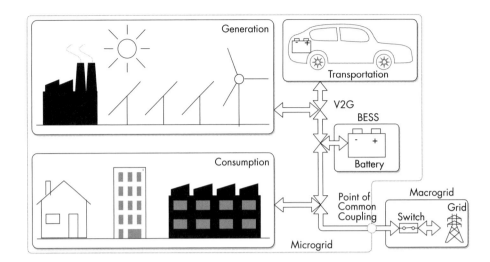

A microgrid provides energy independence and may reduce environmental impact. It helps municipalities break the stranglehold of large electrical utilities. In this application, the full grid is called the macrogrid. A connection between the microgrid and the macrogrid, if present, is called the point of common coupling (PCC).

4.4.1.1 Types

Microgrids can be classified roughly as [13]

- Customer microgrids or true microgrids (μgrids): independently developed for a building, a mine, a community, or a campus.
- Utility microgrids, microgrids or milligrids (mgrids): same as above but within the auspices of a power company.
- Virtual microgrids (vgrids): decentralized but managed as a single unit.
- Remote power systems (rgrids): completely off-grid.

4.4.1.2 Generation

Generation may include renewables (e.g., solar, wind, geothermal, hydro) and nonrenewables (e.g., gas turbines).

4.4.1.3 Transmission

Within some microgrids, the transmission lines may be DC (usually at the battery voltage) or AC. Using AC simplifies the interconnection of various resources considerably, although DC can be slightly more efficient.

4.4.1.4 Loads

Loads may include residential, commercial, and industrial facilities. It may also include mobile loads (transportation).

Loads are classified by priority, to control demand at times of limited energy availability:

- Critical (high priority);
- Adjustable (controllable);
- Sheddable (low priority).

Adjustable and sheddable loads play an important role in the management of the microgrid's balance of generation and consumption. This role is more substantial in a microgrid because each load has a proportionally higher effect in the small microgrid than it would in the larger macrogrid.

4.4.1.5 Storage

BESSs normally include a large stationary battery; typically, the battery voltage is 450V or 900V. Mobile resources may be included[30], as well as mechanical storage. There may be multiple BESSs, using small power or buffer batteries, coupled with large energy batteries. The former would be used for short-term regulation and power quality, while the latter would be used for long-term energy needs.

4.4.1.6 Connection to the Macrogrid

A microgrid may be

- Connected to the macrogrid (connected mode), exchanging energy with it;
- Disconnected from the macrogrid (island mode), or completely off-grid, operating autonomously.

In the connected mode, a BESS provides customer energy management services (see Section 4.3.2) for environmental and economic reasons:

- Power reliability;
- Retail energy time shift;
- Demand charge management;
- Power quality.

The Island mode is useful if the macrogrid is down, unreliable, too expensive, or if it uses dirty sources. Off-grid, or in the Island mode, the BESS uses an energy battery to provide off-grid services (see Section 4.6.1).

4.4.1.7 Control

A combination of centralized and distributed control mechanisms manages the various elements in the microgrid. They do so at the local level (voltage and frequency), the supervisory level (balance of resources and loads, prediction of future supply and demand), and at the grid layer level (in the connected mode, managing the exchange of energy with the macrogrid) [14].

These control mechanisms are divided into three time frames [15]:

1. Primary: at the cycle level;
2. Secondary: at the seconds to minutes level;
3. Tertiary: at the hours level.

4.5 LARGE ENERGY USER APPLICATIONS

Large-scale customers of the power utility (e.g., industrial, commercial, government), serving a large facility or small campus, may own and deploy a BESS to

---

30. Vehicle-to-grid: V2G. See Section 3.9.1.8.

- Provide backup power in case of power failure.
- Reduce the electric bill by not using electricity when it's expensive or by limiting expensive peak demand.
- Clean up the AC voltage and current.

These functions are called customer energy management services. The battery voltage is typically 900 Vdc.

4.5.1 Customer Energy Management Services Overview

Table 4.7 lists the customer energy management services provided by these systems[31] [5].

4.5.1.1 BESS Topology for Each Service

The customer energy management services use a unidirectional BESS. The five unidirectional BESS topologies perform some or all of these four services (Table 4.8) [9].

The difference between a BESS for one service and one for a different service, even if they use the same topology, is the type of battery (see Volume 1, Section 5.1.4), energy, power, or buffer, and the control algorithm.

Group	Class	Description	Time Frame	Battery Type	Power or Energy
Customer energy management services	Power Reliability	Provides backup power in case of failure of the grid	Hours	Energy	0.5~10 MWh
		Same as above, but only until the generator comes on	Minutes	Power	0.1~3 MW
	Retail energy time shift	With time-of-use pricing, when energy is expensive, the customer uses the battery; when it's cheap, the battery is recharged	Hours	Energy	0.5~10 MWh
	Demand charge management	Used if the electric bill is based on peak power demand; the BESS keeps power within a limit during times of peak load demand	Minutes	Power	0.1~3 MW
	Power quality	Ensures that the grid voltage and frequency remain in specifications; ensures that the load current is in sync with the grid voltage	Milliseconds to minutes	Buffer	0.1~10 MW

TABLE 4.7   Customer Side Energy Management Services

TABLE 4.8
Support for Customer
Energy Management
Services by Each
Unidirectional BESS
Topology

Type	Power Reliability	Retail Energy Time Shift	Demand Charge Management	Power Quality
Standby	✓	✓		
Online	✓	✓	✓	✓
Hybrid	✓	✓	✓	✓
Delta	✓	✓	✓	✓
Grid-interactive	✓	✓	✓	(1)
Note: (1) Current only.				

---

31. See Section 4.3.2 for the top portion of this table.

A single BESS to perform all four services should use a large energy battery, which can work as a power battery thanks to its large size, and even as a buffer battery if the cells have low MPT. It should use one of these recommended topologies:

- If you're building just one system, use the online topology.
- If you're buying a ready-made one for high power, get the delta topology.
- If you're designing a new product, consider the grid-interactive topology.

4.5.2 Power Reliability Service

A BESS that provides power in case of an outage from the grid may be called an uninterruptible power supply (UPS), a battery-powered generator, or simply a backup battery[32]. Regardless, in case of power failure, it powers the customer's facility for up to a few hours. It uses a large energy battery (0.5~10 MWh).

Alternatively, the BESS may power the facility for only a few minutes, until a generator can take over. If so, it uses a smaller power battery (e.g., 0.1~ 3 MW, and 17~500 kWh). In this context, a UPS is a large, high-power product that uses a high-voltage battery (as opposed to a small UPS for consumer products that use 12–48V batteries (see Section 2.10.1) Table 4.9 lists some characteristics of each topology when used as a UPS.

4.5.2.1 UPS Design

If you're going to design your own UPS[33], I suggest you implement the online topology, using a standard charger and a standard inverter. There is no need for any control logic, although some intelligence could be added to report the status or to do periodic tests.

A bypass switch lets the grid power the load during a failure of the UPS. To automate that transition, use a DPDT contactor with an AC coil, and power the coil from the output of the inverter (Figure 4.24). When the inverter works, it powers the contactor coil, and the contactor selects the output of the inverter to power the load. If the inverter stops, or if the battery is empty, the contactor coil is no longer powered, and the contactor powers the load from the grid (if present).

TABLE 4.9
Comparison of
Unidirectional BESS
Topologies When Used in
a UPS

Type	Operation	Notes
Online	The load is always powered by the inverter, regardless of what's happening on the grid	Most common for low power, low voltage battery, especially for server rack UPSs
Standby	Remains idle; in case of a blackout, it transfers the load from the grid to its inverter	Cheapest; with an AVR; it's called line interactive
Hybrid	The load is always powered by the BESS, regardless of what's happening on the grid	Poor value: use a line interactive UPS instead
Delta	The load is always powered by the BESS, regardless of what's happening on the grid	Best for high power
Grid-interactive	Remains idle; when it sees a drop in the grid, it disconnects the grid and starts powering the load	Not commercially available

32. Many thanks to Ajay Bhoge of GPS/Rocket Batteries and Sudhir Gokhale of G E Electrotech.
33. By which I do mean an uninterruptible power supply, not just an inverger.

FIGURE 4.24
Online UPS with
bypass contactor.

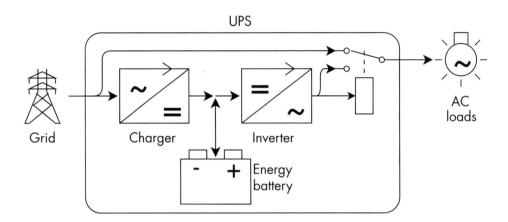

You may be able to find a ready-made package that includes all the power components, including the bypass switch. If you want to implement the off-line topology, add a PLC; you must develop custom code for it to detect when AC power is about to go away and switch the transfer switch (a DPDT contactor) before it does. Then wait for AC power to return and stabilize before transferring back to the grid. Leave the inverter always running, so that it's ready to go in an instant.

4.5.3 Retail Energy Time-Shift Service

Most customers pay for electricity based on how much energy they use, regardless of when they do so. Some power companies charge based on time of use; electricity is more expensive during certain times of the day and the week.

A large energy user may install an ESS to shift the time of use:

- Use the ESS instead of buying electricity when the price is high.
- Recharge the ESS at night, when the price is low.

These savings help offset the cost of the ESS.

This ESS may take many forms, such as making ice at night that later can be used to cool the building. In general, though, a BESS would be used for the purpose.

For example (Figure 4.25(a)):

- On a hot afternoon, when the facility uses a lot of energy and the price for electricity is high, the BESS powers the facility.
- In the evening, when the price for electricity is low, the BESS recharges its battery.

This BESS uses a large energy battery. An analysis of the energy needs of the facility during the high rate periods determines the optimal battery size. Undersizing the battery by a bit is not catastrophic; the company pays the high rate for a few minutes between when the battery is empty and when the rate drops.

This BESS can use any topology. The inverter must be sized to handle the maximum load power, as the BEES takes no power from the grid during the time of high price. With the standby topology (Figure 4.26(a)), a controller controls the charger and the transfer switch.

Just before the rate goes up, the controller disables the charger, turns on the inverter, and switches the transfer switch to select the inverter instead of the grid. This action causes a short interruption in power to the load. Similarly, just after the rate

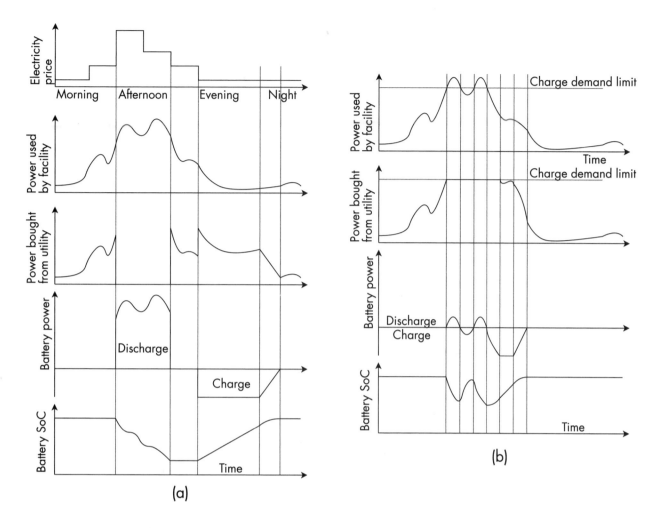

FIGURE 4.25 BESS to reduce the electric bill: (a) time shift, and (b) demand charge.

goes down, the controller switches the transfer switch to select the grid and turns off the inverter. In the evening, the controller turns on the charger.

With the online topology (Figure 4.26(b)), a clock controls the charger. The loads are powered by an inverter, regardless of the time of day.

Just before the rate goes up, a clock disables the charger. The inverter keeps on powering the load from the battery as if nothing happened. Just after the rate goes down, the clock re-enables the charger, but just enough to maintain the battery SoC. In the evening, when the rate is minimum, the clock allows the charger to turn on at full power to recharge the battery fully.

In the delta topology (Figure 4.26(c)), during the period of high rate, a controller opens the switch to the grid and uses the converters to power the loads. When the rate drops, the controller turns the switch back on, to let the grid power the loads, without recharging the battery. In the evening, when the cost of electricity is minimal, it recharges the battery.

In the grid-interactive topology (Figure 4.26(c)), the inverger is permanently connected to the grid, ready to jump in and power the load so that there is no interruption in the power supply during transitions. Just before the cost of electricity

FIGURE 4.26
Time-shift BESS
topologies: (a) standby,
(b) online, (c) delta, and
(d) grid-interactive.

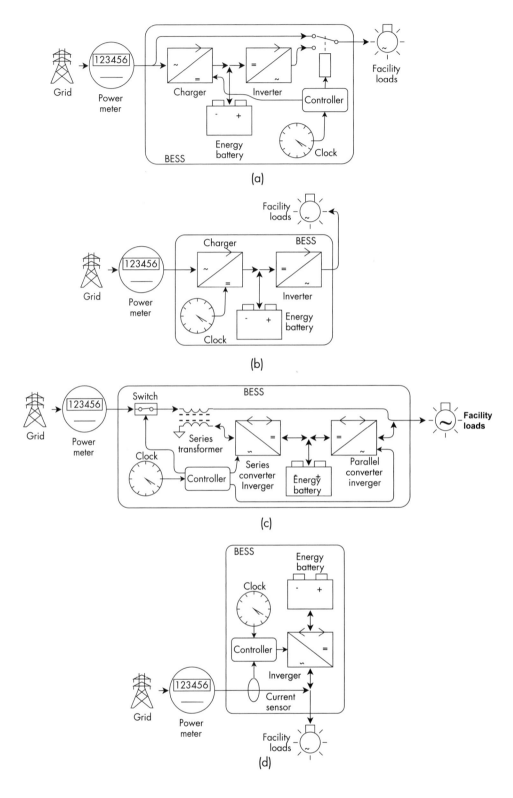

jumps up, the inverger slowly ramps up its output power until it powers the load
completely. The grid remains connected and keeps on providing the voltage reference.
This solution must avoid buying expensive electricity and any attempt to sell electricity
back to the power company. A controller monitors the grid current and adjusts the
output current of the inverger continuously to ensure that the grid power is zero or,

at most, slightly positive. The inverger must supply the facility's power demand, or just slightly less[34].

Similarly, just after the rate cost of electricity jumps down, the inverger slowly ramps down the power, until the grid provides all the power to the load. In the evening, the inverger starts taking power from the grid to charge the battery.

4.5.4 Demand Charge Management Service

With demand charge pricing, the monthly electric bill is based not just on the actual energy used (energy used), but also on the maximum power used at any one time during the month (demand charge)[35]. The power company monitors the peak power consumption over each 15-minute interval. On the most basic level, it charges the customer a demand charge based on that power. It may use a more complex pricing structure, such as using different pricing for on-peak and off-peak times.

Often, the demand charge far exceeds the charge for the energy used. In this example (Table 4.10), the demand charges constitute two thirds of the electrical bill [16]:

If a BESS supplements the power from the grid while the facility requires peak power, the peak power is limited, as is the electrical bill for the month (Figure 4.25(b)). The BESS provides only the power beyond the charge demand limit (e.g., if the limit is 100 kW, and the demand is 120 kW, the BESS only provides 20 kW). After the peak demand is over, the BESS is recharged, at a rate that does not cause the demand from the utility to surpass the charge demand limit (e.g., if the limit is 100 kW, and the demand is 90 kW, the battery only charges at 10 kW). After its battery is charged, the BESS is ready for another peak in demand.

This BESS uses a power battery, large enough to provide the peak energy requirement of the facility, yet not too large, as this would add unnecessary cost. Compared to a battery for retail time-shift energy, this battery is smaller (it only powers the peaks) and more powerful (because it's smaller, its specific current is higher).

A careful analysis of the facility's peak power demand determines the energy required from the battery. This analysis is more critical than with a battery for retail time-shift energy because a single mistake can cost the company the savings for an entire month.

Strategies to avoid demand charges from a single mistake include the following:

TABLE 4.10
Electrical Bill with
Demand Charges

	Units	Cost per Unit	Charge
Basic charge			$23.00
Electricity consumption	12,455 kWh	$0.163	$2,030.17
Delivery charge		$0.00062	$7.72
Transmission charge		$0.00266	$33.13
On-peak demand charge	472 kW	$7.13	$3,565.00
Off-peak demand charge	152 kW	$4.94	$750.88
Total due			$6,210.26

---

34. Selling energy back to the power company is not envisioned, though there is no reason that net metering could not be included as a way to offset the cost of the battery.
35. More precisely, the average power within a 15-minute interval.

- As the battery is nearly empty, reduce the amount of peak shaving (by increasing the power bought from the grid), rather than running out of energy too soon, which would result in a power peak at the end of the high demand period.
- Shut down nonessential loads (e.g., the marketing department [17]).
- As the facility's loads increase and as the battery ages, reduce the amount of peak shaving.

The analysis of the facility's peak power demand is more difficult than for retail time-shift because the estimate is for the excess power over the limit. A 10% error in estimating the peak power can result in a 30% error in the estimation of the excess power. Combine this with a 10% error in estimating its duration can result in a 50% error in the estimation of the battery energy required to power the excess demand.

This BESS can use any topology except for the standby one. In all topologies, the controller senses the grid current and limits the input current from the grid to a maximum, in various ways:

- *Online topology:* by limiting the charger's input power (Figure 4.27(a));
- *Delta topology:* by supplying the extra power with the parallel inverger (Figure 4.27(b));
- *Grid-interactive topology:* by supplying the extra power with the inverger (Figure 4.27(c)).

While all three topologies may be used, the online topology is disadvantageous, as the inverter must be sized to supply the entire power requirement of the load. In the other two topologies, the invergers only need to provide the difference between the power required by the load and the desired maximum power from the grid.

4.5.5 Power Quality Service

A power quality BESS, placed between the grid and the facility, helps both parties:

- It helps the utility by drawing a current that is perfectly matched to the line voltage; this maximizes the power transfer and minimizes the losses in the transmission lines to the facility.
- It helps the facility by providing it with a clean sine wave voltage at constant voltage (and possibly frequency); this allows the equipment in the facility to operate optimally.

The BESS uses a buffer battery to supplement power from the grid when needed (on the order of milliseconds to a few seconds) and may improve the shape of the voltage or the current waveform.

4.5.5.1 *Power Quality BESS*

All topologies can handle power quality issues to different degrees (Table 4.11). Power interruptions are not included here, as they are discussed under Section 4.5.2.

A buffer battery only needs to provide enough energy to power the loads for a short disruption. It provides the power to make up for the small delta between the ideal voltage and current and the real ones.

With the standby topology, an AVR (see Section 4.2.4.1) adjusts the line voltage, attenuates noise, and clamps surges (Figure 4.28(a)). It cannot compensate for the power factor and distortion in the load current. If the AVR cannot handle a line

FIGURE 4.27
Demand charge
BESS topologies: (a)
online, (b) delta, and
(c) grid-interactive.

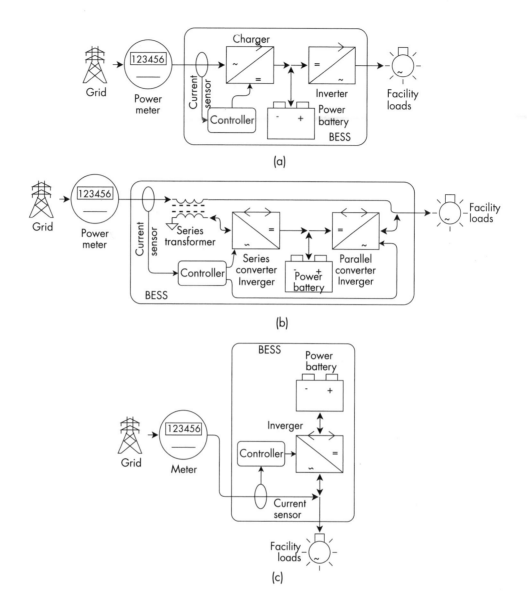

disturbance, the controller transfers the load from the grid to the inverter for just a short time.

This topology is more efficient (the converters only operate when necessary), but takes some time to react. It does nothing for the utility (it does not correct for a distorted current draw) and, indeed, it causes more problems for the utility every time it steps in because the load suddenly disappears from the grid. It can take care of changes in the line frequency for just a few seconds. However, those changes tend to last for many minutes during recovery, so this topology is ineffective in this case.

With the online topology (Figure 4.28(b)), the charger is power factor corrected (PFC) and therefore draws the ideal current from the grid[36]. The inverter generates an entirely new and perfect sine wave voltage for the facility, supporting loads that do not draw current nicely. The load does not see changes in the line frequency.

---

36. With a power factor of 1, and with low distortion.

TABLE 4.11
Handling of Power Quality
Issues by Each BESS
Topology

	Standby	Online	Hybrid	Delta	Grid-Interactive
Transient disturbances	(1)	✓	✓	✓	✓
Voltage disturbances, short		✓	✓	✓	
Voltage disturbances, long	✓	✓	✓	✓	✓
Frequency disturbances		✓	✓		
Harmonic distortion, voltage		✓	✓	✓	
Harmonic distortion, current		✓	✓	✓	✓
Voltage fluctuation		✓	✓	✓	(2)
Noise	(1)	✓	✓	✓	✓
{tfn}Notes: (1) When fitted with an AVR. (2) To a limited extent.					

With the delta topology, the controller monitors the grid voltage and current and adjusts the direction, power, and waveform of the two invergers to compensate for grid disturbances and to correct the waveform of the current to the load (Figure 4.28(c)). It cannot handle changes in line frequency[37].

With the grid-interactive topology, the controller monitors the grid voltage and current. It adjusts the direction, power, and waveform of the invergers to correct the waveform of the current drawn by the load (Figure 4.28(d)). An AVR regulates the line voltage, attenuates noise, and clamps surges; it can help a little with harmonic distortion coming in from the grid. It cannot do anything about other grid disturbances, such as changes in the line frequency. In case of a fault in the BESS, regardless of the topology, a bypass switch (not shown) transfers the loads directly to the grid.

4.5.5.2 BESS Design

If you're going to make your own BESS for power quality, I recommend the online topology, as it is quite simple. If you want to design one for production, consider the grid-interactive using an off-the-shelf grid-tied inverger. In either case, you need a small battery with a very low MPT or a larger battery to be used at lower specific current and only part of the SoC range. Consider ultracapacitors or Li-ion capacitors. If you're going to build one for high power, I recommend the delta topology, as it is quite effective.

4.6   OTHER APPLICATIONS

These applications may use either a low-voltage battery or a high-voltage one. The advantages of using a high-voltage battery instead of a low-voltage one are that the power electronic converters are more efficient and that smaller gauge wires can be used to carry a given amount of power. The disadvantages include that the battery uses a more expensive BMS and that the battery is not isolated from earth.

A ready-made battery includes a BMS. Alternatively, a system can be put together from a high-voltage battery and converters. If building the battery from individual cells, then a BMS is required.

4.6.1   Off-Grid Applications

Off-grid sites include homes in the wilderness, broadcast transmitters, or telecom sites up a mountain, remote research centers, locations in the developing world. The primary power source is renewables (wind, solar, hydro) or fuel (a genset).

---

37.   Well, theoretically, if it were powerful enough, it could supply the other customers, helping the power company's generators recover.

FIGURE 4.28
Power quality device
topology: (a) standby,
(b) online, (c) Delta, and
(d) grid-interactive.

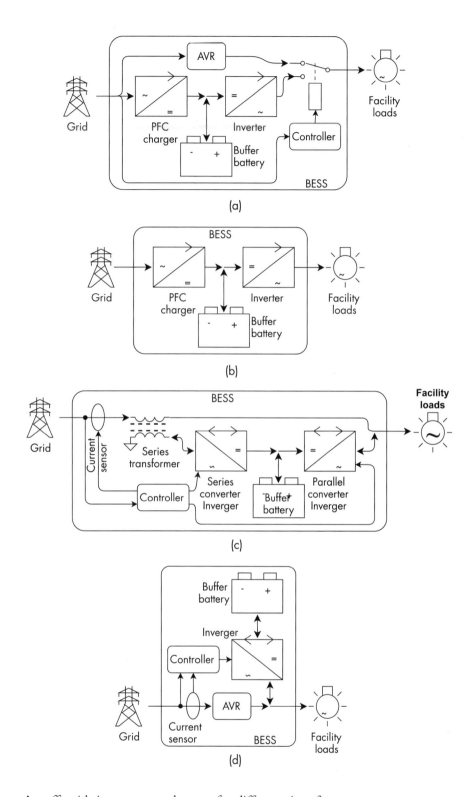

An off-grid site may use a battery for different time frames:

- *Long-term:* Backup power in case the charging source is unavailable (e.g., over-cast skies, broken generator, no wind, dry creek); an energy battery stores energy sufficient to power the site at least two days and as much as one week, giving enough time for relief to reach the site.

- *Short-term:* A power battery powers the site while a generator starts up, while a cloud goes by or as some equipment is being serviced.

These applications are similar to the ones that use a low-voltage battery (see Section 2.4), except that the battery is high-voltage.

4.6.2 Residential Applications

Today, most alternative energy homes use a low-voltage battery: 24V or 48V (see Section 2.5). Yet some homes are now beginning to use a high-voltage battery, as popularized by Tesla's PowerWall. The battery voltage is on the order of 450V. So far, there are only a few choices of invergers for high-voltage batteries. All-in-one AC-coupled storage units integrate a grid-tie inverter and a high-voltage battery; they use a single connection to the AC power line.

4.7 HIGH-VOLTAGE DESIGN

As they operate at such high voltages, these batteries can arc and can be particularly lethal. This section explores ways of avoiding high-voltage arcs and leakages and how to reduce the danger posed by high-voltage applications.

4.7.1 Breakdown Voltage

High-voltage applications require careful design to avoid breakdown (arcing) between two conductors, such as

- Between a cell terminal and a grounded battery enclosure;
- Across an isolation barrier in the electronics;
- Between adjacent tap wires or thermistors;
- Between tap wires and chassis ground.

What matters is not the absolute voltage of a single conductor. What matters is the relative voltage between two conductors; more precisely, what matters is the intensity of the electric field (the voltage gradient) between the conductors.

The voltage gradient between two conductors is affected by their distance and their shape:

- The voltage gradient in the air between two flat conductors spaced 1 mm apart and with 1 kV across them is 1 MV/m (Figure 4.29(a)).
- The voltage gradient increases with a shorter distance: by reducing the distance to 0.5 mm, the voltage gradient is 2 MV/m (Figure 4.29(b)).
- The voltage gradient increases with a sharpness: by restoring the distance to 1 mm but using a pointed conductor instead of a flat plate, the voltage gradient near this point starts at 3 MV/m and then drops rapidly (Figure 4.29(c)).

An arc occurs between two conductors when the voltage gradient in the material between them exceeds the breakdown electric field strength that a material can handle: its dielectric strength. The dielectric strength for air is about 3 MV/m (dry, clean air at sea level pressure); lowering the pressure reduces its dielectric strength. An arc is more likely to occur at higher altitudes than at sea level. However, the dielectric strength increases again for vacuum to 20 to 40 MV/m [18].

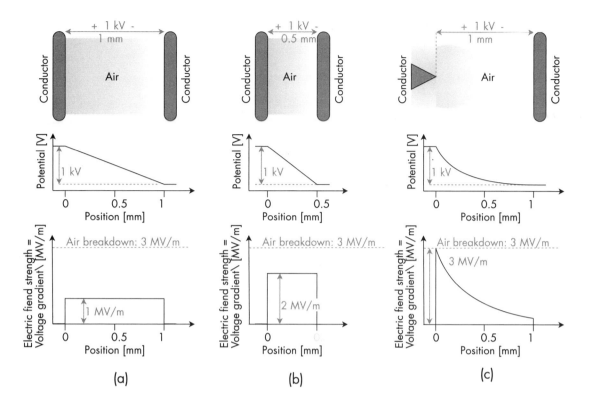

FIGURE 4.29 Electric field strength (voltage gradient): (a) two flat plates, (b) closer plates, and (c) sharp point and a plate.

Always be aware that your body can be one of those two conductors; keep your distance from a high-voltage battery and use high-voltage insulating gloves when coming close to one. Earth ground can also be one of those two conductors; keep grounded metal away from a high-voltage battery or use an effective insulating barrier between the two.

The voltage within a battery module is not affected by its absolute voltage; the voltage between two adjacent cells in a module is 3V, whether the module is used in a high-voltage battery, in a low-voltage one, or is sitting by itself on a shelf. The only concern is once such a module is placed in proximity to a conductor that has or could have a significantly different voltage, such as a grounded metal enclosure.

For a low-voltage battery in a metal enclosure, the voltage between the cells and the enclosure can be considered about 0V (Figure 4.30(a)). If the enclosure is grounded to earth, the voltage between the cells and the case can be as high as the battery voltage (Figure 4.30(b)).

For a high-voltage battery composed of modules in metal enclosures, the voltage between the cells and their case can be considered about 0V. The voltage between adjacent cases is, at most, the voltage of the module. The voltage between the cases and the earth ground is as high as the total battery voltage (Figure 4.30(c)). If the cases are grounded to earth, the voltage between the cells and the case can be as high as the voltage of the battery (Figure 4.30(d)).

Therefore, either the cases must be nonmetallic or they must be isolated from earth ground, or, if grounded, there must be sufficient isolation between the cells and their case.

FIGURE 4.30
Voltage differential:
(a) low-voltage battery
in floating metal case,
(b) same, grounded case,
(c) high-voltage battery
modules in floating
metal cases; and (d)
same, grounded cases.

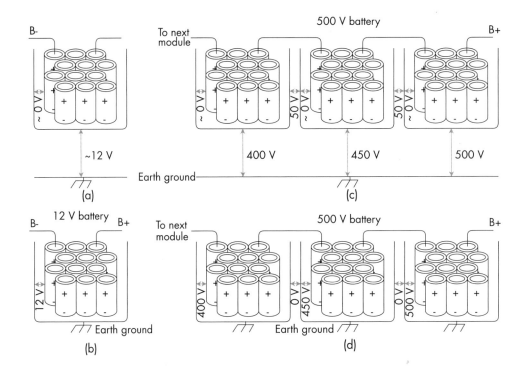

4.7.2 Clearance and Creepage

The discussion in the previous section assumed that the isolator was dry, clean air at sea level. In reality, air can be humid, dirty, and low pressure. That section also ignored the path along the surface of an insulator. We'll consider both in this section.

4.7.2.1 Minimum Distance Between Conductors

Current may flow, or even arc, if the spacing is too short for the voltage (Figure 4.31); directly through the air, if the clearance distance is too short; along a surface, if the creepage distance is too short.

Table 4.12 compares clearance and creepage. The breakdown voltage in a clearance is reduced when the pressure drops (i.e., as the altitude increases). This means that a battery is more likely to arc as it's brought to higher elevations[38]. Sharper radii also reduce it: contacts with sharp copper points tend to arc; big brass balls do not.

Pollution reduces the breakdown voltage of both clearance and creepage, it increases the chance of an arc, and it forms a conductive layer on the surface between the conductors.

Pollution degree is classified as [19]:

1. No pollution or dry pollution only: inside a cleanroom;
2. Nonconductive pollution: commercial buildings, laboratories;
3. Conductive pollution: industrial, manufacturing, agricultural;
4. Persistent conductivity: outdoors.

For creepage, tracking resistance is a measure of the propensity of a material to breakdown when exposed to contaminants. It ranges from Material Group I (best) to Material Group IIIb (worse) [20].

---

38. Paschen's Law.

FIGURE 4.31
Clearance and creepage:
(a) in general, and (b) and
connector cross-section.

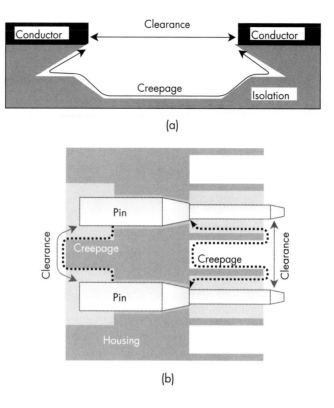

(a)

(b)

TABLE 4.12
Comparison of Clearance
and Creepage

	Clearance	**Creepage**
Path	Through the air (or other gas or vacuum)	Along the surface of an insulator
Affected by	Path length, peak voltage, pollution (humidity), pressure (altitude), sharpness	Path length, RMS voltage, pollution (salts), surface material (and its tracking resistance)
Consequences	Fast and furious: if the electric field strength exceeds the breakdown of the air between two conductors, an arc occurs immediately	Slow and mild: it may take months for leakage to develop along a surface between two conductors at differing voltages

Minimum clearance and creepage for a given voltage are specified in a great variety of mutually inconsistent standards [21, 22]. Table 4.13 is one such table; it lists creepage distance for a given operating voltage.

Table 4.14 lists clearance distances for a given peak voltage. Again, standards vary a lot; use these tables as a rule of thumb and then consult the applicable standards for your application to calculate appropriate distances.

4.7.2.2 Solutions

There are many ways to address clearance and creepage issues, such as by adding insulation[39]. We saw that the dielectric strength for clean, dry air at sea level is about 3 MV/m. The dielectric strength of most other materials is higher (10 MV/m for

---

39. Insulator is a material (e.g., Teflon), while isolator is a component (e.g., an old glass insulator on a power pole) made of insulating material. Insulation and insulating are the act of placing isolation (e.g., after soldering the wires, insulate them with tape), while isolation is the resulting level of withstanding a voltage (e.g., check that the isolation of the battery is at least 1,500V). Insulation is thermal, not electrical.

TABLE 4.13
Creepage Distances [mm]
for a Given Operating
Voltage

Pollution degree	1	2			3		
TCI material group	NA	I	II	III	I	II	III
0.5~0.8 kV	1.8	3.2	4.5	6.2	8	9	10
0.8~1 kV	2.4	4	5.6	8.8	10	11	12.5
1~1.25 kV	3.2	5.0	7.1	10	12.5	14	16
1.25~1.6 kV	4.2	6.3	9	12.5	16	18	20
1.6~2 kV	5.6	8	11	16	20	22	25

TABLE 4.14
Clearance Distances [mm]
for a Given Peak Voltage

	Pollution Degree		
	1	2	3
530V	0.04	0.2	0.8
700V	0.1	0.2	0.8
960 V	0.5	0.5	0.8
1.6 kV	1.5	1.5	1.5
2.6 kV	3	3	3

glass) or much higher (20–170 MV/m for most plastics, 65 MV/m for distilled water), except for helium, for which it's much lower (0.15 MV/m). That means that just about any material provides better insulation than air.

For clearance, the breakdown voltage is increased by (Figure 4.32(a))

- Operating in vacuum;
- Replacing air with a sealed gas or a solid material;
- Adding an insulating barrier between the conductors: plastics such as Teflon, ceramics, fish-paper;
- Insulating the conductors: coating them (e.g., silicone gel), adding heat shrink tubing, rubber boots;
- Completely separating the conductors with isolators;
- Completely enclosing the volume within an insulating material such as potting.

For creepage, the breakdown voltage is increased by (Figure 4.32(b))

- Adding an insulating barrier between the conductors, which lengthens the path;
- Completely enclosing one of the conductors within an insulating material;
- Opening a gap between the conductors;
- Completely separating the conductors with isolators;
- Completely enclosing the volume within an insulating material: potting.

The pollution degree is reduced by manufacturing in a clean environment and placing the high-voltage circuits in a sealed enclosure.

Electrical isolation is classified into five categories:

1. *Functional:* good enough to work, but no safety measures;
2. *Basic:* some barrier, by itself, provides single level protection against electric shock;
3. *Supplementary:* an additional barrier, by itself, at least 0.4 mm thick;

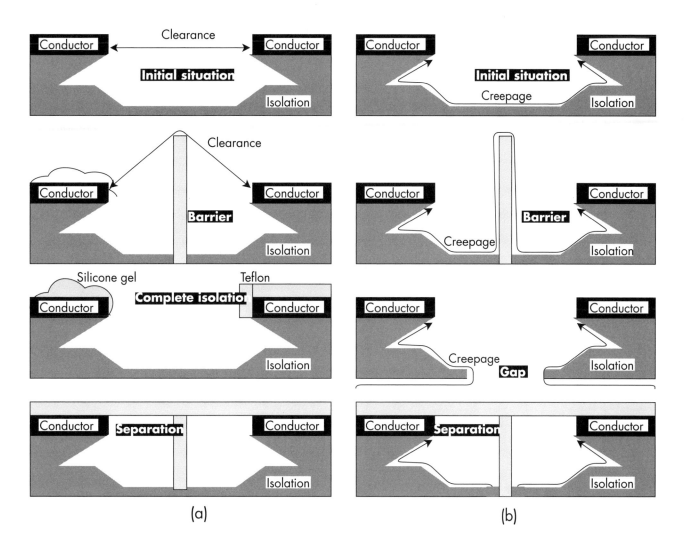

FIGURE 4.32 Remediation: (a) clearance, and (b) creepage.

4. *Double:* the combination of basic insulation plus supplementary insulation (in case basic isolation fails);

5. *Reinforced:* a single barrier that is as effective as double insulation.

4.7.3 Subdividing the Voltage

To increase safety, the voltage of these batteries may be broken up when not in use during manufacture, service, or active operation.

4.7.3.1 Subdivision into Low-Voltage Modules

Subdividing a high-voltage battery into lower-voltage modules may increase safety. For example, nineteen 48V modules may each be connected in series to form a 900V battery (Figure 4.33(a)).

Low-voltage modules can be manufactured, handled, and installed safely. There is no high voltage until the battery is turned on. During service, the battery is turned off, disconnecting the modules and removing the high voltage. A fuse in each module

FIGURE 4.33
Breaking up the voltage:
(a) into low-voltage
modules,(b) fuse and
disconnect inside each
module, (c) safety measures
in a high-voltage battery,
and (d) safety measures in
each half of a split battery.

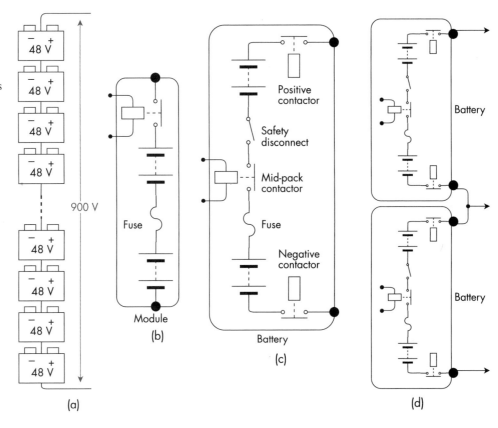

blows in case of an accidental short circuit across its output or parallel connection to a module at a different voltage (Figure 4.33(b)).

Regardless, include some way to prevent accidental connection to its terminals while the module is not yet in use, such as

- *A recessed connector:* a stray conductor cannot contact the terminals accidentally;

- *A contactor:* turned on once the module is placed in service;

- *A safety cutoff:* installed only after the module is placed into service;

- *A circuit breaker:* it is turned on only after the module is placed in service; this can also replace a fuse in the module. It must be rated for the total voltage of the battery, not just the voltage of the individual module.

4.7.3.2   Battery Disconnect

Provide a means that disconnects a high-voltage battery when not in use so that no voltage appears across its terminals. This way, even in case of an accidental connection to one of its terminals, there is no continuous path for any current (Figure 4.33(c)):

- Include two contactors, one at each end of the battery, to galvanically isolate it when off.

- Include at least one mid-pack contactor (or even multiple contactors dividing the battery into lower-voltage sections); this reduces the voltages within the battery when off.

- Include a safety disconnect: either a jumper or an emergency push-button switch that cuts off power to the contactors.
- For a split battery, include all these safety measures in each battery half (Figure 4.33(d)).

4.7.3.3 Isolation Testing with Mid-Pack Contactors

Note that a mid-pack contactor defeats a ground fault detection. The detector only works if the string is complete, and the battery is galvanically isolated from a grounded load. Therefore, the battery must have three states:

1. *Off:* all contactors off (Figure 4.34(a));
2. *Isolation test:* output contactors off, all mid-pack contactors on (Figure 4.34(b));
3. *Operating:* all contactors on (Figure 4.34(c)).

The standard circuits and off-the-shelf products for isolation testing are not designed to work at the highest battery voltages for these applications.

4.7.4 Sensing

For battery voltages up to 450V, sensing the pack voltage, current, and isolation is no more challenging than it is for automotive applications. Any BMS that is designed to work with traction batteries has no problem when used with stationary batteries up to this voltage. However, for higher battery voltages, sensing the pack has to be done with more considerable attention to isolation.

4.7.4.1 Cell Voltage Sensing

For a wired master/slave BMS, ensure that each slave is reliably isolated from the master (isolated communication line, powered by the cells it measures).

4.7.4.2 Pack Voltage Sensing

In low-voltage batteries and traction batteries, the BMS may measure the total battery voltage directly. Doing so in a high-voltage battery could be unsafe. Voltage divider resistors should be placed close to the high voltage terminals; place a fuse next to each of the two battery terminals, followed by a resistor (high resistance, rated for the high voltage) and then a wire to the BMS (Figure 4.35).

Each of the two resistors forms one leg of a voltage divider (the other leg is inside the BMS), so that the voltage that reaches the BMS is down to safer levels, and is current-limited. Configure the BMS to account for the voltage divider. For batteries up to 1 kV, use voltmeter fuses (rated for 1 kV DC, 440 mA), which are readily available.

4.7.4.3 Current Sensing

A cable-mounted Hall effect current sensor is safer than a shunt, because the amplifier for a shunt current sensor may not be rated for high-voltage isolation.

4.7.4.4 Temperature Sensing

In a wired BMS, pay close attention to thermistors and their wiring: should a thermistor touch a live conductor in the battery, bad things would happen.

FIGURE 4.34
Three states of a high-
voltage battery:
(a) off, (b) isolation test,
and (c) operating.

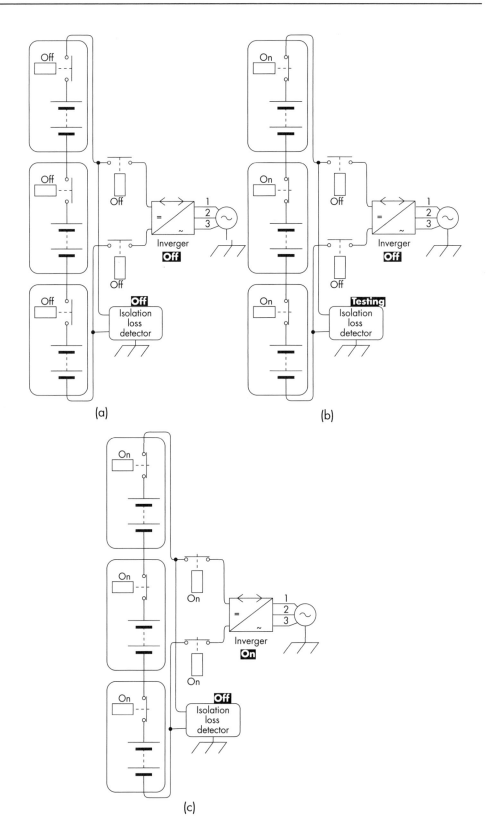

4.7.5 Hipot Testing

As a high voltage battery is built, and then throughout its life, it must be tested to
ensure that its isolation is adequate. This test is called hipot, an abbreviation of high

FIGURE 4.35
Pack voltage sensing.

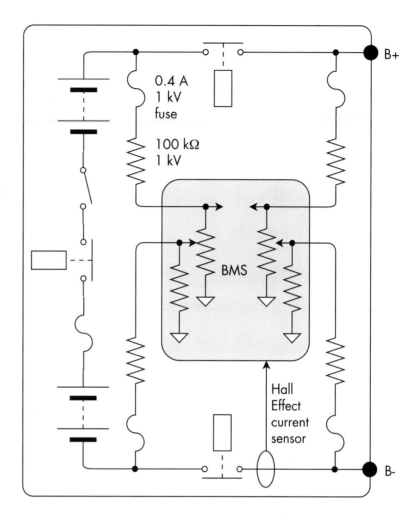

potential (as in high voltage), or dielectric withstand test[40]. This test helps find issues with spacing (creepage and clearance), conductors (incorrect wire voltage rating, crushed insulation, stray strands in wire or braided shields), and contamination (corrosive or conductive).

This test is performed with a hipot tester. The tester is configured for a test voltage that is twice the battery voltage plus 1 kV and the appropriate current threshold, such as 0.5 mA. A test prod is connected to a battery terminal, while the enclosure of the battery is connected to earth ground. The tester generates the high voltage (either AC or DC) between a test prod and earth ground and reports whether the current flowing through earth ground exceeds the configured threshold, indicating that the battery failed the test. The test lasts for 1 or 2 seconds.

4.8 BATTERY TECHNOLOGY

As they operate at such high voltages, these batteries are particularly lethal; being capable of such high current, they can be particularly damaging. The electrical design of the battery can improve these batteries' safety through grounding and isolation, fusing, interlocking, and anti-islanding. All of these are covered below. This section

---

40. Wikipedia, "Dielectric withstand test" https://en.wikipedia.org/wiki/Dielectric_withstand_test

discusses the technology inside a high-voltage stationary battery (just the battery portion, not the power converters).

4.8.1 Cell Selection

These applications may use all three battery classes: energy, power, or buffer (see Volume 1, Section 5.1.4). Each should use the appropriate cell format, type, and chemistry (Table 4.15).

4.8.1.1 Energy Batteries

For energy batteries, use large prismatic cells; 100 Ah are commonly available, but also 200 Ah, 400 Ah, and 1,000 Ah. These are standard Li-ion cells (as opposed to power cells). Select LCO or NMC cells for high energy content. If operated at cold temperatures, select LTO cells.

4.8.1.2 Power Batteries

For power batteries, use power cells or cells that offer a good compromise of energy and power density (see Volume 1, Section 2.2.3). These are available in the small cylindrical or pouch formats (not large prismatic). Small prismatic cells designed for HEV use are starting to become available (see Section 3.9.1.2). Select NMC or LFP cells with a low MPT. For very short charge and discharge times, look into LTO cells. Multiple cells are connected in parallel to achieve the desired capacity; use the parallel-first arrangement (see Volume 1, Section 3.4).

4.8.1.3 Buffer Batteries

For buffer batteries (for the 15-minute market) select LFP and NMC power cells. Again, these come only in the small cylindrical or pouch formats. For markets with periods of 1 and 5 minutes, select LTO cells with extremely short maximum power time (see Volume 1, Section 1.5.2) (less than 40 seconds). If this is not possible, use a much larger battery with LFP or NMC power cells (see Volume, Section 2.5.3). For example, for a battery used in the 1-minute market that must receive or deliver 30 Ah over 1 minute, design a 300-Ah battery so that the current is 6 C. Yes, all the extra capacity is expensive, but the battery will last much longer.

4.8.1.4 Very High Power

For markets with periods of 0.1 minute, and for regulation service, use super-capacitors or Li-ion capacitors (see Section A.3.4). If this is not possible, use a larger-than-needed battery using LTO cells.

Service	Market	Battery Type	Cell Format	Cell Type	Cell Chemistry
Energy	2~10 hours	Energy	Large prismatic	Energy	LCO, NMC (LTO)
Energy	30~60 minutes	Power	Pouch or small cylindrical	Compromise	NMC, LFP (LTO)
Arbitrage	15 minutes	Buffer		Power	NMC, LFP
	1~5 minutes	Buffer	Small prismatic	Very high power	LTO (NMC, LFP)
	0.1 minute	Very high power	Large prismatic	NA	Li-ion capacitor (LTO)
Regulation	10 ms		Large prismatic	NA	Li-ion capacitor

TABLE 4.15   Battery and Cell Types for Various Services

4.8.2 BMS Selection

Many off-the-shelf BMSs (Figure 4.36) can support the 120 or so cells in series in a 450V battery. A few can support the 240 cells in a 900V battery. As most BMSs are limited to 256 cells[41], very few can go up to the 480 cells in series in a 1.8-kV battery and have the isolation required to withstand this voltage. Fewer still can handle arrays of batteries. That is why the largest stationary batteries use custom BMSs, designed by or for the manufacturer of the battery modules.

Table 4.16 lists some of the off-the-shelf BMSs that may be considered for a high-voltage station array battery of 900V and above.

4.8.3 High-Power Circuits

The design of the high-power circuits and components in a high-voltage battery requires some care.

4.8.3.1 Fusing

Protect the battery and the wires with fuses:

- *Main fuse:* Place the main fuse in the middle of the high-voltage string; a fuse placed on an end terminal does not protect against a complete ground fault in most locations within the battery; a mid-pack fuse, while not perfect, is more likely to blow no matter where within the battery a ground fault occurs.
- *Side circuits:* If the battery terminals are also connected to a low-power circuit (such as a small DC-DC converter or a sensor of total battery voltage) place a low current fuse near each terminal, before feeding the small gauge wires going to the low-power circuit.
- Voltage sense wires: With a centralized BMS[42], place a fuse in series with each tap wire, close to the cell; resistors won't work for the purpose because they affect the reading and prevent balancing.

Fuses must be rated for DC operation and the voltage of the entire battery; pay particular attention to the breaking current rating (see Volume 1, Section 5.12.6). For low current, consider fuses for a meter: 0.4A, 1 kV.

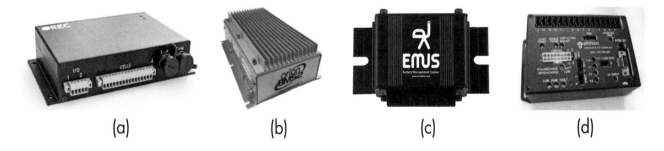

(a)   (b)   (c)   (d)

FIGURE 4.36   High-voltage BMSs: (a) REC SI (courtesy of REC BMS), (b) Ewert Orion BMS (courtesy of Ewert Energy), (c) EMUS (courtesy of EMUS), and (d) Lithiumate Pro.

---

41. That's so that the cell number can be expressed in a single byte. A byte has 8 bits, which allows counting from 0 to 255.
42. This is not a concern in a master/slave BMS that is reliably insulated from the master because a given slave only senses a few cells in series. This is not a concern with a distributed BMS either.

Brand	Model	Voltage [kV]	Power (1)	Array (2)	Additional Features
123 Electric	123 BMS	0.93			USB
Batrium	WatchMon	0.9			CAN, WiFi, USB
Elithion	Vinci HV	1.6	✓	✓	GUI, CAN, isolated, dedicated lines (RS-485 ModBus, Ethernet ModBus, WiFi, XanBus), contactor drive, isolation detection, Hall effect current sensor
Elithion	Lithiumate	0.93			CAN, RS-232, isolation detection, Hall effect current sensor
EMUS	GI	0.93			CAN, USB, RS-232
Ewert	Orion	0.65			CAN, Hall effect current sensor, ground fault detection
GWL Power	RT-BMS	0.7			CAN, USB, RS-232, contactor drive
JTT	X-Series	0.87			2 × CAN, contactor drive
Jon Elis	BMS ver.2	0.93			CAN, USB, Bluetooth, contactor drive
Lithium Balance	LiBAL sBMS	0.93			CAN
Nuvation Energy	High voltage	1.25	✓	✓	Ethernet, power interface: monitor current, voltage, contactor driver, shunt current sensor
O'Cell Technology	BMU LV-150	0.8			CAN, contactor drive
REC	BMS	0.87			CAN, RS-485, contactor drive, 380-A current sensor
Volrad	V-ACT	1.4			

{tfn}Notes: (1) The BMS can evaluate SoC of power battery that is never fully charged. (2) Compatible with arrays: the BMS connects to a common DC bus only if safe and/or includes an array master.

TABLE 4.16   BMSs for High–Voltage Batteries

4.8.3.2 Protector Switch

The only acceptable power switch component for a high–voltage battery is a contactor (see Volume 1, Section 5.12.3):

· Rated for high voltage and high current;
· Rugged;
· Provides galvanic isolation.

Economizer contactors are not recommended due to reliability and noise issues. Instead, a dual coil contactor is preferable as it is more efficient and reliable.

4.8.3.3 Precharge

The DC port of the inverger includes large capacitors. Direct connection of the battery to the inverger results in a large and damaging current inrush as the battery charges this capacitance (see Volume 1, Section 5.13.2). Therefore, before making a direct connection, the inverger's capacitance must be charged slowly, through precharge (see Volume 1, Section 5.13).

The battery connects to the inverger when ready. Therefore, it should be the battery's responsibility to implement precharge. Better invergers include a smart precharge circuit[43]:

---

43. Powertronix (Italy).

- In standby, the inverger's capacitors are discharged, and its main contactors are off.
- When a battery is connected, the inverger sees it and starts precharge.
- At the end of precharge, the inverger turns on its main contactors.

For lower-voltage batteries, precharge can use a simple relay. For a high-voltage battery, it's better to use a contactor because it can handle the high voltage across its contacts when the battery is off.

A single precharge resistor may not be rated for the high voltage; use a set of resistors in series so that each resistor sees a lower voltage. The resistance is not critical; $10 \ \Omega$ total works well.

4.8.3.4 Wire insulation

A typical hook-up wire has PVC insulation rated for 150V, 300V, or 600V. These ratings are too low for high-voltage batteries. Either additional insulation is required, or the wires must be routed so that it cannot come into proximity with the opposite polarity. Otherwise, wires with high-voltage insulation must be selected, such as silicone rubber (for up to 40 kV), fluorinated ethylene-propylene (FEP) (up to 20 kV), rubber (up to 5 kV), or Teflon (PTFE, up to 1 kV).

4.8.4 BMS Power Supply

The BMS is powered by a low-voltage supply (e.g., 12V) that is not supplied by the string of cells. Therefore, there is no fear of overdischarging the cells (see Volume 1, Section 5.8.1).

4.8.5 Environmental

A high-voltage battery is likely to be installed in a clean environment, such as inside a container. Yet, large prismatic cells may be exposed to the environment. In the rare case that the battery is exposed to the rigors of an industrial environment, the battery should be protected from the dust or even metal filings that may float into it.

4.8.6 Modular Battery Design

There are at least a couple of approaches to the implementation of a high-voltage battery:

- Rack of modules;
- Shelves of cells.

Ready-made solutions for high-voltage stationary batteries may use ready-made modules. Each module has low voltage and is therefore considered safe on its own. The modules are fully enclosed for safety. Several modules are placed in a rack to achieve the desired voltage and energy.

4.8.6.1 Front Versus Rear Connectivity

Some modules have the power terminals on the front panel, and some have them on the rear. If on the front panel, the rack provides only mechanical support. All the connections are made on the front panels of the modules. Connections are made manually, after the modules are in place (Figure 4.37(a)). This solution is somewhat unsafe, labor-intensive, mistake-prone, and unfriendly to servicing the modules. If on the rear panel, sliding a module into the rack connects it to buses mounted to the rear of the rack (Figure 4.37(b)). For power and communications, these modules use blind–mate connectors. These connectors self-align when a module slides into place into the rack, thanks to their wedge-shaped guides and to the fact that they are free to float slightly (Figure 4.37(c)). The rear of the rack has bus bars that connect modules in series; modules in parallel are not allowed, for safety reasons. Communication connectors mate to each module.

4.8.6.2   Isolation and Capacitance

The modules may use a plastic enclosure or a metal one. If metal, the modules must be designed to withstand high voltage between the cell and the enclosure and to have low capacitance between them. The reason that capacitance must be kept low is that many applications "bounce the cells up and down" at the PWM frequency of the inverger and possibly also at the AC line frequency. That capacitance presents a burden to the power electronics, which must charge and discharge this capacitance 20,000 times a second. It also introduces differential mode noise into the BMS sensing, which some BMSs cannot handle, and it slows down the operation of a ground fault detector.

Spacing between the cells and the metal enclosure provide this isolation and minimize this capacitance. Isolating barriers on their own only provide isolation; they do not decrease capacitance.

4.8.7   Cells on Shelves Design

A battery may be built in its final location by placing large prismatic cells on shelves, either horizontally (Figure 4.38(a)) or vertically (Figure 4.38(b)). This method is appropriate for one-off sites or when a larger capacity is required than is practical with rack-mounted modules.

First, cells are placed on the shelf. Then, cells are connected in parallel, to achieve the capacity, and then in series to achieve the voltage (see Volume 1, Section 3.4). Each shelf includes a fuse and a safety disconnect: ideally a contactor.

The cells must be actively retained with brackets because relying just on gravity is no help during an earthquake. Once completed, large clear plastic plates are placed vertically in front of the shelves, for protection.

FIGURE 4.37
Modular batteries: (a) front connected (courtesy of AllCell Technologies), (b) rear connected, and (c) blind–mate power connector..

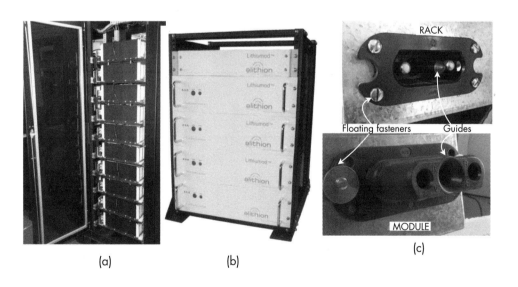

(a)   (b)   (c)

FIGURE 4.38
High-voltage battery
implementation with cells
on shelves: (a) horizontal
cells (courtesy of HNU
Energy), and (b) vertical
cells on shelves (courtesy
of IPS, Integrated Power
Systems, Canada).

(a)   (b)

For this solution, nonconductive shelves would provide isolation and minimize capacitance to ground. However, in most cases, metal shelves are used instead. If so, lifting cells from the metal shelf with a thick insulating layer provides isolation and reduces capacitance to ground.

4.9 SYSTEM INTEGRATION

This section covers the interface between a high-voltage battery pack and the system. It discusses some technologies in the system that relate to the battery pack.

4.9.1 System Grounding and Fault Testing

For safety, the batteries themselves should be floating: isolated from earth ground (see Volume 1, Section 5.14). The only time that one may perform an isolation test on these batteries is when both ends of the battery are galvanically isolated from a grounded power converter through two open contactors (see Section A.6.1).

Once connected to a grounded power converter, the battery is no longer floating because this converter generally does not isolate the battery from the AC line. At this point, an isolation loss test would fail; still, a ground fault current test would at least be able to detect a soft short. A hard short should blow a mid-pack fuse; a fuse at the end may not be affected.

A converter between the battery and the AC line may use any of a variety of configurations (Figure 4.39). Each grounds a battery terminal by connecting it to the AC neutral. Doing so defeats any battery isolation and makes it unsafe. An exception is the three-phase Δ line circuit, which does not have a neutral (Figure 4.39(e)). Yet even this is considered unsafe because any connection to the grid is potentially unsafe.

In any of these grounded systems, a test for ground fault current would detect a soft short inside the battery. It does so by noticing that the two wires from a battery are carrying different levels of current (see Volume 1, Section 5.12.6). To detect a soft short outside the battery, one could monitor the current in the single connection to earth ground. This test cannot be too sensitive because the current in the earth connection of a high power installation can be relatively high due to parasitic capacitance.

For ground fault detection, we use two detectors (Figure 4.40(a)). If there are no shorts, neither detector reports a fault. These detectors are in two places:

FIGURE 4.39
Grounding on AC side:
(a) single-phase, split
battery, (b) single-phase,
single battery, (c) split-
phase; (d) three-phase
Y, (e) three-phase Δ,
and (f) three-phase
Δ with a split leg.

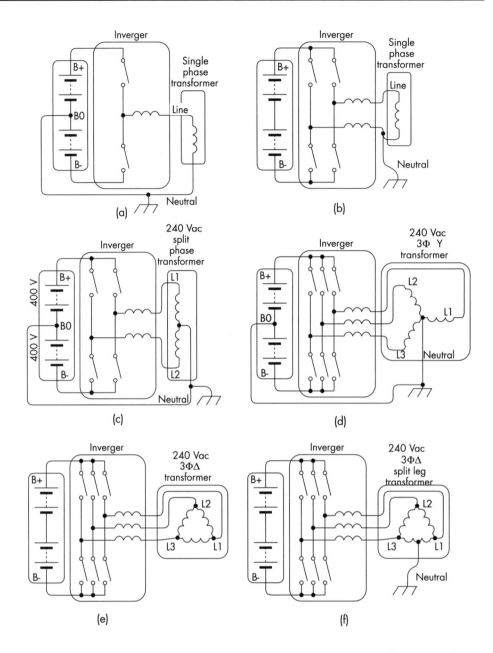

- *Inside the battery:* A battery fault current sensor detects ground faults inside the battery; this is a magnetic field sensor that encloses all the wires from the battery and measures the net current through both wires, which normally should add up to zero.

- *Outside the battery:* An earth fault current sensor detects ground faults outside the battery; this is a magnetic field sensor that encloses the wire between neutral and earth ground, which usually has some current due to leakage and parasitic capacitance; to avoid a false positive due to the usual earth ground current, this sensor is purposely not as sensitive.

Detecting a ground fault inside a split battery requires running all three of its wires through the battery fault current sensor (Figure 4.40(b)). Even if the two battery halves experience a different current, the sum of the current in all three wires adds up to zero, and therefore the current sensor sees no net magnetic field.

FIGURE 4.40
Ground fault current
detection: (a) single
battery, and (b)
split battery.

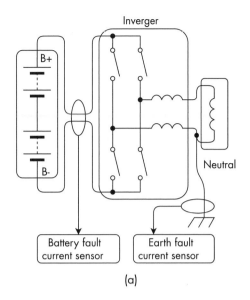

(a)

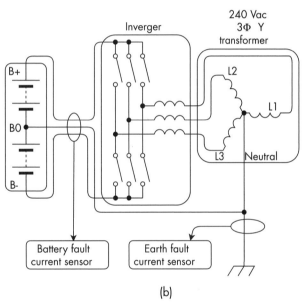

(b)

If there is a soft short (see Volume 1, Section 1.2.2.9) inside a battery, there is a difference in the current in the two battery wires. The battery fault current sensor notices this and reports a fault (Figure 4.41(a)). The fault current also flows through the earth fault current sensor, but this sensor is not as sensitive, so it doesn't report a fault. The earth fault current sensor does see the current due to a hard short outside the battery (Figure 4.41(b)). The battery fault current sensor won't see this current because it does not flow through the battery.

Similarly, for a split battery, the battery fault current sensor senses a soft short inside the battery because the fault current flows through it (Figure 4.42(a)), and the earth fault current sensor detects a hard short outside the battery (Figure 4.42(b)).

4.9.2 Communications

High-voltage stationary plants may communicate using a variety of physical layers and protocols (Table 4.17) (see Volume 1, Section 5.10).

FIGURE 4.41
Ground fault current
detection, single battery
system: (a) inside
the battery, and (b)
outside the battery.

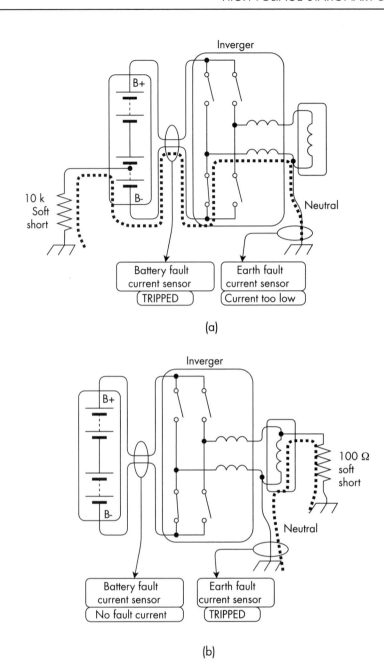

(a)

(b)

A BMS is not likely to cover more than one of these physical layers (probably CAN or RS-485) and may use a proprietary protocol. It is common for a system controller to handle the communications with the plant using one of these protocols and to be configured to communicate with the BMS with whatever method the BMS provides.

4.9.3 Control System Architecture

A larger plant that uses a BESS is likely to be managed through SCADA (supervisory control and data acquisition). SCADA is the standard control system architecture in the power and energy industry. A SCADA system consists of five levels. Table 4.18 shows the levels, how they are typically implemented, and how a high-voltage battery fits in this scheme.

FIGURE 4.42
Ground fault current
detection, split battery
system: (a) inside
the battery, and (b)
outside the battery.

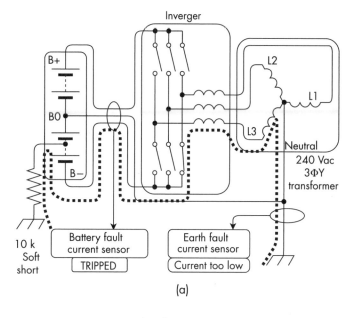

(a)

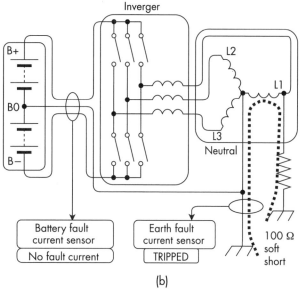

(b)

TABLE 4.17
Physical Layer of
Communication Protocols

Physical Layer	Protocol
RS-485	ModBus RTU, Profibus, Proprietary
Ethernet, TCP/IP	ModBus TCP, RP-570
CAN	CANopen, DeviceNet, Proprietary
MBP	Profibus
Optic fiber	Synchronous optical networking (SONET), synchronous digital hierarchy (SDH), Profibus

4.9.4   Grid Back-Feed

A high-voltage battery connected to the grid and able to back-feed to it uses a grid-tied inverger, which operates as a current source that follows the line frequency from the grid. A grid-tied inverger belonging to a customer of the power company cannot operate without the grid. It mustn't, for the safety of personnel working on

TABLE 4.18
SCADA Control System
Architecture

Level	Description	Device	HV battery
Level 4	Production scheduling		
Level 3	Production control		
Level 2	Plant supervisory	Desktop computer	System controller
Level 1	Direct control	Programmable logic controllers (PLCs), Remote terminal unit/remote telemetry unit (RTU)	BMS
Level 0	Field level	Sensors	Battery sensors

the transmission lines during a power outage (see Section 2.5.2.2). However, a backup system commissioned by the power utility itself uses grid-interactive invergers that can switch from grid-tied mode to off-grid mode the instant the main AC power source drops, to feed the grid.

4.10   BATTERY ARRAYS

A high-voltage system may be subdivided into several smaller sections

- To achieve the desired voltage and capacity;
- To work with an inverger that requires a split battery;
- For safety (each battery has a lower voltage);
- For redundancy and to allow servicing part of the BESS without downtime.

4.10.1   Arrangement

The batteries in an array may be arranged in various ways.

4.10.1.1   Parallel Arrangement

The system may consist of

- Battery array: several batteries connected in parallel to a shared DC bus (DC-coupled) (Figure 4.43(a));
- BESS array: several BESS connected in parallel to a shared AC bus (AC-coupled) (Figure 4.43(b)).

DC coupling has the advantage of flexibility; more batteries may be connected in parallel to increase the size. However, it also the disadvantage that direct parallel connection of batteries at different SoC results in an inrush current between them (see Volume 1, Section 3.3.6). Solutions include using a DC-DC converter for each battery (see Volume 1, Section 6.3.3), using array-compatible BMSs, or an array master (see Volume 1, Section 6.3.2).

AC coupling in a BESS array has many advantages:

- Allows the use of smaller, lower-power invergers;
- Redundancy: should one inverger go down, the system continues to operate;
- Completely avoids the issue of inrush current between parallel batteries.

4.10.1.2   Series Arrangement

Do not confuse a single battery with multiple strings connected permanently in parallel (see Volume 1, Section 6.1.4.1). If the BMS cannot handle more than 256

FIGURE 4.43
Array: (a) DC-coupled
battery array, and (b) AC-
coupled BESS array.

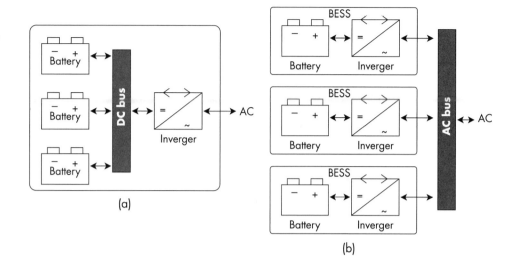

(a)

(b)

cells, yet the battery needs 400 cells in series to achieve the desired voltage, the battery can be divided into two smaller batteries, each with 200 cells in series.

The batteries are ganged so that they may operate as a single unit (see Volume 1, Section 6.4.4). The external system may gather data from each BMS individually (Figure 4.44(a)) or an array master manages the two batteries as one (Figure 4.44(b)) (see Volume 1, Section 6.3.2).

4.10.1.3   Parallel and Series Arrangement

Several batteries may be connected both in the parallel and series to achieve the desired capacity and voltage (see Volume 1, Section 3.1.5). They are managed as an array (see Volume 1, Section 6.3).

4.10.1.4   Split Battery

Some invergers require a split battery: two separate batteries that are connected in series with a tap between them (see Volume 1, Section 6.5). They are ganged so that they may operate together (Figure 4.44(c)).

4.10.2   BMS Topology

In an array, the BMS topology depends on the particular BMS used. Two off-the-shelf BMSs support DC-coupled battery arrays directly.

In a Vinci HV system (Figure 4.45), the BMS in each battery consists of a battery master, an application module, and one or more fiber optic slaves connected to cell boards on the cells. The cells are divided into banks of up to 30 cells in series; a cell board is mounted on each cell; up to four banks are connected to a slave through fiber optic cables. The application module powers the battery master and the slaves, includes contactor drivers, supports current and stack voltage sensors, and tests for isolation.

A battery master manages several batteries, controlling and coordinating their state, and compiling their data into a single set of data for a system controller. Isolation occurs in two places: in the optic fibers and inside the application module. In a Nuvation Energy High Voltage system (Figure 4.46), the BMS in each battery consists of a stack controller, a power interface, several cell interfaces connected to up to 16

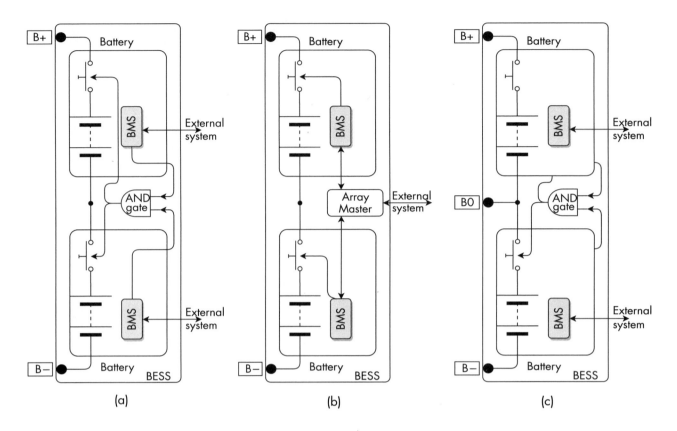

FIGURE 4.44   (a) Series battery array: (a) ganged, (b) with array master, and (c) split battery.

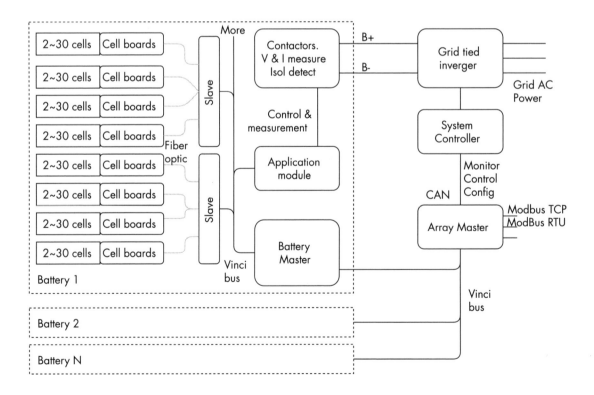

FIGURE 4.45   Vinci HV array topology.

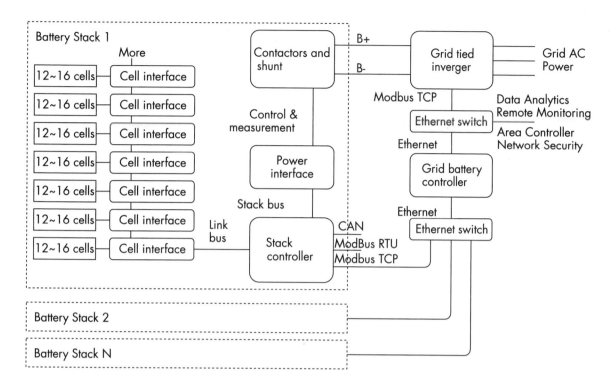

FIGURE 4.46 Nuvation Energy High Voltage array topology.

cells in series. The power interface powers the stack controller and the cell interfaces, includes contactor drivers, supports current and stack voltage sensors, and tests for isolation.

A battery master manages several batteries, controlling and coordinating their state, and compiling their data into a single set of data for a system controller. Isolation is provided in two places: inside the cell interfaces and the power interface.

REFERENCES

[1] "Power Quality Assessment Procedures," EPRI CU-7529, December 1991.

[2] Eto, J. H., "Use of Frequency Response Metrics to Assess the Planning and Operating Requirements for Reliable Integration of Variable Renewable Generation," Joseph H. Eto, Berkeley National Laboratory.

[3] Energy Storage Association, "Frequency Regulation," energystorage.org/energy-storage/technology-applications/frequency-regulation.

[4] Pacific Gas and Electric Company "Voltage Tolerance Boundary," https://www.pge.com/includes/docs/pdfs/mybusiness/customerservice/energystatus/powerquality/voltage_tolerance.pdf.

[5] American Power Conversion Corp., "The Different Types of UPS Systems," EE Times, eetimes.com/document.asp?doc_id=1272971.

[6] AEG, "Uninterruptible Power Supply," aegps.com/en/technology/uninterruptible-power-supply-ups/.

[7] Jülch, V., "Comparison of Electricity Storage Options Using Levelized Cost of Storage (LCOS) Method," Applied Energy, Vol. 183, December 1, 2016.

[8] Schmidt, O., et al., "Projecting the Future Levelized Cost of Electricity Storage Technologies," Joule, January 9, 2019, pp. 1–20.

[9] Sandia National Laboratories, "DOE/EPRI Electricity Storage Handbook in Collaboration with NRECA," January 2015.

[10] Reneweconomy, "Explainer: What the Tesla Big Battery Can and Cannot Do," reneweconomy.com.au/explainer-what-the-tesla-big-battery-can-and-cannot-do-42387/.

[11] Henning, B. B., "The Impact of Renewable Electric Generation on Natural Gas Infrastructure and Operations," ICF International, https://www.ingaa.org/File.aspx?id=11011

[12] Microgrids at Berkeley Lab, "Fort Collins," building-microgrid.lbl.gov/fort-collins.

[13] Microgrids at Berkeley Lab, "Types of Microgrid," building-microgrid.lbl.gov/types-microgrids.

[14] IEEE 2030.7 Standard, standards.ieee.org/standard/2030_7-2017.html.

[15] Dudiak, J., et al., "Hierarchical Control of Microgrid with Renewable Energy Sources and Energy Storage," 2015, pp. 568–571.

[16] "Making Sense of Demand Charges: What Are They and How Do They Work?" https://blog.aurorasolar.com/making-sense-of-demand-charges-what-are-they-and-how-do-they-work/.

[17] Adams, D., Hitchhiker's Guide to the Galaxy, planet Golgafrincham.

[18] Wikipedia, "Dielectric Strength," en.wikipedia.org/wiki/Dielectric_strength.

[19] IEC 60664-1 Standard, "Insulation Coordination for Equipment Within Low-Voltage Systems - Part 1: Principles, Requirements and Tests."

[20] "Comparative Tracking Index (CTI)," Underwriter Laboratories.

[21] IEC 61857, "Electrical Insulation Systems," IEC 60601-1.

[22] "Medical Electrical Equipment," IPC2221A, UL60950-1.

CHAPTER 5

ACCIDENTS

5.1 INTRODUCTION

Despite best intentions, sometimes things go wrong. With Li-ion batteries, sometimes they can go very wrong. Being aware of other people's accidents may help us prevent having an accident ourselves.

5.1.1 Tidbits

Some interesting items in this chapter include:

- Li-ion doesn't kill people; bad design and procedures kill people (Section 5.2).
- Despite warnings to the contrary, water is effective in a Li-ion fire (Section 5.3).

5.1.2 Orientation

This chapter starts with reports of actual accidents in cars, yachts, and airplanes. It then discusses emergency procedures.

5.2 CASE STUDIES

Here are some actual events that resulted in property damage. No one was hurt in any of these cases.

Li-ion can start fires, some of which result in loss of valuable property. When this happens, experts may be called to investigate; I was personally involved in the investigations of most of the following accidents. I tried to give as much detail as necessary for our learning experience, limiting any identifying details.

5.2.1 Automotive Accidents

Poorly executed conversions to Li-ion have resulted in a few cars going up in flames.

5.2.1.1 PHEV Conversion Company

Hybrids Plus converted Toyota Prius cars to plug-in-hybrids, using A123 cells; I designed the BMS and the other electronics for it.

This new technology elicited much excitement. People of varying degrees of expertise volunteered to help to design the battery. One of these volunteers decided that standard 4/0 welding cable was too stiff. He replaced it with tinned copper braid, the kind used for shielding, in a braided cable sleeve wrap (loom). The metal conductor was visible through the loose weave of the loom; I told the boss that such insulation was insufficient. He overruled me because I was not in charge of that aspect of the design.

As we were placing a module in the traction battery enclosure, the module's two cables touched each other and made contact through the weave in the loom. All hell broke loose (Figure 5.1(a)). Wisely, we were wearing safety goggles, and we had placed the battery on a cart next to a sliding door. Within 10 seconds, we opened the door and pushed the cart to in the parking lot, as a thousand 26650 cells were going off like firecrackers. To this day, some cells are still on the roof of the car wash you see on the left of the picture[1].

Afterward, we designed the battery somewhat better, but we still had significant problems with quality control. A battery module used a set of washers on a terminal. Without those washers, bolting down the terminal would put pressure on a plastic enclosure for the module, not on the hard metal of the terminal. Once, the assembly technician omitted those washers in one of the batteries. On a hot summer day in South Carolina, the plastic softened, and the ring terminal at the end of a cable became loose. The battery current, flowing through the high resistance of that contact, generated high heat, which started a fire inside the module (Figure 5.1(b)). The BMS issued an alarm, lighting a Red Triangle on the dashboard. The driver parked on the side of the highway, got out of the car, and watched as the car burned down. The cells next to the loose connection overheated and ignited, emitting a fountain of molten material that cut through the steel battery enclosure like a blow torch. Then the upholstery ignited, burning down the car (Figure 5.1(b)).

Lessons learned

- Most fires in low production volume Li-ion batteries are due to plain old battery design mistakes, which would be bad with any other battery technology, but are made worse by Li-ion's high power capability.
- If the design seems unsafe to you, as an engineer, you must speak up, even if it's not your area.
- Put together quality assurance (QA) procedures, follow them, and have someone use them to check the production process.

Conclusion: I prepared a set of QA procedures, hired a QA inspector, and left the company[2] in good hands. Four months later, the remaining people dissolved Hybrids Plus, formed a new company (Eetrex), which lasted 1 year, during which it lost one more vehicle to fire.

FIGURE 5.1
PHEV conversion company: (a) development battery fire, and (b) converted Prius after the fire.

(a)   (b)

---

1. Northeast corner of 28th Street and Valmont, in Boulder, Colorado.
2. I started Elithion, now in its eleventh year.

5.2.1.2 *The Rockstar*

The Rockstar loves his big old 1959 American car. The car has a Wikipedia page and is featured in a documentary about the Rockstar. But the car is a gas guzzler, so the Rockstar hires the EV Guy to convert it to an HEV.

On November 9, 2010, in the evening, a powerful charger was sitting on a bench, connected to the battery but not yet adjusted for the correct voltage; the BMS was connected just to the cells, not to the charger. The EV Guy left for the evening, forgetting that the charger was turned on. Quickly, the battery was topped, cells started to overcharge, and the BMS was screaming "Stop!" but the charger did not know because it was not connected to the BMS. Cells ignited; a fire burned in the warehouse full of the Rockstar's irreplaceable memorabilia.

EV aficionados blamed the BMS: "Balancing starts fires. BMSs are the problem, not the solution[3]." A tech blogger reported rumors of arsonists. The EV Guy said, "We are still investigating the exact cause although it appears to be an operator error that occurred in an untested part of the charging system."

Lessons learned

- Adjust the CV of the charge to the correct voltage before ever using it to charge a battery.
- Always, always, always let the BMS shut off the battery current if it needs to.

Conclusion: the car was rebuilt and is now using a traction battery from A123.

5.2.2 Marine Accidents

I saw the aftermath of Li-ion battery fires on beautiful and expensive yachts on three occasions when I was hired to investigate the causes.

5.2.2.1 *Racing Yacht T*

This 75-foot Grand Prix racing yacht was built in 2009 in New England in the United States. It used custom made 24V Li-ion batteries from a small manufacturer of a distributed digital protector BMSs offered as a marine BMS. The batteries used seven large prismatic NMC cells in series from International Batteries. The vessel had two batteries connected to a shared bus. The design does not include a way to ensure the safe connection of one battery to the shared bus if it is already connected to the other battery at a different voltage.

On April 26, 2011, the yacht was docked in Antigua with no one on board, charging the battery from shore power. A child in a nearby yacht noticed smoke and alerted others. It took two hours to extinguish the flames. There was little left of the battery that burned (Figure 5.2).

What we know is that

- In 2009, there was a problem with the battery; A.M., the battery manufacturer, visited the yacht. He noticed that the BMS reported that the voltage of a particular cell is low; he did not measure the actual voltage with a meter to see if the voltage was truly low, or if the BMS was reporting the wrong voltage. A.M. did not check the cell board to see if the balancing load was stuck on, which would have discharged just the one cell. Instead, he assumed that the cell was

---

3. A renowned blogger in the EV enthusiast world.

FIGURE 5.2
Burned out BMS from
yacht T; note the remains
of two EV-200 contactors,
one on each end.

low due to high self-discharge current. A.M. replaced the low voltage cell, but did not balance the new cell (at about 50% SoC) to match the other cells (at 100% SoC) because he was not equipped to do manual balancing and because he decided that BMS balancing (at 60 mA) would take too long.

- On April 11, 2011, the yacht was docked in Florida. The captain noticed that the batteries were off and would not take a charge; over the next few days, he hired a local electrician to restore the system with help from A.M., who was back in the United States. The electrician measured the cell voltages, which indicated that both batteries were slightly overdischarged; the minimum cell voltage was 2.745V in one battery and 2.9V in the other battery, both below the manufacturer's specification of 3.0V. Because the cells should not have dropped that low just from self-discharge in a short time, A.M. considered this to be an indication that a load was connected directly to the battery cells, bypassing the BMS. A.M. decided that the cells were not too discharged; he told the electrician how to bypass the charging contactor to enable charging until the voltage was high enough to power the BMS, after which the BMS can take over controlling the charging contactor. The principal instructed the electrician to do that on both batteries. Afterwards, the batteries were recharged from shore power.

- The yacht sailed to Antigua. During the voyage, the batteries were charged from the on-board generator. Once there, the crew took the yacht for a run.

- The yacht came back to the dock and was plugged into shore power. This was the first time since Florida and the second time since the electrician bypassed the BMS. The crew went to dinner, and the captain received a call that the yacht was on fire.

A capable team was hired to do a forensic analysis. During a 4-year inquest, some crucial questions were never asked or, by the time they were asked, people could no longer remember the answer. The inquest was ended abruptly and before the team was able to come to conclusions.

Although we will never be able to prove the cause, I can report the following issues with the battery manufacturer, in order of severity:

- For using NMC cells on a vessel that are not as safe as LFP cells;

- For paralleling two batteries with BMSs that cannot ensure safe connection to a shared bus;

- For replacing a cell without gross balancing it first with the rest of the cells in the batteries;

- For instructing the electrician to bypass the BMS without making sure that the bypass is removed afterward;

- For having reason to suspect that there were loads connected directly to the battery cells and not following through with the yacht manufacturer to correct the issue;

- For not training the yacht manufacturer on the proper design of a Li-ion battery and not providing instructions that for the crew.

Lessons learned

- Choose an experienced company to design a battery for a vessel.

- Then hire a Li-ion expert to review the design.

Conclusion: The insurance company for the battery manufacturer settled immediately. The insurance company of the vessel sued the insurance company of the defunct cell manufacturer. Legal and forensic work lasted for 4 years until someone noticed that the same big corporation owned both insurance companies, so they stopped the inquest and quietly settled. The battery manufacturer has since switched to using LFP cells.

5.2.2.2 Catamaran M.F.

This beautiful 62-foot family cruising catamaran was built in 2009 in California, United States. The battery is an array of several 12V Li-ion unprotected battery modules from Valence, using 26650 LFP cells. Several modules in a row are connected in parallel, and then two rows are connected in series to achieve 24V. Two identical sets, one on each hull, are connected in parallel with cables. The batteries include monitoring and balance electronics, but no protector switches. Indeed, there is nothing to prevent overcharge or overdischarge of the batteries.

On November 3, 2011, while docked in Curaçao, the vessel was hit by lightning, damaging the electronic equipment. The vessel was dry-docked for a long time, with no shore power. The batteries were discharged down to 0V.

The Valence manual has no warnings on the special requirements of Li-ion batteries. The captain called Valence and was told to disconnect the batteries and recharge them individually. The captain did so with the batteries in the starboard battery compartment and then reconnected them.

Rather than charging the battery banks individually, the captain decided to speed up the process by reconnecting the cables between the two hulls and let the starboard array charge up the port batteries; as he learned to do with lead-acid batteries. The current was too low to blow the fuses between batteries due to the high resistance of the damaged cells; although it was high enough to overheat the port batteries.

White smoke started wafting from the port hull. The captain did not know what to do. Eventually, he used cable cutters in the burning battery, but that took too long. Soon the vessel was on fire (Figure 5.3).

In order of severity, the issues are

- Valence:

  - For not understanding the proper use of Li-ion batteries;

  - For not training the designers of the vessels on the proper way to design Li-ion batteries (e.g., safety disconnects and a protector switch controlled by the BMS);

FIGURE 5.3
Catamaran M.F.:
(a) salon, and (b) portside
battery compartment.

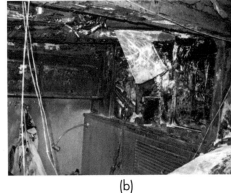

(a)   (b)

- For not ensuring that the battery modules were used properly and that a protector switch was installed;
- For not training the boat builder on how to build Li-ion batteries out of those modules;
- For not explaining clearly in the user manual the proper handling of Li-ion batteries;
- For not properly training the technical support person who gave the captain directions.
- The boatbuilder:
  - For not installing safety disconnects and a way for the BMS to shut off the battery current;
  - For not training the captain on Li-ion batteries.
- The captain:
  - For treating Li-ion batteries as if they were lead-acid.

Lessons learned

- Always let the BMS disconnect the battery if required to protect the cells; this would have prevented the battery from discharging completely.
- Never recharge a Li-ion battery that was overdischarged; some cells were probably reversed when the total voltage reached 0V.
- Include warnings on the unique requirements when reconnecting Li-ion batteries.
- Never connect batteries in parallel without making sure the voltages are the same.
- Train captains on Li-ion handling and safety.
- Place safety disconnects in each battery; make them hard to re-close, to keep untrained people from closing them without checking the voltage first.

Conclusion: The parties chose to blame the lightning strike even though it was not directly related to the fire. The matter was settled without legal action. The catamaran was repaired.

5.2.2.3 Luxury Yacht L

This 100-foot luxury racing yacht was built in 2009 in the United Kingdom. It uses Li–ion batteries from a renowned European manufacturer of marine electronic power products. The batteries use Thundersky type, large prismatic LFP cells. They include charge transfer balancing, but do not include a protector switch. They communicate with the rest of the vessel through RJ45 cables.

On May 2, 2015, while docked in Antigua, one of the batteries ignited spontaneously. The fire engulfed both the top of the cells and the electronics on top of the cells (Figure 5.4).

The actual cause of the fire and the order of events are unclear.
What started the fire?

- A cell went into a thermal runaway spontaneously.

- A cell was overcharged and went into thermal runaway.

- A cell was overdischarged, reversed, shorted, and went into a thermal runaway at the next charge.

- An electronic component in the BMS overheated and the electronic board caught fire.

Once the fire started, how did it propagate?

- The fire started in a cell, and then:
  - Spread to the rest of the cells and then to the board.
  - Spread to the board and then the board ignited the rest of the cells.
- The fire started in the board and spread to the cells.

The logs showed that this particular battery had reported a fault a few days before the accident; it was disabled and disconnected from the rest of the system.

We analyzed the components in the burned battery. All the cells except one were bloated, which is an indication of overcharging. Note how the fifth cell from the left is the only one that did not burn. It is the only cell that was not bloated; note that the board was most consumed by fire above that fifth cell.

FIGURE 5.4
Burned battery from Yacht L: (a) BMS electronic assembly, and (b) cells.

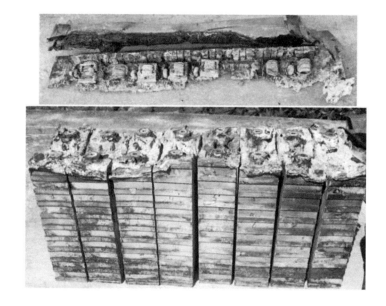

I also opened a good battery and reverse-engineered the circuit in the electronic assembly on top of the cells. It is a monitor that implements charge transfer balancing. There is no protector switch inside the battery, although the vessel had a contactor in series with that battery that the BMS could control.

Yes, a cell may go into thermal-runaway spontaneously; however, our experience with LFP cells tells us that this is highly unlikely:

- A cell that undergoes a thermal runaway is consumed entirely; yet none of these cells were; all the cells appear to be damaged more or less equally.

- LFP cells are not known to self-combust, not without an external cause.

- All the cells appear to be damaged from the fire in the PCB assembly above them or by burning electrolytes escaping the cells at the top.

Therefore, we all agreed that the fire must have started from the board.

My analysis of the evidence pointed to the hypothesis that the charge transfer balancing circuit misbehaved and overcharged one cell using energy from the other cells.

The battery manufacturer engineers, who know the design much better than I do, strongly disagreed. They stated that it's physically impossible for the charge transfer circuit to overcharge any one cell. They hypothesize that the board ignited first due to a component failure[4].

If it is true that it's physically impossible for the charge transfer circuit to overcharge a cell, then my next hypothesis is that a transistor in the board shorted, placing a transformer directly across a cell. This discharged the cell and overheated the transformer until it caught on fire.

The problem with this hypothesis is that, usually, a small transformer that is burning cannot ignite an entire PCB assembly. The most that it could do is to scorch the PCB. In any case, the transformer wire would fuse; regardless, I don't believe that a large prismatic cell is likely to catch fire from a PCB burning above it. Therefore, I still am not sure what happened.

Lessons learned

- Charge transfer balancing is not worthwhile in an LFP battery that is turned on at all times, (see Volume 1, sections 2.7.3 and 4.7.2.1) and, if it breaks down, it could start a fire.

- A vessel must have procedures in place to remove a battery on fire and throw it overboard within 10 seconds (see Section 5.3).

Conclusion: The parties settled amicably, the damage was repaired.

5.2.3 Aviation Accidents

5.2.3.1 *Dreamliner*

The Boeing 787 Dreamliner uses a 32V LCO Li-ion battery with eight cells in series.

On January 7, 2013, a battery overheated and started a fire in an empty plane in Boston (Figure 5.5) [1]. I have no direct experience with this event; I believe it deserves to be mentioned as the most infamous Li-ion fire to date.

---

4. There are significant legal ramifications with this difference in the interpretation of the cause.

When the fire first occurred, I was among those self-appointed experts who pontificated on the matter [2]. I was (wrongly) theorizing that the problem must have been that the designer of the airplane would not let the designers of the battery incorporate a protector switch in the battery, which resulted in abuse of the Li–ion cells and eventually the fire. The NTSB report [3] ruled out the first part by stating:

"The ... battery design incorporated provisions to ... inhibit discharging ... to protect the cells against overdischarge."

It ruled out the second part by stating:

"The battery failure did not result from overcharging, overdischarging, external short circuiting, external heating, installation factors, or environmental conditions of the airplane."

The NTSB concluded that impurities in the cells were to blame, which resulted in internal shorts inside the cells.

Lessons learned

- Don't jump to conclusions about an event until you have all the data.
- No BMS can prevent an internal cell short; however, a BMS can warn of an overtemperature and do so as soon as possible. The 787 battery had only two thermistors; they measured bus bar temperature, not cell temperature. Even voltage data, measured at high resolution, can give an early indication of impending problems within a cell.
- Design batteries to minimize fire propagation and contain any fire.
- Protect the nearby equipment from such a fire.

Conclusion: The 787 still uses a Li–ion battery, although in a sturdier case designed to contain a fire. Unfortunately, its weight counteracts the advantage of a lighter Li–ion battery. As recently as December 1, 2017, a cell overheated and leaked in a 787 approaching landing in Paris; no other damage was reported.

FIGURE 5.5
Burned Dreamliner battery (National Transportation Safety Board).

5.3 EMERGENCY PROCEDURES

In case of direct short across an unprotected battery, focus immediately on removing the short circuit, before thermal runaway starts[5]. Then be ready with the following measures, in case a fire does occur anyway.

In case of fire, cell manufacturers specify to use an ABC-type fire extinguisher, which is rated for electrical fires. They advise against using water because lithium metal reacts with water; however, there is no lithium metal in a good Li-ion cell. In any case, lithium represents only 1%~2% of the mass of the cells. What burns is the electrolyte.

When a battery is undergoing thermal runaway, the benefits of using water outweigh the risk due to the conductivity of water and the reaction of water with the little lithium present in the battery. Water absorbs heat to keep the battery cool and prevent propagating to the other cells. The steam dissipates the energy in the runaway cell, keeping the battery cool. Water also dilutes the organics in the electrolyte to a point below their flammability limit.

For small batteries in consumer products

- At home throw the battery outside so that it may burn safely; use a garden hose to douse it with water.
- In an airplane: dunk immediately a consumer product on fire in a bucket of water, and add more water as it steams away.

For large batteries

- At the factory: build a battery on a cart, next to a wide door to the outside; in case of an event, quickly open the door and roll out the cart (see Volume 1, Section 7.2.3).
- In a vehicle: pull over and exit the vehicle immediately; don't try to put out the fire because you can't.
- At sea: cut the cables, leaving enough cable to use as a way to pull the battery; break off the battery from its mounts with an ax; and throw the battery overboard.
- At home: don't try to put out the fire with a fire extinguisher because you can't. If available, use a water hose to douse the battery from a distance.
- In the air: there is not much you can do other than trying to land.

REFERENCES

[1] Wikipedia, "Boeing 787 Dreamliner Battery Problems," en.wikipedia.org/wiki/Boeing_787_Dreamliner_battery_problems.

[2] Rushe, D., "Boeing Chief Defends Use of Lithium Batteries in Grounded 787 Planes," The Guardian, January 30, 2013, www.theguardian.com/business/2013/jan/30/boeing-chief-defends-battery-787s.

[3] NTSB, "Auxiliary Power Unit Battery Fire Japan Airlines Boeing," 787-8, JA829J, www.ntsb.gov/investigations/AccidentReports/Reports/AIR1401.pdf

---

5. Caveat: I am not an expert on fire suppression.

APPENDIX A

BATTERIES

A.1 INTRODUCTION

This appendix collects details that would bog down the discussion if they were included in the main body of Volume 1:

- Deeper technical discussions that are of interest only to some readers;
- Subjects that are only tangentially related to the main subject of the book;
- Directories of off-the-shelf products; this does not constitute an endorsement of any of these companies and products[1].

A.2 FUNDAMENTAL CONCEPTS

These items expand on Volume 1, Chapter 1.

A.2.1 Short Discharge Time

In 2013, I proposed the concept of short discharge time[2]. Theoretically, short discharge time is the time required to discharge a full cell or battery through a short circuit, hence the name.

Given a cell's or battery's DC resistance, capacity, and voltage, the short discharge time is:

$$\text{short\_discharge\_time [h]} = \text{capacity [Ah]} \star \text{DC\_resistance } [\Omega] \text{ / nominal\_voltage [V]}$$
$$= \text{total\_energy [Wh] / short\_circuit\_internal\_power [W])} \tag{A.1]}$$

The short discharge time for Li-ion cells ranges from 0.004 to 0.06 hours (15 to 220 seconds). Therefore, seconds is a more practical measure of short discharge time than hours:

$$\text{short\_discharge\_time [s]} = 3600 \text{ [s/h]} \star \text{capacity [Ah]} \star \text{DC\_resistance } [\Omega] \text{ / nominal\_voltage [V]} \tag{A.2}$$

I discovered that many were turned off by the word "short" in short discharge time and by the theoretical nature of the definition; I was not too successful in promoting the concept. Therefore, five years later, I introduced a slightly different

---

1. While the information presented in a directory is believed to be accurate, I take no responsibility for any errors.
2. It appears that the short discharge time concept was developed independently by others who did not publicize it; I discovered this after I first made the public aware of the concept in the fall of 2012.

concept, maximum power time, which has a more practical definition (see Volume 1, Section 1.5.2). The maximum power time value is simply twice the short discharge time value.

A.2.2   But, But, But... Electron Flow

Students fresh from a Physics 101 course may "correct" us when we say that current flows from the more positive voltage to the more negative voltage[3], sneering knowingly at that fool Benjamin Franklin for picking the "wrong" direction for current.

Dear student[4]:

1. Current is not just electrons:
   - Positive ions and semiconductor holes travel in the same direction as conventional current.
   - Negative ions and electrons travel in the opposite direction.
2. Franklin could have picked the direction of either positive or negative ions and neither would have been wrong; he was not wrong when he picked the direction of positive ions.
3. The math and the circuits work the same, regardless of whichever direction we use for current.
4. The sooner you forget about electron flow and accept conventional current without reservation as your personal savior, the sooner you will start understanding electricity because you'll be speaking the same language as those who can help you understand it.
5. In any case, if you want to be precise about the physics involved, the latest understanding of current flow in a conductor involves photons, rather than the flow of electrons[5].

A.3   LI-ION CELLS

These items expand on Volume 1, Chapter 2.

A.3.1   Li-Ion Cell Operation at a Molecular Level

Knowing how a Li-ion cell operates on a molecular level is not required to use it, although having a sense of its mechanism may be instructive.

A.3.1.1   Intercalation

In a Li-ion cell, individual lithium ions or atoms are intercalated inside the crystal lattice in the active material in an electrode; they are nested in the empty spaces between the atoms in the active material (Figure A.1(a)).

Having asked various experts in Li-ion technology, I received multiple and inconsistent explanations of whether intercalated lithium is in the form of Li+ ions or neutral lithium elemental atoms:

---

3. This is true inside a load. Inside a source, the current flows in the opposite direction.
4. Including you, Davide Andrea, Liceo Cannizzaro, Palermo, Sicily, 1974.
5. Gestalt Aether Theory.

- Some understand that ions are intercalated in both electrodes.
- Others understand that neutral atoms are intercalated in both electrodes.
- Others understand that ions are intercalated in the positive electrode and neutral atoms in the negative electrode.
- Finally, some studies suggest that the atoms may share their electrons loosely with the surrounding material, making them slightly positive on the average.

In the following discussion, I assume the first understanding: Li+ ions are intercalated in both electrodes.

A.3.1.2 Rocking Chair Processes

While charging, lithium ions transfer from the positive electrode to the negative electrode. While discharging, ions transfer back the other way. This process is called a rocking chair due to the way the ions go back and forth during a charging and discharging cycle.

To be consistent with the good practice in electrical schematic diagrams, I have drawn the cell with its positive terminal at the top and negative one at the bottom (Figure A.1)[6]. I placed the external circuit on the right. This sets the direction of the current in the following discussion as clockwise while discharging. Remember that current flows in the same direction as positive ion flow, and in the opposite direction of electron flow.

The four numbers in the drawings match the numbers in the following discussion. We can divide the Li-ion process into four simultaneous processes:

1. Deintercalation;

FIGURE A.1
Cell: (a) lithium ions intercalated in the active material, (b) charging process, and (c) discharging process.

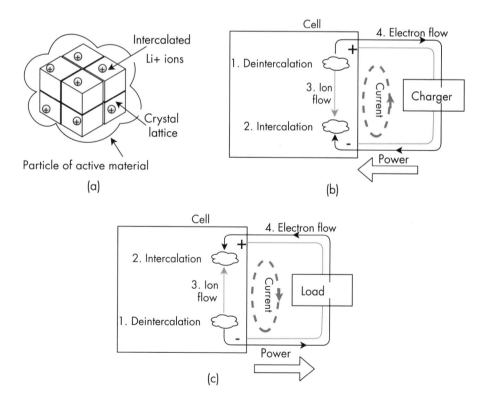

6. This is rotated 90° from the typical representation of a cell's inner workings.

2. Intercalation;

3. Positive lithium ion flow;

4. Electron flow.

While charging, the charger is the primary force; it causes current to flow counterclockwise (Figure A.1(b)):

1. A lithium ion deintercalates from the active material in the positive electrode (top).

2. Simultaneously, a lithium ion intercalates into the active material in the negative electrode (bottom).

3. Ions move downward inside the cell.

4. Electrons move clockwise within the cell and in the external circuit, though the charger.

While discharging, the cell is the primary force; it causes current to flow clockwise (Figure A.1(c)):

1. A lithium ion deintercalates from the active material in the negative electrode (bottom).

2. Simultaneously, a lithium ion intercalates into the active material in the positive electrode (top).

3. Ions move upward inside the cell.

4. Electrons move counterclockwise within the cell and in the external circuit, through the load.

A.3.1.3   The Process in Detail

Let's examine the charging process in some detail (Figure A.2(a)):

1. *Deintercalation:* In positive electrode (top), two items exit a particle of lithium oxide active material simultaneously: a Li+ ion and an electron; this leaves an empty spot in this first particle, which remains electrically neutral; we will follow the Li+ ion in part three and the electron in part four.

2. *Intercalation:* In negative electrode (bottom), two items enter a graphite particle simultaneously: a Li+ ion and an electron (these are not the same ones as the ones in part one); this fills an empty spot in this second particle, which also remains electrically neutral.

3. *Ion flow:* Inside the cell, the cloud of Li+ ions moves one step down (current flows one step counterclockwise):

   - A Li+ ion deintercalates from the first particle (discussed in item 1), leaving an empty spot.
   - In the positive electrode, that Li+ ion swims a bit down in the electrolyte mixed with the active material.
   - In the electrolyte in the positive electrode, Li+ ions move one step down.
   - A Li+ ion enters the separator from the top side.
   - Inside the separator, Li+ ions move one step down.
   - A Li+ ion exits the separator from the bottom side.
   - In the electrolyte in the negative electrode, Li+ ions move one step down.

FIGURE A.2
Li–ion processes:
(a) charging, and
(b) discharging.

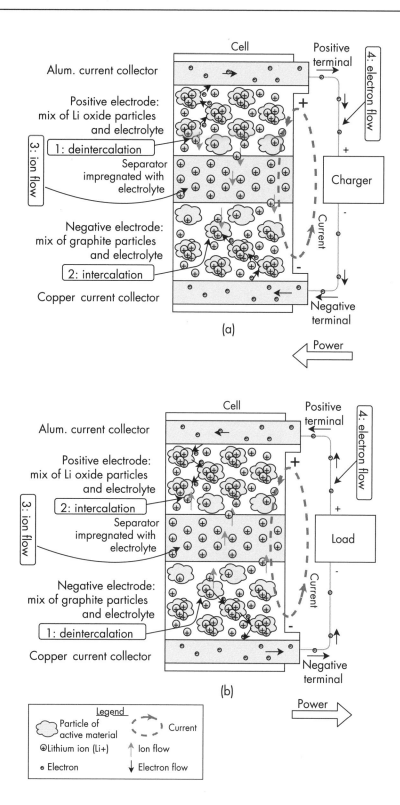

• A Li+ ion enters the second particle (discussed in item 2), the one with an empty spot.

4. *Electron flow:* Simultaneously, the cloud of electrons moves one step clockwise (current still flows one step counterclockwise):

• An electron exits the first particle (discussed in item 1).

- The released electron jumps upward and into the next particle of active material (the particles are conductive and are electrically connected).
- The particle receives that electron and releases a different electron, which jumps into the next particle.
- And so forth, upward, up to a particle sitting against the positive current collector.
- An electron jumps from that particle into the positive current collector (top).
- The electrons in the positive current collector move one step to the right.
- An electron exits the positive current collector and enters the cells' positive terminal.
- Another electron exits the cells' positive terminal and enters a wire.
- In the wire, the electrons move one step clockwise (current moves counterclockwise).
- Another electron exits the wire and enters the charger's positive terminal.
- Another electron exits the charger's negative terminal and enters a wire.
- In the wire, electrons move one step clockwise (current moves counterclockwise).
- An electron exits the wire and enters the cell's negative terminal.
- Another electron exits the cell's negative terminal and enters the negative current collector (bottom).
- The electrons in the negative current collector move one step to the left.
- Another electron exits the negative current collector and jumps into a particle of active material.
- Electrons jump from particle to particle, as described earlier.
- An electron enters to the second particle (discussed in item 2).

The discharging process is equal and opposite and uses a load instead of a charger (Figure A.2(b)).

The electrodes and current collectors in LTO cells are different, though the general process is the same.

An electron microscope video of the inside of a Li-ion half-cell [1] shows a shade growing inside the carbon electrode, starting from the separator and growing towards the back. This indicates that, while charging, a lithium ion swims through the electrolyte mixed with the active material and intercalates inside the first unoccupied spot it encounters; therefore, the spots closest to the separator are filled first. Similarly, while discharging, lithium deintercalates first from the active material closest to the separator.

A.3.2 Cell Chemistry

Please understand that I am not a chemist. I am well aware of what can happen when neophytes attempt to explain something they do not fully grasp. If you find the following discussion wildly preposterous, I apologize. Still, please, humor me and hear me out.

There is plenty of chemistry going on in the Li-ion cell electrolyte and the making of a Li-ion cell. Electrochemistry explains why the lithium ions behave the way they do in a Li-ion cell. However, the process of charging or discharging a Li-ion cell does not involve any chemical reaction, not in the neophyte's narrow understanding that a chemical reaction involves a material changing into a different material, and that ionization is a physical phenomenon more than a chemical one. Understandably, chemists bristle at this statement.

Charging and discharging a Li-ion cell involves the flow and settling of ions, which is something that chemists consider an electrochemical process. Chemists do see an electrochemical reaction in the electrodes because one electrode gains an electron while the other one loses one. When a Li+ deintercalates from an electrode, an electron leaves the electrode, which an electrochemist calls oxidation. Simultaneously, at the other end of the cell, when a Li+ intercalates the opposite electrode, an electron enters it, which an electrochemist calls reduction. The two, taken together, form a redox reaction. The increase in stored energy during charging and the corresponding decrease during discharging is due to the difference in the electrochemical potential of the two different materials in the two electrodes.

So, do charging and discharging involve a chemical reaction? There certainly is no chemical reaction in the lithium: positive lithium ions simply move from one side to the other, without ever being converted to anything other than Li+. According to one understanding of how Li-ion cells work, what starts as Li+ as it pops out of one side, remains Li+ as it crosses over, and remains Li+ as it settles on the other side. The only question is whether there is a chemical reaction inside the electrodes.

Chemists use the tools at their disposal[7] and consider the charging and discharging of a Li-ion cell as an electrochemical reaction[8]. As a nonchemist, I use the tools at my disposal and see this process simply as the flow of one material within other materials, all of which remain chemically unchanged: plain physics.

According to another understanding, the lithium ion gains an electron and becomes an elemental lithium atom inside an electrode; even so, this is still a simple physical process, akin to a person discharging a buildup of static electricity by touching grounded metal.

Nonchemists may not see much difference between the process inside a Li-ion cell and the one inside a forward-biased silicon diode[9]:

- In both cases, operation relies on the two terminals having different potentials.

- In both cases, one terminal loses electrons, and the other one gains electrons.

- In neither case is there any material change.

To a nonchemist, both a diode and a Li-ion cell operate on just plain physics. Yet, to a chemist, the operation of a Li-ion cell is electrochemistry, and the operation of a diode is physics.

Chemists tell us that the electrode composition changes during charge or discharge, such as $LiC_6 \rightleftarrows C_6 + Li^+ + e^-$ in a graphite negative electrode. I see lithium entering or exiting an electrode whose chemical composition remains unchanged, in the same way that adding sugar to water is not a chemical reaction because it does not change water to a different compound.

By analogy, charging a Li-ion cell is akin to me leaving my work cubicle, going up the stairs, and sitting at the break room's table. Discharging is like me getting up from the break room's table, going down the stairs, and sitting back in my cubicle.

---

7.   Abraham Maslow said: "…it is tempting, if the only tool you have is a hammer, to treat everything as if it were a nail."
8.   For example, $LiC6 + CoO2 \rightarrow C6 + LiCoO2$.
9.   I suppose that the difference between a Li-ion cell and a diode is that lithium ions travel in a Li-ion cell, while silicon nuclei do not move in a diode.

A chemist might say that the cubicle chair changed into a Cubicle_ChairMan compound when I enter it and back to Cubicle_Chair when I exit it; that the break room changed into a BreakRoom_ChairMan compound when I enter it and back to Breakroom_Chair when I exit it. I say that I am the same man in both cases; I just happen to be sitting in a different chair and have more potential energy when I am upstairs. Also, each chair is the same chair, whether or not I am sitting in it. It just happens to be compressed or relaxed, but it is still a chair.

Just like I sit in a chair without becoming part of that chair, the lithium ion fits in a nook in the electrode's lattice without becoming part of that electrode. The chair I'm sitting in compresses and loses some air to accommodate my fat butt, but it's still a chair. The electrode expands a bit and gains an electron to accommodate the Li+, but it's still the same material.

That is not to say that no chemical reactions take place in a Li-ion cell: on the contrary, they do. They are reactions that degrade the electrolyte and electrodes, resulting in cell aging, which manifests itself as a reduction in capacity and increase in resistance. These include:

- While charging: lithium plating over the SEI layer;
- While discharging: electrochemical stripping of any lithium plating;
- At all times, especially at high voltages and high temperatures: disassociation of the electrolyte.

A.3.3 Lithium-Metal/Solid-State Secondary Cells

Besides Li-ion, other rechargeable cells use lithium, although they do not meet the definition of a Li-ion cell (see Volume 1, Section 2.1.3) (Figure A.3(a)) in one or more respects:

- At least one electrode does not use lithium intercalation.
- They contain bulk lithium metal.
- They use a solid electrolyte.

Unlike a Li-ion cell (Figure A.3(a)) these cells use bulk lithium metal on the negative terminal (Figure A.3(b)) may use a solid electrolyte (Figure A.3(c)), or both (Figure A.3(d))[10].

At the time of this writing, these cells are not quite ready for commercial adoption due to cost, performance, and sometimes safety issues. Some manufacturers have ceased to exist or at least stopped offering these cells. That is why I only discuss them in this appendix.

Just like Li-ion cells, lithium metal cells do require the protection of a BMS to monitor each cell to ensure that it's operated within its safe operating area.

A.3.3.1 Lithium Metal

A healthy Li-ion cell does not contain any bulk lithium metal; it only contains loose lithium ions. Conversely, a lithium metal cell does contain bulk lithium metal.

In a lithium metal cell, intercalation occurs only in the positive electrode. In the negative electrode, the process is different:

- *Discharge*: Lithium is stripped from the lithium metal.

---

10. Thank you to Dean Frankel of Solid Power for help in this section.

FIGURE A.3
Cell composition:
(a) Li-ion, (b) lithium
metal, (c) solid electrolyte,
and (d) all solid-state.

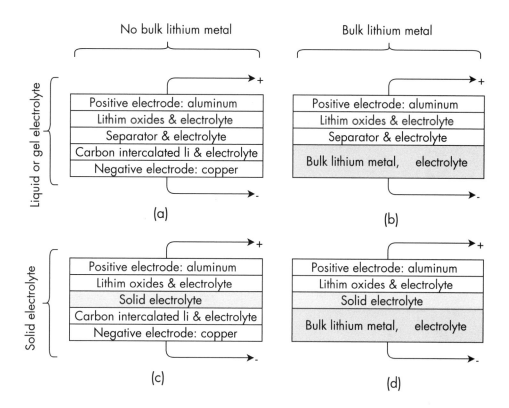

- *Charge*: Lithium is plated back onto the lithium metal.

The lithium metal is the current collector; there is no separate current collector.

Compared to Li–ion, a lithium metal cell promises to increase the active material (about twice as much), and therefore twice the specific energy (2~4 MJ/kg) (see Volume 1, Section 2.5.6). However, the volumetric energy density is not impressive.

The limiting factor is that these cells still use the same positive electrode as Li–ion cells. Cell scientists talk about specific capacity rather than specific energy, so let's do the same here:

- Specific capacity of lithium metal: 3.8 Ah/g;
- Specific capacity of NMC positive electrode: 0.2 Ah/g typical, up to 0.4 Ah/g maximum.

As you see, the capacity of a solid lithium cell is severely limited by the NMC electrode. In the future, more exotic materials for the positive electrode may result in a higher specific capacity.

There are serious safety issues with lithium metal cells, including:

- Lithium metal melts at only 180°C.
- Lithium may ignite spontaneously in air.
- Lithium may reignite after a fire is extinguished as lithium reacts with water to form hydrogen and lithium hydroxide [2].

Solving these issues has proven very challenging, especially with liquid electrolytes, which is one reason that lithium metal cells are still under development.

Lithium metal rechargeable cells include Li–S (lithium-sulfur) cells with a liquid electrolyte. Compared to Li–ion, these cells have very high specific energy, which makes them very promising. However, they have low power density, terrible cycle life, low voltage, and instability at high temperatures. Winston Batteries briefly offered Li-S cells.

A.3.3.2   Solid Electrolyte

An electrolyte is a material that has high ionic conductivity yet low electronic conductivity. The electrolyte in a Li-ion cell is what burns in case of a thermal runaway. It decomposes rapidly when the cell is charged to a high voltage.

Li-ion cells use an organic liquid or a gel electrolyte. By this definition, cells that use a solid electrolyte are not Li-ion cells. Among other materials, a solid electrolyte may be a polymer.

A solid electrolyte (also known as fast ion conductor) can have many advantages:

- It promises to eliminate the possibility of a thermal runaway.
- Depending on the chemistry, it may be explosion-proof.
- It may minimize cell degradation in a cell kept at a fully charged state.
- Compared to a Li-ion cell, it has a much lower self-discharge current.

A separator is required in a Li-ion cell. Inherently, a cell with a solid electrolyte does not need a separator because the solid electrolyte is the separator.

A.3.3.3 All Solid-State Cells

These cells use both a lithium metal negative electrode and a solid electrolyte. Solid Power is developing an all-solid-state pouch cell with a lithium alloy negative electrode and an NMC positive electrode. Compared to a Li-ion NMC cell, the manufacturer claims that these cells have:

- Similar voltage range and Coulombic efficiency;
- Much lower self-discharge current
- Significantly lower power density: 1 C maximum, and an MPT (see Volume 1, Section 1.5.2) on the order of 300 seconds;
- Operation limited to higher temperatures:
  - Poor performance under 25°C;
  - Sweet spot temperature of 70°C;
  - Max temperature of 150°C.
- No hazard from overcharging (though damage does occur);
- Inherently no thermal runaway.

While standard Li-ion cells use an organic electrolyte, which is flammable, these all-solid-state cells do not, and are therefore safer, at least in this one respect.

On the other side, unlike Li-ion cells, these all-solid-state cells contain lithium metal. The manufacturer claims that lithiated carbon, which may be found in a Li-ion cell other than LTO, is more reactive than their lithium metal electrode.

These cells pack even more energy than standard Li-ion cells. If their terminals are shorted together, all the energy is discharged, heating the cell to a high temperature. However, these cells have significantly higher internal resistance, higher than even the worse Li-ion cell, which limits the short circuit current; yes, all their energy ends up heating the cells, but over a longer period. Therefore these cells, when short-circuited, reach a lower temperature than an unprotected Li-ion cell; although a standard Li-ion cell is protected[11] (unlike an all-solid-state cell) and won't fully discharge through a short circuit.

In conclusion, these all-solid-state cells appear to be safer in some respects, not in other respects.

Lithium metal polymer (LMP) cells are also all-solid-state cells. Avestor in Quebec, Canada, developed a lithium metal polymer cell that used a lithium polymer solid electrolyte. French EV manufacturer Bolloré acquired the assets of defunct Avestor; now its BatScap subsidiary manufactures those batteries for its BlueCar [3]. Other automotive manufacturers are researching this technology. Other companies developing such cells include Pellion Technologies, Sion Power, PolyPlus, Solid Energy Systems, and Ion Storage Systems [4].

Lithium-sulfur cells with a solid electrolyte are another example of an all-solid-state cell.

A.3.3.4 Lithium-Air

This cell uses a reduction of oxygen in the positive terminal. This cell is under development and promises energy densities comparable to petrol fuel. The problem to overcome is that the electrode tears itself apart when lithium ions intercalate inside of the silicon.

A.3.4 Li-Ion Capacitors (LIC)

A Li-ion capacitor (LIC) is a high-voltage Electric Double Layer Capacitor (EDLC, ultracapacitor). Doping one of the electrodes with lithium ions increases the operating voltage, compared to standard EDLCs; the higher voltage increases the energy density. This component is a capacitor, not a battery cell, because charge is stored electrostatically, not electrochemically. LICs are useful in applications that require high-speed charging and discharging (on the order of 1 second[12]), such as for voltage support in a high-voltage transmission line (see Section 4.3). The power characteristics of an LIC equal or exceed those of either a Li-ion cell or an EDLC (Table A.1).

As in any capacitor, the voltage of an LIC drops linearly as it's discharged (Figure A.4(a)). Therefore, a DC-DC converter is required between an LIC and a constant voltage bus (Figure A.4(b)).

Unlike a capacitor (and like a Li-ion cell), an LIC has a minimum voltage; note how the linear curve for the LIC stops at around 2.8V. Therefore, the majority of the charge in an LIC is not accessible. However, this is OK, because:

- Most of the energy in a capacitor is at the highest voltages; not much energy is unused by not having access to the lower voltages in an LIC.

- An LIC is used in buffer applications; what matters is the power, not the energy.

LICs are available in pouch and small prismatic formats[13].

A.3.5 EIS and Nyquist Plots

Researchers use electrochemical impedance spectroscopy (EIS) to create a Nyquist plot (Figure A.5), which is a representation of the amplitude and phase of the cell

---

11. . The separator is engineered to melt and stop ion flow at a high temperature.
12. Note that we can't use "C" for specific current because an LIC is not a battery and does not have a capacity (in Ah).
13. JSR Micro Ultimo series; jsrmicro.be.

TABLE A.1
Comparison of Li-Ion
Capacitor to a Li-Ion Cell
and to a Super-Capacitor

	Li-Ion Cell	Li-Ion Capacitor	EDL Capacitor
Power density	Low to medium	Similar to EDLC	Very high
Energy density	High	Slightly more than EDLC	Very low
Capacity or capacitance	1~1,000 Ah	1~3 kF	1F~1kF
Voltage [V]	~2.5~4.2	2.2~3.8	0~2.5
Voltage vs. SoC	Relatively flat	Linear decrease	Linear decrease
Temperature [°C]	5~45 (charging)	–30~70	–40~70
Self-discharge	Very low	High at 3.8V, very low at 3V	High
Thermal runaway	Yes	No	No

FIGURE A.4
LIC: (a) voltage,
and (b) usage.

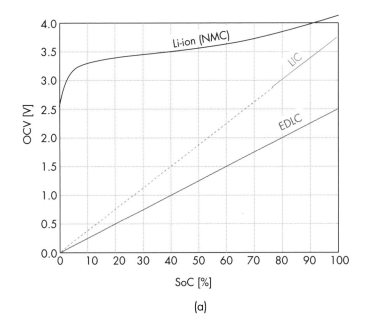

(a)

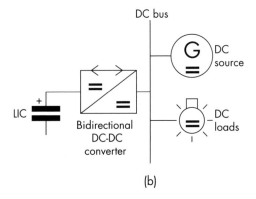

(b)

impedance over a wide range of frequencies. The highest frequency is on the left and the lowest frequency is on the right. DC cannot be shown in a Nyquist plot because it is an infinite distance past the top right corner of the plot. The X-axis is the real part: resistance. The Y-axis is the imaginary impedance: reactance; it uses negative values because cell designers mirror the Nyquist plot vertically.

The plot has two or three semicircular humps and a long upward slope on the top right. The shapes mean different things to different scientists:

FIGURE A.5
Nyquist plot of a Li–ion
cell's AC impedance.

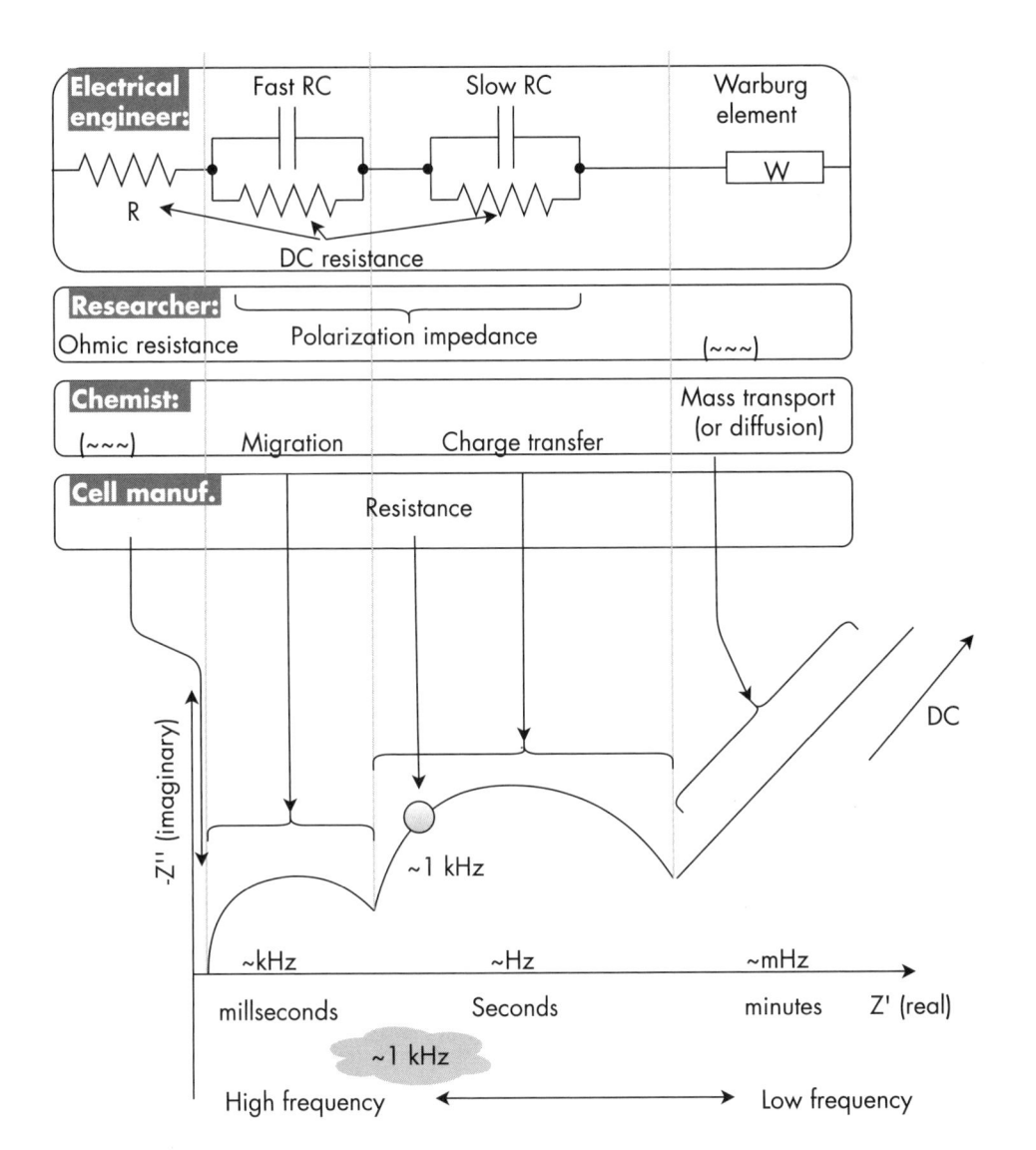

- To a chemist, these indicate particular processes in the cell; specifically, the slope indicates the relatively slow diffusion of ions.
- To a researcher, the impedance elements are divided into:
  - R: Ohmic resistance; affects the behavior at DC; due to:
    - Resistance in electrodes and separator;
    - Effective resistance in the electrolyte, excluding diffusion processes.
  - All others: polarization impedance; affects the behavior at AC; due to:
    - Effects on the surface of the electrodes: double-layer capacitance, reaction kinetics;
    - Diffusion processes in the electrolyte.
- To an electrical engineer, these indicate circuit elements in the expanded Randles circuit [5] equivalent model:
  - R for the horizontal displacement from the zero point;
  - Two RC tanks for the humps; R = resistor, C = capacitor (more accurately, a constant phase element (CPE) rather than a capacitor);

• A Warburg element for the slope on the right.

For a standard LCO cell[14], comparing the ohmic resistance and real part of the polarization impedance (polarization resistance) reveals that [6]:

• The ohmic resistance is dominant, except for a degraded cell, where the polarization resistance dominates at low SoC.
• Both increase with cell degradation, especially rapidly at the end of life, although ohmic resistance increases far more.
• Ohmic resistance is relatively constant over SoC, while polarization resistance increases significantly at the low and high ends of the SoC.

Chemically, a Warburg element[15] models semi-infinite linear diffusion. Electrically, it is a constant phase element (CPE) that adds a 45° phase shift independent of frequency. It appears as a line with a slope of $-1/2$ in a Bode plot. A Warburg element cannot be implemented using standard passive electronic components. It can only be emulated in software.

A.3.5.1 *Resistance*

This term means different things to different people:

• Cell manufacturers: AC impedance at 1 kHz; this could be inside the big hump, or in the dip to the left of it, or in the small hump. Therefore, the impedance at 1 kHz could be anything; it certainly is not DC resistance; indeed, the impedance at 1 kHz is entirely unrelated to the DC resistance.
• Researchers: The ohmic resistance is the short distance between the Y-axis and the beginning of the first hump.
• Electrical engineers: The "almost" DC resistance, in the time scale of seconds to 1 minute.

A.3.6   COULOMBIC EFFICIENCY

The Coulombic efficiency of a Li-ion cell is nearly 100%. One theory[16] is that Coulombic efficiency and capacity fade are tightly related. In rough terms, each time a cell is charged or discharged, not all lithium ions are converted to a stored charge. The ones that aren't, block a site in an electrode, making it unavailable to store charge, decreasing the capacity. Based on this understanding, over a charge cycle, if 1% of the ions fail to intercalate, then the Coulombic efficiency is 99%, and, after every cycle, the capacity is reduced by 1%.

Let's use this understanding to estimate Coulombic efficiency based on the capacity fade that is reported by the cell manufacturer. In the long term, we see that the capacity of a cell is reduced by 10% to 20% over a large number of cycles. Let's take a 20% reduction and 3,000 cycles as typical numbers. Then the reduction in capacity each cycle is:

$$\eta = e^{\wedge}\left(\ln(80\%/3{,}000)\right) = 0.9999256$$

14. Moli Energy ICR18650H.
15. Named after Emil Gabriel Warburg.
16. Based on the work of Professor Jeff Dahn of Dalhousie University.

For every 10,000 electrons that move in and out of a battery, 10,000 lithium ions cross the cell and reach the opposite electrode. Of those, 9,999 ions intercalate and 1 fails to do so, reducing the capacity to 9,999/10,000, or 99.99%.

Therefore, the typical Coulombic efficiency of this Li-ion cell would be 99.99%. In reality, people have measured a Coulombic efficiency as low as 98% [7], which would imply that those cells have a cycle life of only 10 cycles. Otherwise, the understanding that Coulombic efficiency and capacity fade are intimately connected is incomplete: there's more to Coulombic efficiency than capacity loss.

A.3.7   Self-Discharge Current Measurement

It's impossible to measure the self-discharge current directly because it occurs inside the cell, although it can be evaluated indirectly:

• The cell voltage is monitored accurately and for a very long time.
• The voltage drop over that time is measured.
• It is translated to a loss of charge by using the cell's OCV versus the SoC curve.
• Given the cell capacity, this is converted to self-discharge current.

Alternatively, the cell is charged at a current that matches the self-discharge current, adjusted by a servo loop to keep the cell voltage constant. The test must last for a month or so to overcome limitations in the measurements.

A.3.8   ENERGY MEASUREMENT

For an accurate derivation of the energy stored in a cell, the power during a full discharge cycle is integrated. Discrete integration is used, either empirically or graphically.

Empirical measurement of energy requires a sample of the battery and a data acquisition system:

• Charge the cell fully.
• Clear a running total for the energy.
• Discharge the cell slowly (say, at about 0.1 C, over about 10 hours).
• Once a second, measure the voltage and the current, calculate the power (in W), and add it to the running total.
• When the cell is empty, stop.
• The running total is the energy, in W-s.
• Divide by 3,600 (seconds in an hour) to convert to Wh.

Graphic estimation of energy requires an accurate plot of voltage versus the SoC at a low discharge current (Figure A.6):

• Measure the area under the curve, from 100% SoC to 0% SoC, and down to 0V.
• The plot has units of V (vertically) and Ah (horizontally), so the area has units of V-Ah or Wh.
• The area under the curve is the energy in the battery, in Wh.

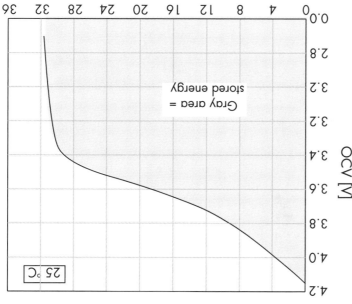

FIGURE A.6
Estimating energy
graphically.

A.3.9   WHAT SHOULD BE IN THE SPECIFICATIONS

From the battery designer's standpoint, the amount of information in today's typical specification sheet is somewhat useful but not quite sufficient for an accurate analysis when choosing a cell and an accurate model when using it.

If the cell manufacturer provided more extensive and useful data in the first place, the battery designer would have a much better starting point. Such data could include:

- Maximum power time (or DC resistance) at 50% SoC, 25°C, at 1 C charge and discharge;
- Maximum power time (or DC resistance) over SoC, temperature, current, calendar time;
- Thermal resistance to case and terminals, from the hottest point inside the cell;
- Self-discharge current versus temperature.
- Permanent DC resistance gain and capacity loss from extreme conditions or events at various temperatures, SoC levels;
- For pouch and prismatic cells, information on mechanical cell retention, including required pressure and resilience of retaining plates;
- Low-frequency electrical model and equivalent component values over SoC, temperature, and current;
- Suggested maximum charging current over internal cell temperature (including cold charging) for various levels of permanent degradation of the cell (resistance gain, capacity loss);
- Suggested maximum discharging current over internal cell temperature for various levels of permanent degradation of the cell (resistance gain, capacity loss);
- Static OCV versus chemical SoC.

The data should be presented in tables as well as in charts.

A.3.9.1 The Ideal Specification Sheet

The ideal specification sheet for a Li-ion cell should contain the items shown in Figure A.7.

- Manufacturer name and model number;
- Chemistry and discharge curves at various rates;
- Charging and discharging conditions, recommended, fast and cold;
- Cycle and calendar life and self-discharge rate;
- Mechanical dimensions and mass, including thread sizes and terminal spacing.

Separately, the manufacturer should also provide:

FIGURE A.7 Ideal specification sheet.

BestCell
McMurdo Station, Antarctica

High energy density
LiFePO4
large prismatic cell

BC80AH

Recommended CCCV charging		
Maximum current	-50	A
Constant voltage	3.6	V
Temperature	15~45	°C
Termination current	5	A

Fast charging		
Maximum current	200	A
Maximum voltage	3.4	V
Temperature	15~45	°C

Cold charging		
Maximum current	2	A
Constant voltage	3.4	V
Temperature	-10~15	°C

Life		
Cycles (1) (2)	2500	
Calendar (2) (3)	60	mo
Self-discharge (3)	< 3 %	mo

Recommended discharging		
Maximum current	100	A
Cut-off voltage	2.7	V
Temperature	0~60	°C

Fast discharging		
Maximum current	400	A
Cut-off voltage	2.7	V
Temperature	15~60	°C

Cold discharging		
Maximum current	50	A
Cut-off voltage	2.0	V
Temperature	-20~15	°C

Resistance		
DC resistance (3)	12	mΩ
Maximum Power Time	245	s

Mechanical		
Dimensions	250 × 132 × 64	mm
Mass	3.2	kg

1) Recommended charge discharge
2) To 90 % Ah
3) At 50% SoC, 25 °C

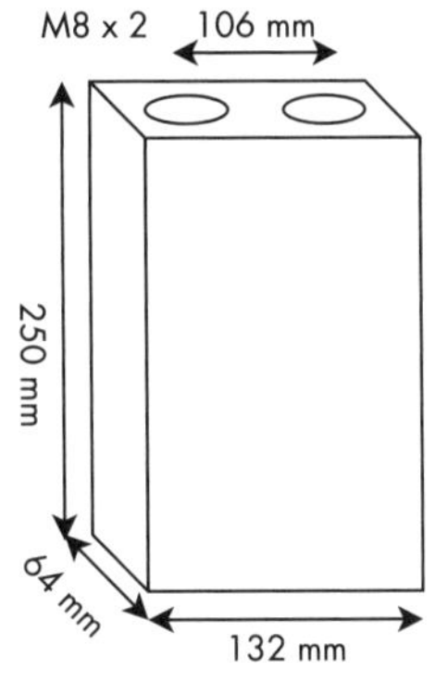

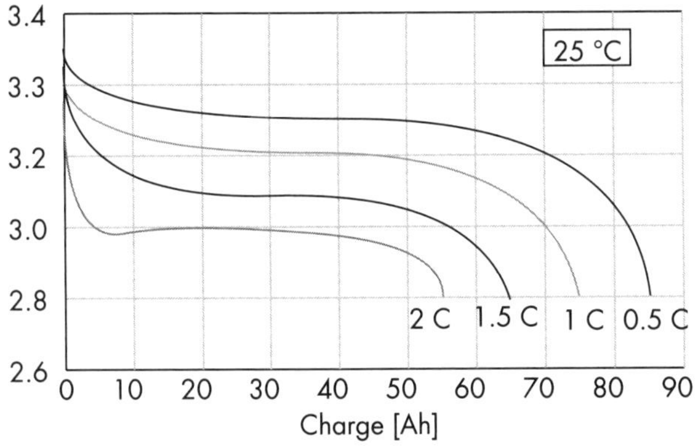

- Safety (MSDS) and transportation certifications;
- Discharge plots at low temperatures;
- Temperature plots during discharge at various discharge rates.

A.3.10 Li-Ion Cell Manufacturer Directory

Table A.2 has a partial list of manufacturers of Li-ion cell manufacturers.

A.4 CELL ARRANGEMENT

These items expand on Volume 1, Chapter 3.

A.4.1 Self-Balancing Li-Ion Myth

There is a myth that some Li-ion chemistries are self-balancing, and therefore a series string using such cells does not require any external balancing. One source of this myth is a famous man behind the Chinese LFP industry. In the early years, he would insist that his company's cells needed no balancing because of their "secret sauce" of rare-earth elements. He would say that if these LiFePO4 cells were charged to 4.0V, they would self-balance. Later, the story changed: "balance" meant the chemical balance inside an individual cell, not the SoC balance among cells in series. At the same time, the maximum voltage in the specs for those cells changed from 4.0V down to 3.8V and eventually down to 3.6V.

A later source of this myth is a hobbyist who noticed that some batteries for power tools did not do balancing. He assumed that this proved that the cells must be self-balancing. For example, older battery packs for Makita power tools reportedly did not do any balancing. I imagine that, by the time they got significantly out of balance, their internal resistance may have increased to the point that the power tool was too weak. It is interesting to note that newer Makita packs do have balancing wires[17].

A rumor is that Makita batteries used LMO cells[18]; given that the battery did not do any balancing, some concluded that LMO cells must be self-balancing. Some have even proposed pairing LMO cells with LCO cells so that the LMO cells would balance not only themselves but also the LCO cells [8].

The next step of the rumor was that the modules in the traction battery for a Nissan Leaf electric vehicle must also be self-balancing because they use LMO[19]. A test was performed, which showed that cells that started top-balanced, were discharged, and then recharged, had their voltages converge back [9]. This is perfectly normal (see Volume 1, Section 3.2.6.1). That was seen instead as proof that LMO cells are self-balancing[20]. It isn't. Indeed, the Nissan Leaf battery does do balancing.

The rumor mill suggests that the Coulombic efficiency (see Volume 1, Section 2.5.4) of LMO cells drops at high voltage, which would result in a low level of self-balancing for LMO cells while charging. If it is true that, at high voltage, the CE drops, then the most charged cell would waste more power into heat, and would not charge as fast, allowing the others to catch up. Again, as far as I can tell, this is just an internet rumor [10]. I cannot find any scientific source stating that the CE of an LMO cell is voltage-dependent.

---

17. Lesson learned?
18. It doesn't. It appears to be using Sony US18650V3 cells, which are not LMO.
19. They do: they use a dual chemistry that included LMO.
20. It only proved that the cells did not get noticeably out of balance after only one cycle.

Company	Location	Chemistry	Small cyl.	Large cyl.	Small pr.	Large pr.	Pouch
A123	China	LFP	✓				✓
Advanced Battery Factory	China	LCO					✓
Advanced Electronics Energy	China	LCO					✓
Altair Nanomaterials	China	LTO					✓
Amperex (ATL)	China	LCO		✓		✓	✓
Automotive Energy Supply (AESC)	Japan	NMC/NMO					✓
B&K Technology	China	LCO					✓
BAK	China	LCO	✓				✓
Bestgo	China	LFP, NMC				✓	✓
BYD	China	LCO	✓				
CALB	China	LFP				✓	
EAS	Germany	NCA		✓		✓	
EiG	South Korea	NMC LMO LFP	✓				✓
Electrovaya	Canada	LCO					✓
Enerdel	USA						✓
Enertech	China	LCO, LFP					✓
EVPST	China	LFP				✓	
Full river	China	LCO	✓				✓
GBS	China	LFP				✓	
GEB	China	LCO	✓				✓
Great power	China	LCO, LMO	✓		✓		✓
GS Yuasa (see Lithium Energy)	Japan						
Harding Energy	USA	LCO, LFP	✓				✓
Hangzhou LIAO Technology	China	LFP			✓		
Headway	China	LFP		✓			
Heter	China	LFP, LCO	✓				
Hitachi	Japan	LCO	✓				
HECO	China	LCO					✓
HYB	China	LCO	✓		✓		
Hyper Battery	China	LCO, LFP					✓
Jiangsu Frey Battery Technology	China						
K2 Energy	USA	LFP	✓				
Kokam	China	NMC, LCO					✓
LG	So Korea	LCO	✓				
Lishen	China		✓		✓		✓
Liotech	Russia	LFP				✓	
Lithium Energy	Japan				✓		
Lyno Power	China	LFP			✓		
Lumos	China	LFP, NMC	✓				✓
Maxwell	Japan	LCO			✓		
Molicel	Canada	LCO	✓		✓		
Panasonic	Japan	LCO	✓		✓		
PSI	China						

TABLE A.2   Li-Ion Cell Manufacturers

TABLE A.2 (CONTINUED)

Company	Location	Chemistry	Small cyl.	Large cyl.	Small pr.	Large pr.	Pouch
RealForce	China	LFP NMC LCO	✓				
Renata	Switzerland	LCO					✓
SAFT	France	LFP			✓		
Samsung	So Korea						
Sanyo (now Panasonic)	Japan		✓				
Sequence	Taiwan	LCO	✓				✓
Sincpower	China	LFP, NMC	✓				✓
Sinopoly	China	LFP				✓	
Sony (now Murata)	Japan						
Tenergy	USA	LCO	✓				✓
Toshiba	Japan	LTO		✓			
Winston	China	LFP				✓	
Xiamen	China	LFP				✓	
Xinchi (See Syncpower)							
Yok Energy	China	LCO					✓
With smaller Chinese companies, it is hard to distinguish actual manufacturers from resellers that rebrand cells manufactured by others.							

For a cell to be self-balancing, rather than having a voltage-dependent CE, it would need a voltage-dependent self-discharge current (see Volume 1, Section 2.7.3). Well, as it turns out, it is. Self-discharge is higher at high SOC levels.

You could fully charge an unbalanced string and then leave it sitting on a shelf. After a year, the string will be at a lower SOC (due to self-discharge) and will still be imbalanced, but the degree of imbalance will be somewhat less. In this sense, the string balanced itself, but not top-balanced, not completely, and it took a year.

A.4.2 Current in Each Cell in a Parallel Block

If the cells in a parallel block have different internal resistance, the current is divided among them according to their resistances.

A.4.2.1 Constant Voltage

If a block of cells in parallel is charged at a constant voltage, the current is divided unequally among the cells. Knowing the voltage applied to the block, and the OCV and resistance of each cell, you may calculate the current in each cell:

$$In = (Vb - Vn) / Rn \qquad (A.3)$$

where n = number of a cell, In = current in cell n [A], Vb = bank voltage [V], Vn = Open Circuit Voltage of cell n [V], and Rn = resistance of cell n [Ω].

A.4.2.2 Constant Current

If a block is charged at constant current, then we must first calculate the block voltage:

$$RP = \Pi \, \text{IN} \, R \, (n) \tag{A.4}$$

and

$$Vb = (1 * \textstyle\sum_i^N Vn * RP / Rn) / \textstyle\sum_i^N RP / Rn \tag{A.5}$$

where Rp = product of all cell resistances [Ω], N = number of cells in the block, n = number of a cell, Rn = resistance of cell n [Ω], Vb = bank voltage [V], I = bank current [A], and Vn = open circuit voltage of cell n [V].

For example, for four cells:

$$Vb = (1 * R1 * R2 * R3 * R4 + V1 * R2 * R3 * R4 + V2 * R1 * R3 * R4 + V3 * R1 * R2 * R4 + V4 * R1 * R2 * R4) \, / \, (R2 * R3 * R4 + R1 * R3 * R4 + R1 * R2 * R4 + R1 * R2 * R3) \tag{A.6}$$

or, equivalently:

$$Vb = (1/R1 + 1/R2 + 1/R3 + 1/R4) + V1/(1/(R2 + 1/R3 + 1/R4) + V2 \, / \\ (1 + R2* (1/R1) + 1/R3 + 1/R4) + V3 \, / \, (1 + R3*(1/R1 + 1/R2 + 1/R4)) \\ + V4 \, / \, (1 + R4* (1/R1 + 1/R2 + 1/R3)) \tag{A.7}$$

Now that we have the block voltage, we can calculate the individual currents as shown in the previous section.

A.4.3 Kim Arrangement

The series-first arrangement has the real disadvantage of requiring many more BMS sense inputs (see Volume 1, Section 3.5.1). The parallel-first arrangement has the debatable disadvantage of needing a fuse in series with each cell (see Volume 1, Section 3.3.12).

A group of researchers [11–13] proposed one of the strangest circuits I have seen to try to overcome these real and debatable disadvantages. The circuit is similar to a parallel-first arrangement, except that it uses resistors (Figure A.8(a)) or fuses (Figure A.8(b)) between columns of cells. Although these circuits are presented as fail-safe, they can actually be downright dangerous.

These arrangements have three goals:

1. If a cell shorts out, it does not draw excessive current.
2. Allow cells in a row to equalize (in case they have different internal resistance or different SoC).
3. Avoid the need for a BMS with $S \times P$ sense inputs.

These arrangements do achieve goal 1: a shorted cell won't discharge the cells in the same row, although we saw that, in some cases, this may be a solution in search of a problem.

Goal 2 is not quite achieved in the resistor version: cells are equalized under steady conditions. However, when the battery current is variable, it takes too long for the cells to equalize. Therefore, a weak cell in a string is not supported by its buddy; instead the whole string acts weak. This arrangement is barely better than the series-first arrangement. The fuse arrangement does achieve goal 2, but the size of the fuses is an issue:

FIGURE A.8
Kim arrangement: (a) with
resistors, and (b) with fuses.

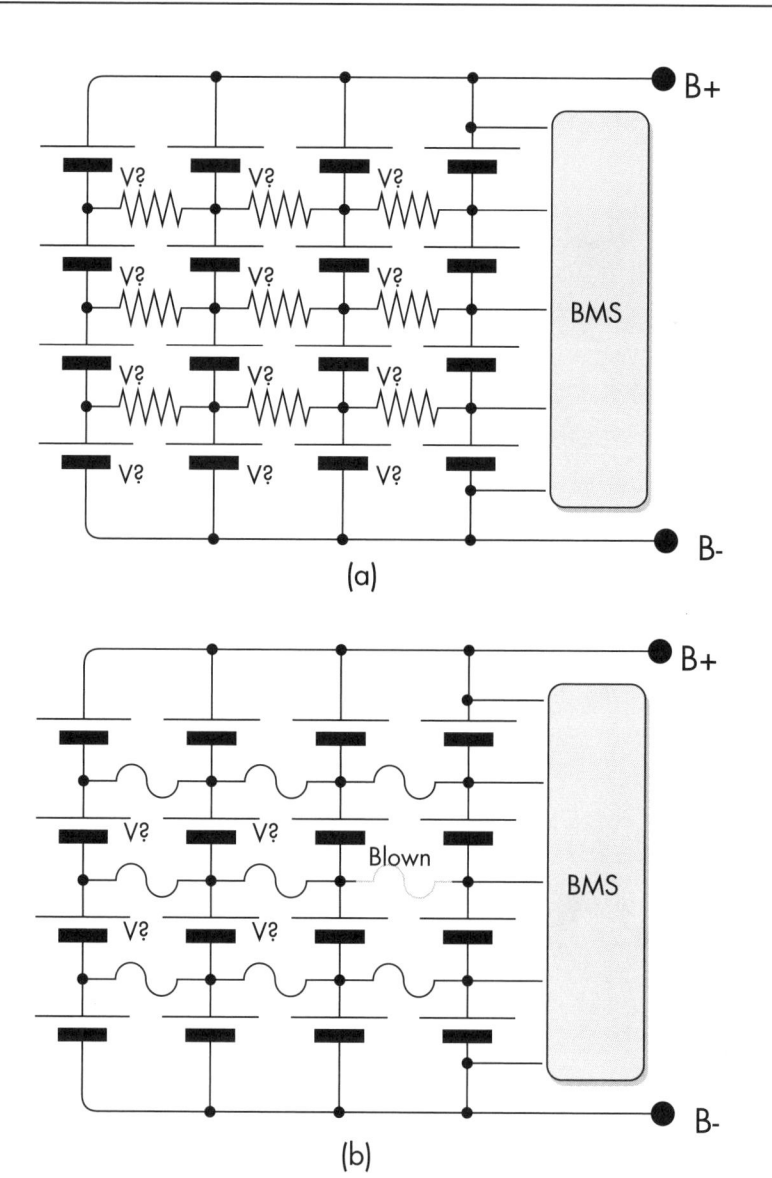

(a)

(b)

- If the fuses are rated for less than the total load current, if each string has a weak cell but in different rows, the total load current blows up the fuses; this makes this circuit no better than the series-first arrangement.

- If the fuses are rated for more than the total load current, they blow if a cell becomes a hard short; however, it won't if the cell develops a soft short; this makes this circuit no better than the parallel-first arrangement.

Use of a BMS with S taps monitoring only one string achieves goal 3. However, it is dangerous because the BMS doesn't know the voltage of many cells:

- Resistor version (Figure A.9(a)): The BMS doesn't know the voltage of the other strings; using resistors from each junction between cells to average the voltage of all the cell doesn't work either (see Section 3.5.1).

- Fuse version (Figure A.9(b)): If a fuse blows, the BMS doesn't know the voltage of the cells past that fuse.

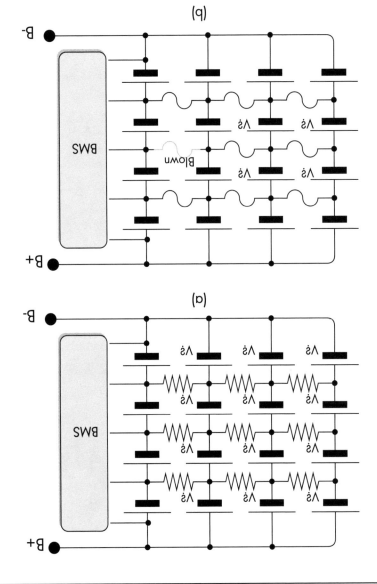

FIGURE A.9
BMS with Kim circuit:
(a) resistor version,
and (b) fuse version.

Ironically, this circuit's attempt at making a battery fail-safe is actually a safety hazard.

The complete version of the fuse circuit includes contactors at the end of each string (Figure A.10(a)).

Assuming that a cell failing as a complete short is a problem that needs a solution, then this circuit can certainly isolate the shorted cell:

- If the bottom cell, the BMS opens the switch in series with it (Figure A.10(b)).
- If the next cell up, the bottom fuse blows, and the BMS opens the switch (Figure A.10(c)).
- In the next cell up, all the fuses beneath it blow, and the BMS opens the switch (Figure A.10(d)).

Therefore, regardless of where the cell is, the fuses between the cell and the nearest battery terminal blow and the BMS opens the last connection, where there is no fuse.

For P parallel strings of S cells in series, the fuse version uses $(P-1) \star (S-1)$ fuses (Figure A.11(a)). Contrast this to the parallel-first arrangement, which uses $P \star S$ fuses (Figure A.11(b)). For example, in a 4P6S circuit, the Kim circuit uses 15 fuses while parallel-first uses 24 fuses. Yet the parallel-first arrangement is safe because when a fuse blows, it completely isolates the shorted cell all by itself. Also, a row in the parallel-first arrangement is completely symmetrical (each cell is treated exactly like any other cell in the row). In my view, this is well worth a few extra fuses.

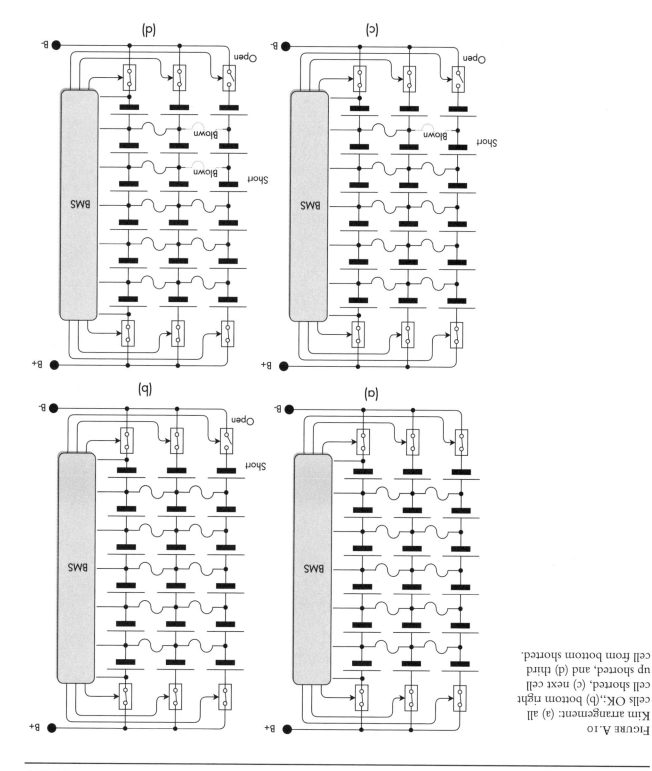

FIGURE A.10
Kim arrangement: (a) all cells OK.;(b) bottom right cell shorted, (c) next cell up shorted, and (d) third cell from bottom shorted.

FIGURE A.11
4P6S arrangement: (a) Kim
circuit, and (b) parallel-
first with fuse-per-cell.

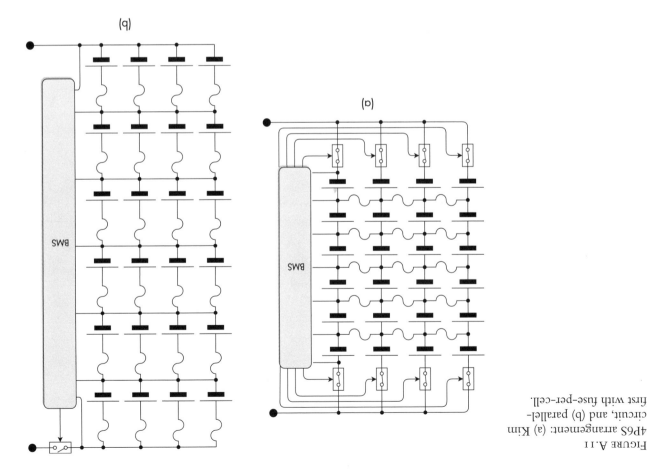

(a)

(b)

Of course, the way to make the Kim circuits safe is to use a BMS with S * P taps,
which is as expensive as the series-first arrangement. At this point, you might as well
use a series-first arrangement.

A.5 LI-ION BMS

These items expand on Volume 1, Chapter 4.

A.5.1 Charge Transfer Balancing Electronics

Charge transfer balancing uses electronic circuits to transfer energy among cells. Here
we go into some detail on how it is done and how it should not be done.

A.5.1.1 Magnetics

The DC-DC converters in balancing circuits use magnetics (inductors or transformers)
to convert the voltage of a cell to a current and then back to a different voltage, to be
applied to another cell, a string, or a bus. They may operate in the reverse direction.
This is the proper technique for transferring energy between two voltage sources
(remember that cells and strings are voltage sources). It is appropriate because:

- It can control the current in or out of a cell to the desired level; there are no
  current surges.
- It can transfer energy from one voltage to another voltage.
- It may provide isolation between the two voltage sources if so required.

For example, for cell-to-cell balancing:

- An inductor is connected across the most charged cell, for a specific duration.
- The current in the inductor increases linearly, until it is disconnected; an amount of energy is slowly transferred from the cell into the inductor and stored there.
- The inductor is disconnected from the first cell and connected to a lower voltage cell, in the opposite direction.
- The current in the inductor decreases linearly, until it drops to 0, while energy is slowly released from the inductor into the second cell.

A.5.1.2 Shared Transformer

A simple implementation of the string-to-cell balancing topology uses a single transformer with as many secondary windings as there are cells; these windings power DC-DC converters, one for each cell; the output voltage of all converters is the same. Only the more discharged cells take current from their DC-DC converter. As those cells are charged, the output of all the DC-DC converters rises. When the output of a DC-DC converter reaches the voltage of its cell, it starts charging it as well. In the end, all the cells are fully charged.

This approach works well enough for top balancing using the cell voltage algorithm.

A.5.2 Ineffective Charge Transfer Balancing Techniques

The following approaches to charge transfer balancing do not work as intended.

A.5.2.1 Single DC-DC Converter

Some solutions use a single DC-DC converter and $2 \times N$ relays to connect its output to one of N cells. This circuit is terrifying because it's all too easy for a relay to misbehave and short out an entire section of the battery. In any case, $2 \times N$ relays suited for the job cost more than N DC-DC converters.

A.5.2.2 Capacitors

Many have suggested using just capacitors to balance a string: connect a capacitor to a cell and then to another cell; repeat. Eventually, the voltages will match. This works to some degree, although it harms the cells, is at best 50% efficient, and requires infinite time to balance completely.

Both cells and capacitors are voltage sources. Connecting two voltage sources in parallel is the electrical equivalent of dividing by zero; bad things happen. An ideal cell, an ideal capacitor, and an ideal switch all have zero series resistance. When first connecting the two, the current is infinite:

$$\text{Current} = \text{Voltage difference} / \text{series resistance} = 1V / 0\Omega = \infty \text{ A} \qquad (A.8)$$

In reality, cells, capacitors, and switches do have nonzero series resistance, so the current is large but finite. That current generates heat in the series resistances; the energy in this heat is equal to the energy transferred, regardless of the value of the series resistance. In other words, half the energy goes into heating the cell, the capacitor, and the switch. Half is transferred between cells. That's in the best case. In reality, the transfer is even less efficient.

The current may be 100 C (though it lasts for a very short time). This high current exceeds the specified maximum charge current for the cells and degrades them. During this pulse of high current, the terminal voltage changes significantly, in proportion to the cell's internal resistance. Initially, the capacitive balancing circuit balances to this skewed voltage, not to the actual OCV of the cell. Eventually, as the two cell voltages start getting closer to each other, this effect is reduced, and capacitive balancing will balance the OCV. However, this takes time, making capacitive balancing slower than balancing using magnetic components.

One more problem with this technology is that the high current pulses produce significant EMI radiation.

Despite these limitations, Chinese companies sell stand-alone capacitor-based charge transfer balancers under the label of "equalizers" or similar names.

A.5.2.3 Relays

Regularly, someone invents a way to reconnect cells in parallel to balance them, and then in series during regular operation; relays would be used to reconfigure the cell arrangement.

Yes, this would work; however:

- The efficiency would be 50% at best, for the same reasons as for capacitor balancing.
- The relays would have to carry the full load current, making them large and expensive.
- The timing would have to be perfect; otherwise, the relay would risk shorting across a cell as it switches over, with dire consequences, or the load would remain unpowered for a split second.

When the cost and complexity of this solution are considered, regardless of the inefficiency, it quickly becomes clear how impractical it is.

A.5.3 BMS Isolation

To ensure that a battery is isolated from the control circuit, the BMU itself must not defeat this isolation: it must have two isolated sections, high voltage and low voltage (Figure A.12).

The detail of how a particular BMS achieves isolation is not of interest to the battery designer. All that the designer cares for is that the BMU is isolated somehow.

A.5.4 Designing Your Own BMS

Before deciding to build your own BMS, consider the advantages and disadvantages (Table A.3).

Entities that may choose to design a custom BMS include:

- High-volume manufacturers: for reasons of cost, manufacturing control, intellectual property;
- Automotive manufacturers;
- Companies that offer components for the battery industry;
- Companies that wish to increase their value through intellectual property;
- Students who need to complete a course requirement;

• Hobbyists who want the learning experience.

Otherwise, an off-the-shelf BMS is best for most applications. I strongly recommend that you seriously explore such solutions before you embark on a BMS design project. For a large battery, consider an off-the-shelf or semi-custom BMS. For a small battery, consider a ready-made design from an IC manufacturer, offered as an evaluation board.

If you absolutely must design your own BMS, you should find this book helpful in guiding your selection of features to implement. However, this book does not discuss the actual mechanics of designing a BMS. Other books are better suited (see Section C.2.1.3).

I maintain a list of integrated circuits for BMS design [14]. Designing a reliable, fully featured BMS takes anywhere between 2 weeks (for a small, low voltage battery, starting from an evaluation board from Texas Instruments) to 2 years (for a large, high voltage battery); the latter takes 2 to 10 engineers, and $500,000 to $2 million.

Table A.3
Advantages and Disadvantages of Off-the-Shelf Versus Custom BMS

	Off-the-Shelf BMS	Custom BMS
Design costs	None	$50,000~$250,000, 3~15 people
Design time	None	0.5~2 years
Production costs	None	$10,000~$100,000 to set up manufacturing, 2~15 factory staff
Per-unit cost	Set by supplier	High volume: 50% to 80% of off-the-shelf, since only required features are implemented, low volume: 1.5 to 3 times more than off the shelf
Control	If a large customer, some control over supplier's priorities; supplier may be open to make changes	100% control of production process and costs; requested features will be implemented
Ownership	None	Owns design, may own patents
Risks	Supplier stops offering product, supplier vanishes or is acquired, supplier is unwilling to make changes to product	In-house engineers do not design a reliable BMS

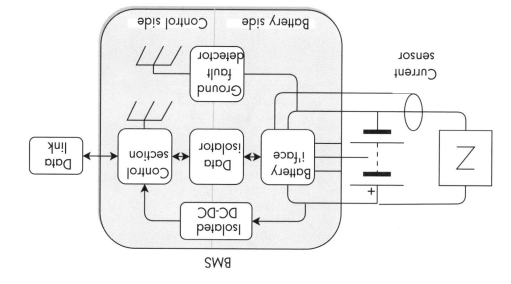

Figure A.12
BMS isolation.

I can hear the hobbyist state:"Heck, I can do it with a Raspberry Pi and $20 worth of parts." My reply is:"More power to you; go for it! It will be a great learning experience." However, I must also warn you that, in my experience, despite having heard so many people announce such a project, not once have I seen a reliable BMS resulting from a Raspberry Pi. May yours be the first.

A.5.5 "I Don't Need No Stinking BMS"

Early this century, some hobbyists venturing into Li-ion joined a bandwagon of BMS skeptics who refused to use a BMS based on rationalizations such as these:

- BMSs balance, balancing causes heat, heat causes fires, and therefore BMSs cause fires.
- Bottom balancing ensures that all the cells are at the same voltage, so a low-voltage cutoff based on battery voltage is sufficient; therefore, a BMS is not needed.
- BMS manufacturers are pushing you to buy a BMS because of their profit motive.
- I have run my EV without a BMS for a year and I only had to change a few bad cells.

I believe that the actual reasons were lack of funds, shortsightedness, and a limited understanding of Li-ion safety issues.

Top-balancing is best for energy batteries, such as traction batteries for BEVs (see Volume 1, Section 3.2.6). On the contrary some BMS skeptics believe that it is better to use a bottom-balanced battery[21] with a load with a low-voltage cutoff. In this narrow respect, they are correct. However, they are assuming that the battery stays bottom balanced on its own (it won't) and neglecting what happens at the top end; a bottom-balanced battery is unbalanced at the top. Without a BMS stopping the charging process, the lowest capacity cells will be overcharged. Overcharging Li-ion cells is immediately more dangerous than undercharging, although undercharging will be dangerous at the next recharge.

Interestingly, this movement evolved through phases that parallel the evolution of the "anti-vaxxer"[22] movement. After spending years debating this issue with BMS skeptics in the EV conversion world, I moved on. Since then, I sense that the movement has lost steam, and that EV hobbyists are now generally using a BMS.

The new batterlsphre is with UAVs; too many UAV users use no BMS. Many use a balancing charger, which top-balances the battery (that's nice), but there's no cell-level undervoltage protection in the UAV. Although, honestly, why worry about voltage protection in a battery that is so severely abused through excessive charging and discharging current? The battery is a mess anyway.

The continuing work is to educate the hobbyist with a single cell to power their Arduino or electronic cigarette. Even a single cell needs a BMS, although a simple one: CCCV charging, low-voltage cutoff, and discharge current limit.

A.5.6 Off-the-Shelf BMS Company Directory

Off-the-shelf BMSs for a given application are listed in the respective sections for that application. Table A.4 lists off-the-shelf BMS companies.

---

21. Although bottom balancing is not ideal, by all means, it is not bad.
22. Someone who opposes vaccinations (despite scientific evidence) fearing that they cause autism.

Company	Type			Application				Focus	Country	Website
	A	D	P	S	L	T	H			
123 Electric								Small automotive	Netherlands	123electric.eu
Apollo								Small battery	China	apollogroup.com.hk
Ayaa			✓	✓	✓			Small battery	China	ayaatech.com
Batrium								Small automotive	Australia	batrium.com
Belktronix								Small automotive	United States	belktronix.com, Bryan Belk
BestTechPower		✓	✓	✓	✓			Small battery	China	bestechpower.com
Elektromotus						✓		General purpose	Lithuania	elektromotus.lt, Gintautas Paluckas
Elithion	✓	✓			✓	✓	✓	General purpose	United States	elithion.com, Davide Andrea
Energus				✓				LV stationary	Lithuania	energusps.com, Sarunas
EVLithium								e-bikes	China	evlithium.com, Alex Chang
EV Power					✓	✓		Small automotive	Australia	ev-power.com.au
EVPST					✓	✓		General purpose	China	evpst.com
Ewert Energy Systems		✓			✓	✓		Multiple	United States	orionbms.com, Chris Ewert
GWL Power					✓	✓		LV stationary	Czechia	www.ev-power.eu
I + ME ACTIA								Automotive	Germany	www.ime-actia.de/
ION Energy								General purpose	India	www.ionenergy.co
Jon Elis					✓			Automotive	Lithuania	jonelis.eu, Jon Elis
JTT					✓			e-bikes	Canada	jttelectronics.com
Lipotech					✓			Low voltage	Italy	lipotech.net, Francesco
Lithium Balance					✓			Multiple	Denmark	lithiumbalance.com, Kasper Torpe
Low Carbon Idea					✓			General purpose	China	lowcarbon-idea.com
Micro Vehicle Lab								General purpose	Japan	www.mvl.co.jp
Nuvation	✓	✓		✓			✓	Multiple	USA	nuvation.com, Joseph Xavier
O'Cell			✓		✓			General purpose	China	ocelltech.com
Powerlogics								Small battery	Korea, so	powerlogics.co.kr
REC		✓			✓			Multiple	Slovenia	rec-bms.com, Maja Pozar
RoboteQ				✓				Robots	United States	roboteq.com, Cosma Pabouctsidis
Shenzhen Li-ion Bodyguard Technology								Small battery, e-bike	China	ws-pcm.com
Shenzhen Leadyo		✓	✓	✓	✓			small battery, e-bike	China	leadyo-battery.com
Shenzhen Smartec		✓	✓	✓				Small battery	China	szsmartec.com
Thunderstruck								Automotive	United States	thunderstruck-ev.com
Ventec								e-bikes, scooters	France	ventec-ibms.com, Rashel Reguigne
Volrad		✓						Low voltage	Turkey	volrad.com.tr, Gökhan Özçetin
XJR Technology		✓	✓	✓	✓			Small battery	China	Through AliExpress
Zeva						✓		General purpose	Australia	zeva.com.au, Ian Hooper

Type codes: A: array; D: digital; P: protector (includes power switch); application codes: H: high voltage; L: low voltage stationary; M: mobile; S: small; T: traction.

TABLE A.4   BMS Companies

The individual "Application" chapters describe the BMSs themselves. For example, if a manufacturer is listed with an application T code, the "Traction Batteries" chapter describes that manufacturer's BMSs.

A.6   BATTERY DESIGN

These items expand on Volume 1, Chapter 5.

A.6.1 Isolation Loss Detection

Automatic isolation loss detection may use one of three methods: static DC, dynamic DC, and AC.

A.6.1.1 Static DC isolation loss tests

Static DC isolation is the simplest isolation test. However, it doesn't detect a ground fault to the battery's mid-point; plus, the test by itself constitutes a ground fault.

A high-value resistor connects the B+ terminal to ground. An identical resistor connects the B- terminal to ground. The current between the resistors and ground is measured.

If there is no current to ground, the battery may be isolated:

• If the battery is floating, the current is zero (Figure A.13(a)).
• If B- is grounded, the current is positive (Figure A.13(b)).
• If B+ is grounded, the current is negative (Figure A.13(c)).
• If a point in the middle of the battery is grounded, the current is also zero, which is bad because there is a fault (Figure A.13(d)).

A.6.1.2 Dynamic DC Isolation Loss Tests

The test occurs in two phases:

1. Connect a resistor between B+ and the ground, and see if there's current through it.
2. Same, but to B-.

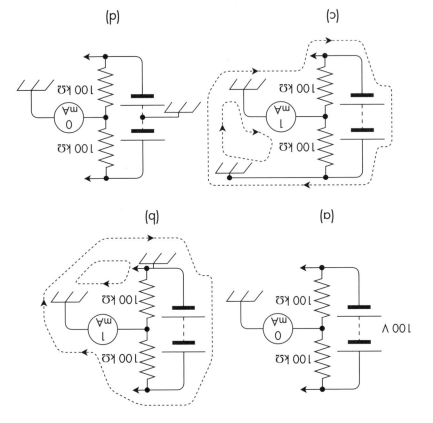

FIGURE A.13
Static DC isolation loss detector: (a) no fault, (b) B- grounded, (c) B+ grounded, and (d) mid-point grounded.

When done, the resistor is disconnected (so there's no effect on the battery and other tests). If there is no current to ground in both phases, the battery is isolated. Otherwise, from the measurements, it is possible to calculate where in the battery the loss of isolation occurred and the resistance between that point and the ground:

- If the battery is floating, the current is zero in both phases (Figure A.14(a)).

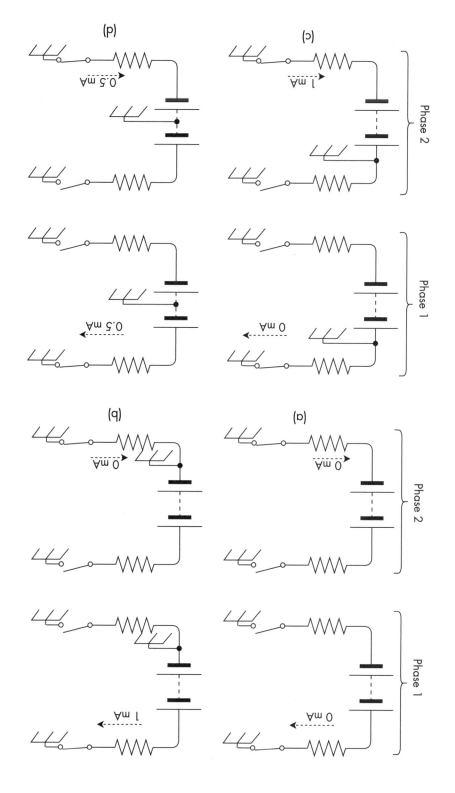

FIGURE A.14
Dynamic DC isolation loss detector: (a) no fault, (b) B− grounded, (c) B+ grounded, and (d) mid-point grounded.

- If B− is grounded, the current is positive and only in the first phase (Figure A.14(b)).
- If B+ is grounded, the current is negative and only in the second phase (Figure A.14(c)).
- If a point in the middle of the battery is grounded, both phases show current (Figure A.14(d)).

In practice, a DC test uses a slightly different method. However, this is a more straightforward explanation.

The DC dynamic test gives more information than the static test. However, it is more prone to false positives due to:

- The electrical noise on the battery: the voltage relative to ground changes rapidly when powering a switching load;
- The capacitance between the battery and ground, either just from the physics of a large battery in a metal enclosure or an actual filter capacitor; it slows down the test.

A.6.1.3 AC Isolation Loss Tests

A very low-frequency (~0.1 Hz) signal relative to earth ground is applied to the battery.

If there is no current to the ground, the battery is isolated:

- If the battery is floating, the current is zero (Figure A.15(a)).
- Otherwise, there is some current:
- If B− is grounded (Figure A.15(b));
- If B+ is grounded (Figure A.15(c));
- If a point in the middle of the battery is grounded (Figure A.15(d)).

A ground fault attenuates the signal. The amount of attenuation may be converted to the resistance of the ground fault. Due to the AC coupling, there is no way of knowing where in the battery the fault occurred.

This test takes about 30 seconds (due to the low frequency used) and operates continuously.

A.6.2 Fault Current Detection

This automated test detects a soft ground fault inside a battery that is grounded externally by the application.

Normally, every bit of current that leaves the B+ terminal of the battery comes back into its B− terminal. If there's a difference between these two currents, it means that some of the current is escaping elsewhere, presumably through a soft ground fault.

As described, this test cannot detect a soft ground fault current that is less than about 1% of the maximum battery current because the accuracy of Hall effect sensors is about 1%. For example, it cannot detect a ground fault that conducts less than 1A in a battery for up to 100A. This is insufficient for most applications.

Instead of measuring the currents separately and then taking the difference in the readings, the solution is to run both currents through the same magnetic current sensor: run both power cables through the same toroidal current sensor. Then the

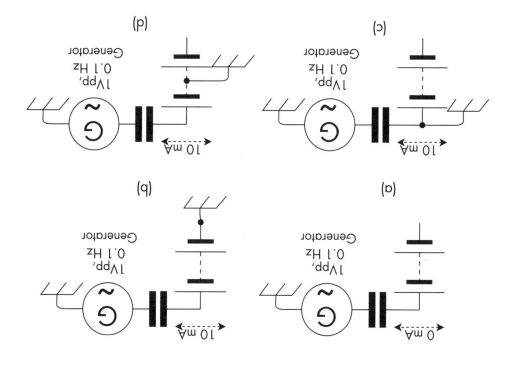

FIGURE A.15
AC isolation loss
detector: (a) no fault,
(b) B− grounded, (c)
B+ grounded, and (d)
mid-point grounded.

difference is taken magnetically and a much more sensitive Hall effect sensor can be used[23].

If this test reports no ground current fault, either there truly is no fault (Figure A.16(a)), or it's a false negative:

- If there is a soft ground fault to B− inside the battery, there's no voltage across the ground fault, therefore, no current through it, so there is no measurable effect and no fault is detected (Figure A.16(b)).
- If there is a soft ground fault outside the battery, whether to B− (Figure A.16(c)) or B+ (Figure A.16(d)), the two currents in the battery terminals are not affected, so no ground fault is detected.

This test does report a ground current fault if there's a soft short inside the battery:

- If there is a soft ground fault to B+ inside the battery, there is a difference in the current in the two battery terminals (Figure A.16(e)).
- That's also the case if there is a soft ground fault to a point in the middle of the battery (Figure A.16(f)).

A.6.3 Engineering Documentation

A battery design must be documented to enable you and others to assemble the battery and to maintain it in the future.

A.6.3.1   Electrical Schematic Diagram, Wiring Diagram

Either a schematic diagram or a wiring diagram represents the way the components in the battery are connected. These two diagrams serve different purposes, are intended for different people and use different conventions (Table A.5).

---

23. This is the same method used in GFI outlets in your bathroom, except that a GFI detects AC and this one detects DC.

The electric diagram helps us understand how the battery works (Figure A.17(a)). Electrical engineers use them primarily; other people may struggle a bit trying to make sense of them.

The components and wiring are laid out according to conventions that enable others to read it the way a musician reads a musical score. Unless it is impractical, follow these rules:

- Higher voltages are placed above lower voltages; for example, the most positive cell is placed at the top of a string, and the most negative one at the bottom.
- Arrange the logic flow of signals and control functions from left to right; for example, inputs are placed on the left edge of the sheet and enter a BMS from the left; contactor drive lines exit the BMS from its right side and go to contactors placed in the right half of the sheet; the main battery terminals are at the right edge of the sheet.
- Reduce clutter by replacing wires for large networks with symbols at each point that is connected to that network; it is implicitly understood that equal

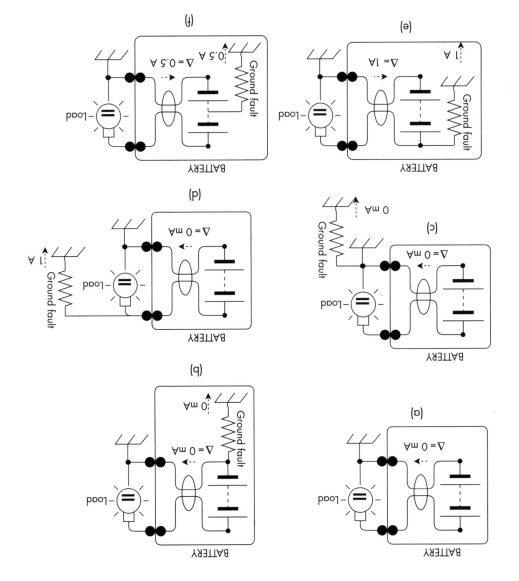

FIGURE A.16
Ground fault current detector: (a) no fault, (b) B- grounded inside, (c) B- grounded outside, (d) B+ grounded outside, (e) B+ grounded inside, and (f) grounded mid-point.

Table A.5
Comparison of Electrical
Schematic Diagram and
Wiring Diagram

	Schematic Diagram	Wiring Diagram
Intended user	Electrical engineer	Assembly and repair people
Component placement	Higher voltages on top; logic flow left-to-right	As physically laid out
Components	Electrical symbols	Boxes or pictures
Pin-outs	In logical order by function	In numerical order
Lines	Indicate networks	Indicate wires
Line color	Conveys no information	Wire color
Wire size	Not shown	Indicated
Connection points	Any nearest point	Where it actually occurs
Ground and power supply networks	Implicit; all identical ground symbols are seen as connected to each other; same for power supply symbols	Explicit: a line is drawn for each wire

network symbols are connected together: signal grounds, power grounds, earth grounds, +12V supply, B+, B−, 230Vac, K1 drive, Interlock.

- However, do not overuse network symbols; use actual wires when a wire joins only two components, and they are next to each other.

The wiring diagram helps us understand how the battery is built (Figure A.17(b)). Assembly and repair people use them primarily; electrical engineers may struggle a bit trying to make sense of them. The components and wiring are laid out according to how they are placed physically. Ideally, both diagrams should be created, to help in both applications.

These diagrams should be complemented by a Bill of Materials, a list of what is in the battery, specifications, and how to buy some more (Table A.6). I assigned the internal part numbers using my free part number assignment utility[24]. Other useful documents include:

- *MSDS*: material safety data sheet for the cell.
- *Mechanical*: drawings and 3-D files of the enclosure.

If you keep a lab book documenting your design process, add it to the package. It may refresh your memory when trying to understand why you made particular design choices.

A.6.3.2 Manuals

Prepare instruction manuals for the factory:

- Incoming QA procedures;
- Assembly instructions;
- Testing procedures;
- Final QA procedures.

Prepare instruction manuals for the user:

24. Partnumber.com, http://partnumber.com/.

FIGURE A.17
Electrical diagrams: (a) schematic diagram, and (b) wiring diagram.

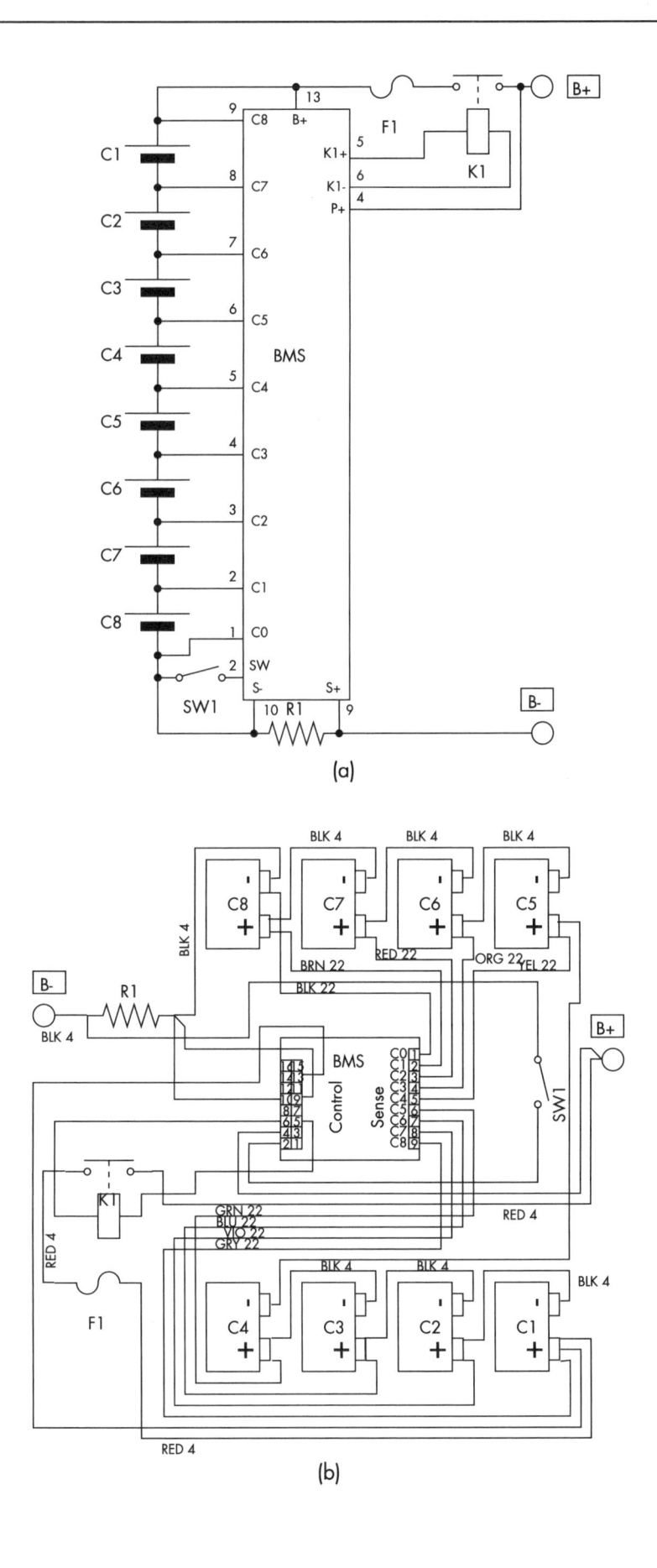

Qty	Our P/N	Description	Label	Manufacturer	Manufacturer P/N	Vendor	Vendor P/N
1	EAZ0120BM	24V BMS	ASM1	Shandong	BMS24	Alibaba	BMS24
1	EFU0104TZ	100-A fuse, 56-Vdc, bolt	F1	Littelfuse	142.5631.6102	Digikey	F6794-ND
1	EKSZ600PJ	Fuse-holder, bolt	FH1	Littelfuse	04980900ZXT	Digikey	F3094-ND
1	ERS5010CP	100-A shunt current sensor	R1	Deltec	MKA-100-50	Deltec	MKA-100-50
1	ESR9311VB	EV contactor, 9~36 V coil	K1	Tyco/Kilovac	EV200AAANA	Online comp.	EV200AAANA
1	EST6012RE	Rocker switch, SPDT	SW1	CW Industries	GRS-4012-0026	Digikey	SW304-ND
20	ETO1921RI	5/16" 4 AWG ring terminal	T1-20	Molex	0190710243	Digikey	WM13723-ND
9	ETO1903RI	5/16" 22 AWG ring terminal	T21-29	3M	MVU18-516R/SK	Digikey	920010-36-ND
1	ETR1938SB	5/16" stud terminal, black	BT1	Eaton	C1938-1	Digikey	283-3909-ND
1	ETR1938SR	5/16" stud terminal, red	BT1	Eaton	C1938-1R	Digikey	283-3910-ND
3 m	EWI0400PA	4 AWG, black THHN	W9	South wire	20499001	Home D	866180
3 m	EWI0400PA	4 AWG, red THHN	W9	South wire	20498201	Home D	405800
1 m	EWI2200PB	22 AWG wire, black	W0	Alpha wire	3051 BK005	Digikey	A2016B-100-ND
1 m	EWI2211PB	22 AWG wire, brown	W1	Alpha wire	3051 BR005	Digikey	A2016N-100-ND
1 m	EWI2222PB	22 AWG wire, red	W2	Alpha wire	3051 RD005	Digikey	A2016R-100-ND
1 m	EWI2233PB	22 AWG wire, orange	W3	Alpha wire	3051 OR005	Digikey	A2016A-100-ND
1 m	EWI2244PB	22 AWG wire, yellow	W4	Alpha wire	3051 YL005	Digikey	A2016Y-100-ND
1 m	EWI2255PB	22 AWG wire, green	W5	Alpha wire	3051 GR005	Digikey	A2016G-100-ND
1 m	EWI2266PB	22 AWG wire, blue	W6	Alpha wire	3051 BL005	Digikey	A2016L-100-ND
1 m	EWI2277PB	22 AWG wire, violet	W7	Alpha wire	3051 VI005	Digikey	A2016V-100-ND
1 m	EWI2288PB	22 AWG wire, gray	W8	Alpha wire	3051 SL005	Digikey	A2016S-100-ND
8	EZC0100LP	100 Ah 3.2 V LiFoPO4 cell	C1-8	CALB	CA100FI	EV Power	CA100FI

TABLE A.6   Bill of Materials

- User's manual;
- Emergency procedures.

A.7 MODULES AND ARRAYS

I don't have any items for Volume 1, Chapter 6.

A.8 PRODUCTION AND DEPLOYMENT

These items expand on Volume 1, Chapter 7.

A.8.1 Test Fixtures

Test fixtures help assembly and improve quality.

A.8.1.1 Cell Voltage Sense Harness

This test fixture tests that a cell voltage harness is connected to the cells correctly: good contact and proper order (Figure A.18(a)):

- If all is OK, all the LEDs light up.
- If a wire is disconnected, one of the two LEDs next to that wire remains off.
- If two wires are reversed, the LED between the two wires remains off and the two LEDs next to it are brighter.

All resistors are 2.2 kΩ. The power is 1W so that they survive if the wires are connected completely backward in a 48V block of cells.

A.8.1.2 Cable Test Fixture

This fixture tests a straight cable that has a connector at each end and the same number of wires and contacts on each end (Figure A.18(b, c)).

- If all is OK, all the LEDs light up.
- If a wire is open, all LEDs remain off.
- If two wires are reversed, the LED between the two wires remains off.

The battery (or power supply) voltage is $2V \times$ number of LEDs and the resistor is 1 kΩ, 1/4W.

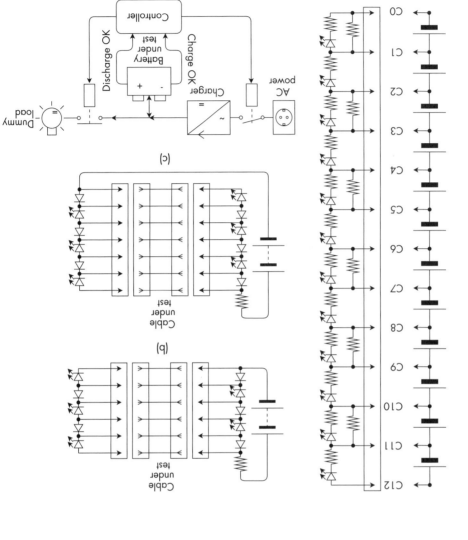

FIGURE A.18
Test fixtures: (a) cell voltages, (b) cable with even number of lines, (c) with odd number of lines, and (d) battery.

A.8.1.3 Battery Test Fixture

This fixture cycles a battery (Figure A.18(d)). The battery tells a controller whether charging or discharging is OK. The controller decides if it's time to charge or discharge and turns on the charger or the load accordingly. If the battery has a communication link, the controller logs the battery performance and calculates its characteristics, such as its capacity.

A.9 DYSFUNCTIONS

These items expand on Volume 1, Chapter 8.

A.9.1 BMS Damage from Excessive Voltages

If a cell voltage is reversed, the typical BMS initially clamps its voltage to about −0.5V. Afterward, the input is damaged. Specifically, two circuits in the BMS do this clamping:

- The voltage sensing circuit has at least one reverse protection diode across its input, one inside the IC that measures the voltage and, in better designs, a Zener diode outside the IC; when the cell voltage is reversed, those diodes become forward biased, clamping the reverse cell voltage to about 0.5V.
- The dissipative balance circuit includes a MOSFET to turn on the balancing load; this MOSFET has an intrinsic diode across its output; when the cell voltage is reversed, that diode becomes forward biased, again clamping the reverse cell voltage to about 0.5V.

Initially, the cell voltage is clamped to −0.5V as the BMS absorbs the discharging current. Eventually, this current fuses the diodes or overheats them, burning up the PCB.

A.9.2 BMS Immunity to Excessive Voltages

We saw that the BMS may be exposed to excessive voltages on the cell sense taps (see Volume 1, Section 3.2.12). It is technically possible to design a BMS that includes a switch that stays open to protect the voltage sense input "until some time has passed"[25]. However, this won't help if excessive voltage occurs at some unknown time after some time has passed. This is not a solution.

It is also technically possible to design a BMS that includes a switch that quickly opens when it sees excessive voltage or a voltage reversal, regardless of how unexpectedly this may happen. However, it is extremely challenging to design such a switch such that:

- The voltage drop across the switch, when closed, is low enough to avoid introducing measurement errors; this excludes a simple diode.
- The breakdown voltage of the switch is very high to allow operation in high-voltage batteries.
- It recovers automatically after the event; this excludes fuses.
- The switch is small, so as not to add bulk to the BMS.

- The cost of the switch and its associated control circuitry is a small portion of the per-cell cost of the BMS, assumed to be ~1$/cell; this means less than $0.20 per cell.

An experienced electrical engineer will tell you that, today, a circuit that meets all these requirements is impossible[26].

REFERENCES

[1]   "ECC-Opto-Std Mode 2: Visualizing Lithium Dendrite Growth in a Graphite vs Lithium Metal Cell," https://www.youtube.com/watch?v=922FeDvw2Rk.

[2]   National Institute for Occupational Safety and Health (NIOSH), www.cdc.gov/niosh/npg/npgd0371.html.

[3]   Green Car Congress, "Bolloré Brings Road-Ready BlueCar EV to Geneva; Plans to Build More," www.greencarcongress.com/2006/03/bollor_brings_r.html.

[4]   Quartz, "Solid Power Raises $20 Million in the Race to Build All-Solid-State Batteries," qz.com/1383884.

[5]   John Edward Brough Randles, "Kinetics of Rapid Electrode Reactions," Discussions of the Faraday Society.

[6]   Professor Kung, Chung-Chun, 306 control lab.

[7]   Plett, G., Battery Management Systems, Volume I, Norwood, MA: Artech House, 2015, p. 48.

[8]   Parabellum, endless-sphere.com, "Balancing LiPo with 'Self Balancing' Parallel LiMn?" May 12, 2012.

[9]   Liveforphysics, endless-sphere.com, "NO Such Thing as SELF-BALANCING 18650V (Makita)," May 6, 2011.

[10] Graves, R., "Syonyk's Project Blog", syonyk.blogspot.com/2016/12/a-tale-of-3-bionx-packs-self-balancing.html.

[11] NREL (National Renewable Energy Laboratory), "Design and Analysis of Large Lithium-Ion Battery Systems."

[12] Santhanagopalan, S., et al.; "Fail-Safe Designs for Large Capacity Battery Systems," U.S. Patent application US20130113495A1.

[13] Kim, G. -H., et al., "Fail-Safe Design for Large Capacity Lithium-Ion Battery Systems," Journal of Power Sources, Vol. 210, 2012, pp. 243–253.

[14] liionbms.com/php/bms_chips_options.php.

---

26. I did design a cell board that meets most of these requirements, but the cell board cost tripled, and no one wants to pay for them.

APPENDIX B

APPLICATIONS

B.1 INTRODUCTION

This appendix collects details on applications that would bog down the discussion if it were included in the main body of this volume:

- More in-depth technical discussions that are of interest only to some readers;
- Subjects that are only tangentially related to the main subject of the book;
- Directories of off-the-shelf products[1].

B.2 BATTERIES WITH CAPACITORS

Having noticed that some batteries are excellent at energy storage and that some large capacitors are excellent at power storage, many have considered devices that combine the two technologies to reap the benefits offered by each. In principle, such an electrical storage system would provide high-energy density as well as high-power density. However, most suggested topologies never reach production, as they turn out to be technically ineffective, or economically nonviable. If you are considering one of the following topologies, please consider the points that I make below before you decide to proceed with the implementation of a battery with capacitors.

B.2.1 Directly in Parallel

In this topology, a large capacitor is connected directly across each cell (Figure B.1(a)), or a single high-voltage capacitor is connected directly across the battery terminals (Figure B.1(b)). This is the easiest way to add capacitors to a battery, although it is the least effective.

For a capacitor to be of benefit, its energy must be used by the system. This means that the capacitor must be allowed to charge and discharge. A capacitor's voltage increases significantly as it charges, and decreases significantly as it discharges. If a capacitor's voltage doesn't change, it is neither charging nor discharging. Conversely, a battery's voltage doesn't change much in the short time frame in which a capacitor is asked to provide power (e.g., accelerating a vehicle) or accept power (e.g., during regenerative braking).

By connecting a capacitor directly across a battery, the voltage of the capacitor is constrained by the battery. If the battery voltage doesn't vary, the capacitor voltage

1. This does not constitute an endorsement of any of these companies and products. While the information presented in a directory is believed to be accurate, I take no responsibility for any errors.

FIGURE B.1
Batteries and capacitors:
(a) capacitor across
each cell, (b) capacitor
across entire battery, (c)
capacitor through DC-DC
converter, and (d) DC
motor application.

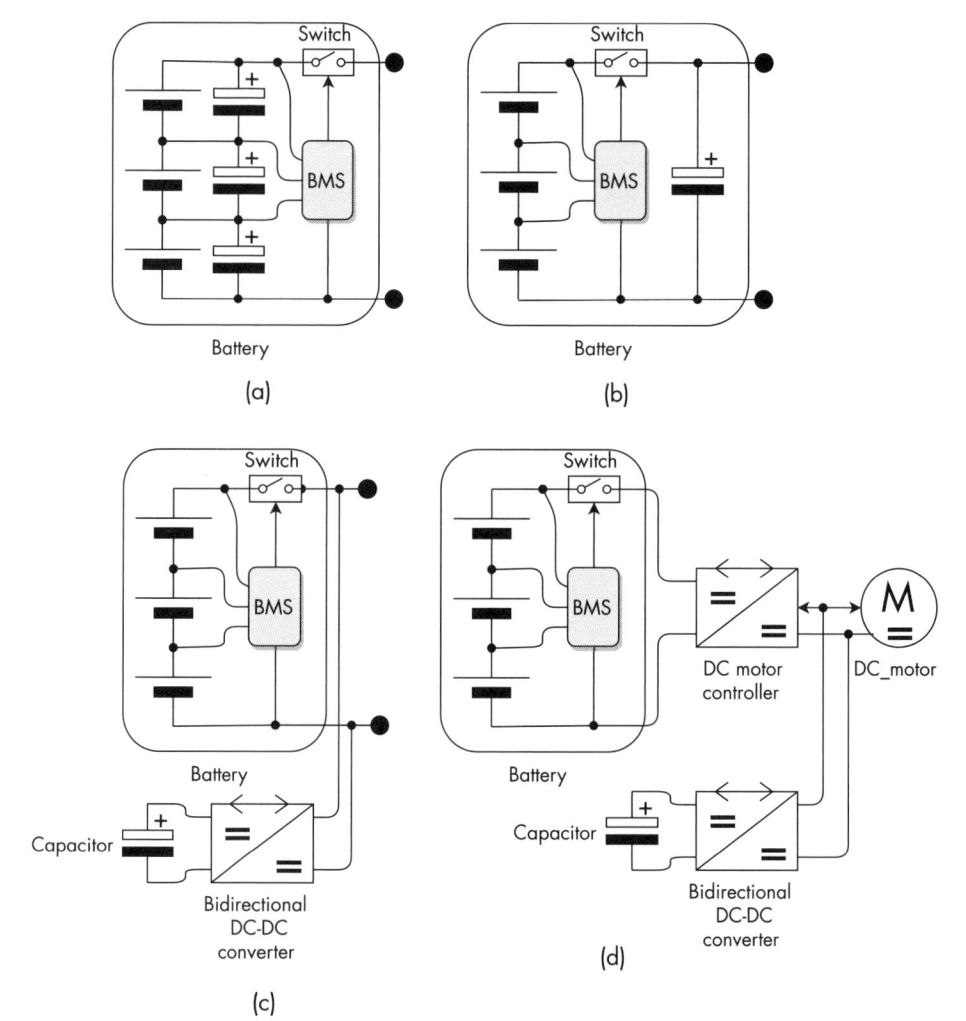

cannot vary either. Therefore, no energy can be extracted from the capacitor or returned to the capacitor. Therefore, the capacitor has no effect.

As we know, the battery has internal resistance, so its voltage does vary as the current changes. Therefore, yes, some charge is moved in and out of the capacitor. How much? Assuming that the battery voltage jumps by 5% under load, 10% of the energy in the capacitor is used[2]. That little amount of energy does not justify the cost and complexity of the large capacitor. Therefore, due to physics, connecting a capacitor directly to a cell or a battery is ineffective.

Still, if you do decide to pursue this topology, you need to know how it affects the BMS. As it turns out, the BMS doesn't care if you use a capacitor across each cell or a high-voltage capacitor across the entire battery:

- *Across each cell*: The BMS doesn't even know if a low-voltage capacitor is placed directly across each cell; as far as the BMS is concerned, the parallel combination of a Li-ion cell and a capacitor is no different from a Li-ion cell with slightly different AC impedance.

---

2. In a capacitor, $E = CV^2/2$. The energy moved is $\Delta E = CV_1^2/2 - CV_2^2/2 = C(V_1^2 - V_2^2)/2 = C(V_1^2 - (0.95 V_1)^2)/2 = C(V_1^2 - 0.95^2 V_1^2)/2 = C(V_1^2 (1 - 0.95^2)/2 = C(V_1^2 (1 - 0.9025)/2 = 0.0975 CV_1^2/2$. So, about 10%.

• *Across the battery:* A high-voltage capacitor placed across the entire battery appears to the BMS the same as any capacitive load; don't forget precharge!

A high-voltage capacitor across the entire battery, large enough to do the job, normally consists of a string of ultracapacitors in series. Ultracapacitors are rated for low voltage, typically 2.7V. Therefore, the string requires a separate capacitor management system to prevent each capacitor from being operated outside its safe operating area. This device functions in a way that is very similar to a BMS for Li-ion cells: undervoltage or overvoltage protection and balancing, except that the minimum voltage is 0V.

B.2.2 Through DC-DC Converter

We saw that placing a capacitor directly across cells or a battery is ineffective because the battery keeps the capacitor voltage from swinging, and therefore from releasing energy. Placing a bidirectional DC-DC converter between the battery and the capacitor remedies that. It isolates the capacitor with a variable voltage from the battery with a mostly constant voltage. This allows the capacitor to charge and discharge (Figure B.1(c)). The load is connected to the battery (rather than to the capacitor) because it prefers to be powered by the stable battery voltage than from the variable capacitor voltage. This is the most effective way to add capacitors to a battery, although it is also the most complex and expensive.

In general, the cost of the high-power capacitors and the cost and complexity of a high-power bidirectional DC-DC converter make this approach impractical. In practice, it's cheaper, smaller, simpler, and more effective to use high-power cells. Indeed, batteries using capacitors and bidirectional DC-DC converters are rare in products that reach the production stage: during development, designers realize that there are better ways of achieving the desired performance; a product with capacitors never leaves the prototype stage.

In most applications, using high-power cells results in equivalent performance at a lower cost than coupling batteries and capacitors.

B.2.3 With DC Motor

With a traction battery for a DC motor, it is possible to connect the DC-DC converter for the capacitor directly to the motor (Figure B.1(d)). If so, the capacitor can be used to absorb regen current, or help acceleration.

One strategy is to keep the capacitor normally full, to help with acceleration, in which case it is available to absorb regen energy only after an acceleration. The opposite strategy is to keep it empty, to help with regen, in which case it is available to help with acceleration only after braking. Or somewhere in between. It's even possible to place the capacitor in series with the DC motor, to absorb regen energy. Still, this is getting off-topic, since the battery is completely unaware that there is a capacitor in the system.

B.3 MOTORS

This section explores all electromagnetic[3] motors, not just the ones used in the drivetrain of electric vehicles.

---

3. A few motors are not electromagnetic: they may use piezo, ultrasound, or electrostatic fields. Their power is too low for traction applications at the power levels covered in this book.

B.3.1   Motor Classification

Electromagnetic motors could be classified based on the drive they require or based on their construction.

B.3.1.1   *Motor Drive Voltage*

Motors can be classified based on the drive voltage that they require

- Straight DC (Figure B.2(a));
- Sinusoidal AC, single-phase (Figure B.2(b)) or three-phase, and fixed frequency (e.g., 50 or 60 Hz), or variable frequency (Figure B.2(c));
- Trapezoidal (Figure B.2(d)), usually three-phase, for BLDC;
- Rectangular, either unipolar (Figure B.2(e)) or bipolar (Figure B.2(f)), such as for stepper motors.

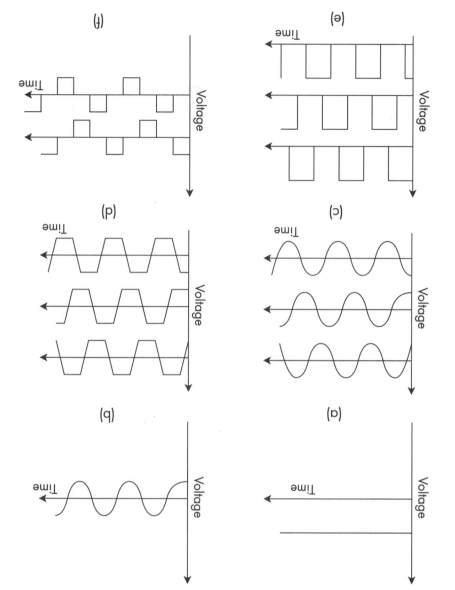

FIGURE B.2
Motor drive waveforms:
(a) straight DC, (b) single-phase sinusoidal AC, (c) three-phase sinusoidal AC, (d) trapezoidal AC, (e) three-phase, unipolar rectangular, and (f) quadrature, bipolar rectangular.

B.3.1.2 Practically All Motors Are AC Motors

At their core, nearly all motors are AC motors. For example:

- AC motor (Figure B.3(a)): driven by the sinusoidal AC waveform of the AC power supply;
- DC motor (Figure B.3(b)): an internal mechanical commutator generates AC for the motor windings;
- BLDC motor (Figure B.3(b)): an electronic commutator generates AC for the motor windings;
- Stepper motor (Figure B.3(c)): an external stepper motor driver generates AC for the motor windings.

However, a homopolar motor and a rail gun are exceptions, as they are true DC motors because the magnetic field does not vary.

B.3.1.3 Synchronization

Motors that use other than straight DC can be classified based on the synchronization (Table B.1).

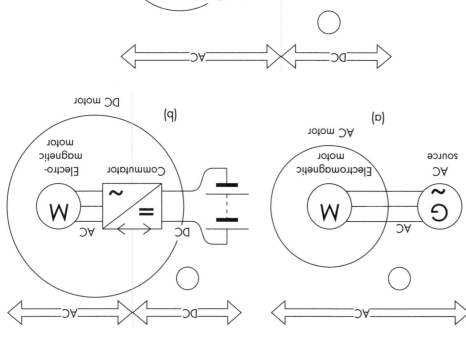

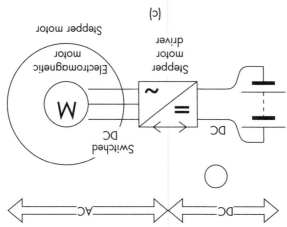

FIGURE B.3
At their core, nearly all motors are AC: (a) AC motor, (b) DC or BLDC motor, and (c) stepper motor.

Table B.1 Motor Synchronization

Industry Term	Term Used in this Book	Start	Run
None	Asynchronous	Rotates at a speed that is completely unrelated to the frequency of the drive	
Asynchronous	Asynchronous start/ subsynchronous run	Operates asynchronously (at low torque), until it reaches the subsynchronous speed	Rotates at a slower speed than the frequency of the drive; the speed decreases at higher torque
	Subsynchronous	Rotates at a slower speed than the frequency of the drive; the speed decreases at higher torque	
Synchronous	Subsynchronous start/synchronous run	Operates subsynchronously (at low torque), until it reaches the synchronous speed	Rotates at the speed set by the frequency of the drive; when loaded, the rotation angle lags, but the speed remains locked to the drive waveform
	Synchronous	Requires either a Variable Frequency Drive or a separate motor; a damper winding makes it a subsynchronous start motor	

The speed difference between the magnetic field (which is synchronous with the drive) and the motor is called the slip. When a motor is loaded, it slows down (the slip increases).

B.3.2 Motor Construction

Motors can be classified by how they are constructed:

- *Motion:* rotary (most common) or linear;
- *Magnetics:* magnets, iron, and copper winding;
- *Brushing:* brushed commutator, brushed slip ring, or brushless.

B.3.2.1 Motion

Motors can be classified based on their motion:

- Rotary:
  - The motor is propelled by a rotating magnetic field and turns a shaft.
  - It produces torque.
  - Rarely, the shaft is stationary, and the outer case of the motor turns.
- Linear:
  - The motor is propelled by a magnetic field that moves linearly and moves a rod or a bar.
  - It produces force.
  - Alternatively, the motor drags itself along a fixed track.

In a standard rotary motor, the stator is just inside the stationary motor enclosure, while the rotor is mounted to the rotating shaft. The following discussion generally

refers to rotary motors because they are much more common. When talking about linear motors, the word "rotor" refers to its slider.

B.3.2.2 Magnetics

Motors can be classified based on the magnetic material in the rotor and stator:

- *Iron rotor:* the rotor is magnetic metal, not permanently magnetized; it has no winding.
- *Magnet rotor (+ iron):* the rotor has a permanent magnet, no winding (hybrid steppers also have iron).
- *Magnet stator (+ iron):* stator has permanent magnets.
- *Copper rotor (+ iron):* the rotor includes a winding; some low inertia motors have no iron.

All motors produce a magnetic field by running current through one or more windings (typically copper magnet wire). Some motors also use permanent magnets to generate a fixed magnetic field. The windings can be classified as

- *Armature:* The windings driven by a variable voltage to generate a variable magnetic field[4].
- *Field:* Optional windings driven by DC to generate a continuous magnetic field, just like a permanent magnet.

Generally, in a Permanent Magnet DC motor, the armature winding is on the rotor[5]. In an induction motor, both the rotor and the stator carry AC; so, technically, they are both armatures; yet the industry uses this term only for the rotor.

A motor may generate a constant magnetic field with a permanent magnet, a field winding, both, or neither. A homopolar motor and a rail gun are an extreme case because the "winding" is the straight line that conducts current; there is no iron. The number of poles in a rotary motor affects the speed: the more poles, the slower the speed. For example, a three-phase AC, two-pole motor, powered at 60 Hz and unloaded, would rotate at about 60 RPS (3,600 RPM). A ceiling fan uses many more poles, so its speed is much slower. However, what really slows down a ceiling fan is that it operates with a high slip (see Section B.3.1.3).

B.3.2.3 Brushed Versus Brushless

Motors may be

- *Brushless:* A time-varying voltage is applied to the stator, not to the rotor, so there's no need for brushes; an electronic commutator may generate the time-varying voltage from straight DC.
- *Brushed:* Current flows through stationary brushes into a sliding contact area in the rotor:
  - *Slip ring:* The drive voltage is applied to the rotor through a slip ring, a continuous conductive ring.

---

4. In a generator, the armature generates an AC voltage.

5. In a small AC generator, the armature is on the rotor. In a large AC generator, it is on the stator.

- *Commutator:* The drive is applied to the rotor through the commutator, a ring divided into isolated sections; this applies the drive to the appropriate winding, given the present position of the rotor.

B.3.2.4 Servo Motors

Servo motors are complete assemblies that include a rotary motor and a matched driver. They receive DC plus a control signal that sets the position, the velocity, or the torque. The rotation of small servo motors for toys (RC motors) is often limited to less than a full turn. What matters in these motors are the position and the torque.

Confusingly, some companies refer to standard motors as "servo motors" because they may be used in a servo system; yet they are not servo motors; they are just plain motors.

B.3.3 Motor Classification Tables

Table B.2 classifies all the types of electromagnetic motors based on the drive voltage that they require and the synchronization of their motion. It also indicates its construction: magnetics and brushing. Table B.3 presents the same information in list form.

Synchronization	Drive				Construction	
	Straight DC	Waveform[2]			Magnetics[4]	Brush[5]
		Sinusoidal AC[3]	Trapezoidal	Rectangular		
Asynchronous or NA	Homopolar[6], rail gun[6,7]				Copper rotor	SR
	ECM BLDC/servo[8]				Magnet rotor	BL
	Permanent magnet				Magnet stator	Commutator
	Wound stator: self-excited: series wound, universal[9], shunt wound, compound wound, SEM/Sepex[10]	Repulsion, universal[9]	Universal[9]	Universal[9]	Copper rotor	
A/S		Repulsion start induction run[11]				
Quasi-synchronous		Wound rotor induction, 1Φ or 3Φ				SR
		Induction, squirrel cage: single-phase: split-phase, capacitor start, capacitor run/permanent split capacitor (PSC), capacitor start-run[12], shaded pole, polyphase (e.g., 3Φ), linear induction motor (LIM)[7]				Brushless

TABLE B.2   Motor Classification by Drive and Synchronization

Synchronous[1]	Drive				Construction	
	Straight DC	Waveform[2]			Magnetics[4]	Brush[5]
		Sinusoidal AC[3]	Trapezoidal	Rectangular		
S/S		Hysteresis			Iron rotor	Brushless
Synchronous		Synchronous reluctance		Variable reluctance stepper/ switched reluctance/ variable reluctance/ switched reluctance motor (SRM)[15]		
		Permanent magnet AC (PMAC)/ permanent magnet synchronous motor (PMSM)/brushless AC (BLAC)[13]/directly excited synchronous motor (DESM)/ internal permanent magnet (IPM)	BLDC[14], externally commutated	Permanent magnet stepper, hybrid stepper, coil gun[7]	Magnet rotor	
		DC-excited synchronous/self-excited synchronous			Copper rotor	SR

1. Synchronization: whether the motor rotation is synchronized with the phase of the drive: NA: not applicable for straight DC motors; asynchronous: speed is completely independent of the drive frequency; A/S: starts asynchronous, runs subsynchronous; subsynchronous: speed is slower than the drive frequency, S/S: starts subsynchronous, runs synchronously; synchronous: speed is locked to drive frequency.
2. Driven by time-varying voltages: sinusoidal, trapezoidal, or rectangular (unipolar or bipolar).
3. Fixed sinusoidal line frequency (50/60 Hz, 1Φ, 3Φ), or variable sinusoidal frequency (1Φ, 3Φ, polyphase).
4. Materials: The material in the stator or rotor (rotor = slider for linear motors). Copper rotor: the rotor includes a winding. Some low inertia motors have no iron, just the winding. Homopolar motors have neither winding nor iron. Magnet rotor: the rotor has a permanent magnet, no winding (hybrid steppers also have iron); the stator has iron. Magnet stator: the stator has permanent magnets, and the rotor has iron. Copper rotor: the rotor includes a winding. Some low inertia motors have no iron, just the winding. Homopolar motors have neither winding nor iron. Iron rotor: the rotor is magnetic metal, not permanently magnetized; it has no winding.
5. Brushing: electrical connection to the rotor, brushless (BL): no brushes. Slip ring (SR): brushes and slip ring; commutator: brushes and commutator.
6. True DC motor; requires no commutation.
7. Linear motor.
8. ECM: electronic commutator motor: transistors/ICs (contrast with externally commutated BLDC: external electronics) and BLDC: brushless direct current. Inside the motor, an electronic commutator converts the applied DC voltage to a trapezoidal wave that is in sync with the rotation of the rotor, to power specific windings. "Servo" is a misnomer, derived from servo-loop, which is something completely different.
9. Universal motor: appears twice, as it may be driven by AC or DC; it's a series wound motor.
10. Sepex: separately excited; a separate source powers the field winding; the field is independent of armature current.
11. Starts asynchronous, runs subsynchronous.
12. Two capacitors. 13. Requires a sinusoidal waveform (contrast with BLDC motor).
14. Brushless DC; requires a trapezoidal waveform (contrast with BLAC motor: sinusoidal; contrast with ECM BLDC: internal electronics).
15. The difference between a stepper motor and a switched reluctance motor is the former has many more poles; the drive can be unipolar or bipolar.
16. Permanent magnet.

TABLE B.2   Motor Classification by Drive and Synchronization (continued)

B.3.4   Specific Motors

Having explored an overview of all motors, let's discuss some specific ones.

B.3.4.1   Straight DC Motors

DC motors receive straight DC power at their terminals and convert it to AC internally[6]. They do so with either an electromechanical commutator on the shaft or with solid-state electronics; the latter are brushless motors.

---

6.   Except for homopolar motors and rail guns.

Name	Drive	Start/Run	Magnetics	Brushing
BLDC, externally commutated	Trapezoidal	Synchronous	Magnet rotor	Brushless
BLDC, internally commutated, Electronic Commutator Motor	Straight DC	Asynchronous	Magnet rotor	Brushless
Brushless AC (BLAC)	Sinusoidal	Synchronous	Magnet rotor	Brushless
Capacitor run, single-phase induction (squirrel cage)	Sinusoidal	Subsynchronous	Copper rotor	Brushless
Capacitor start-run, single-phase induction (squirrel cage)	Sinusoidal	Subsynchronous	Copper rotor	Brushless
Capacitor start, single-phase induction (squirrel cage)	Sinusoidal	Subsynchronous	Copper rotor	Brushless
Coil gun inductive mass driver (linear)	Rectangular	Synchronous	Magnet slider	Brushless
Compound wound, wound stator, self-excited	Straight DC	Asynchronous	Copper rotor	Commutator
DC-excited synchronous	Sinusoidal	Synchronous	Copper rotor	Slip ring
Directly excited synchronous motor (DESM)	Sinusoidal	Synchronous	Magnet rotor	Brushless
Electronic commutator motor brushless DC (ECM BLDC)	Straight DC	Asynchronous	Magnet rotor	Brushless
Externally commutated BLDC	Trapezoidal	Synchronous	Magnet rotor	Brushless
Homopolar	Straight DC	Asynchronous	Copper rotor	Slip ring
Hybrid stepper	Rectangular	Synchronous	Magnet rotor	Brushless
Hysteresis	Sinusoidal	Subsynchronous/synchronous	Iron rotor	Brushless
Induction (squirrel cage), polyphase	Sinusoidal	Subsynchronous	Copper rotor	Brushless
Internal permanent magnet AC (IPM)	Sinusoidal	Synchronous	Magnet rotor	Brushless
Linear induction motor, LIM (linear)	Sinusoidal	Subsynchronous	Copper slider	Brushless
Permanent magnet synchronous motor (PMSM)	Sinusoidal	Synchronous	Magnet rotor	Brushless
Permanent Magnet (PM)	Straight DC	Asynchronous	Magnet stator	Commutator
Permanent magnet AC (PMAC)	Sinusoidal	Synchronous	Magnet rotor	Brushless
Permanent magnet stepper, unipolar or bipolar	Rectangular	Synchronous	Magnet rotor	Brushless
Permanent split capacitor (PSC), single-phase induction (squirrel cage)	Sinusoidal	Subsynchronous	Copper rotor	Brushless
Rail gun (linear)	Straight DC	Asynchronous	Copper slider	Slip ring
Reluctance	Sinusoidal	Synchronous	Iron rotor	Brushless
Repulsion	Sinusoidal	Asynchronous	Copper rotor	Commutator
Repulsion start induction run	Sinusoidal	Asynchronous/slip	Copper rotor	Commutator
Self-excited synchronous	Sinusoidal	Synchronous	Iron rotor	Brushless
Separately excited motor (SEM), Sepex, wound stator	Straight DC	Asynchronous	Copper rotor	Commutator
Series wound, wound stator, self-excited	Straight DC	Asynchronous	Copper rotor	Commutator
Shaded pole, single-phase induction (squirrel cage)	Sinusoidal	Subsynchronous	Copper rotor	Brushless
Shunt wound, wound stator, self-excited	Straight DC	Asynchronous	Copper rotor	Commutator
Split-phase, single-phase induction (squirrel cage)	Sinusoidal	Subsynchronous	Copper rotor	Brushless
Squirrel cage: see individual types	Sinusoidal	Subsynchronous	Copper rotor	Slip ring
Switched reluctance motor (SRM)	Sinusoidal	Synchronous	Iron rotor	Brushless
Synchronous reluctance	Rectangular	Synchronous	Iron rotor	Brushless
Universal: wound stator, self-excited	Any	Asynchronous	Copper rotor	Commutator
Variable reluctance	Rectangular	Synchronous	Iron rotor	Brushless
Variable reluctance stepper, unipolar or bipolar	Rectangular	Synchronous	Iron rotor	Brushless
Wound rotor induction	Sinusoidal	Subsynchronous	Copper rotor	Slip ring
Wound stator: see individual types	Straight DC	Asynchronous	Copper rotor	Commutator

DC motors have these characteristics:

- The torque is proportional to the motor current.
- The speed is roughly proportional to the motor voltage.

There are a variety of DC motors:

- **Brushed commutator:**
  - Permanent magnet motors: Magnets in the stator generate a fixed magnetic field; the only winding is in the rotor.
  - Wound stator motors: Windings in the stator generate a magnetic field that depends on the current:
    - Self-excited: The magnetic field is proportional to the motor drive:
      - Series-wound: The field winding is wired in series with the armature; it operates at low voltage and high current:
      - Universal motor: This series wound motor is designed to work either at DC or AC.
      - Shunt-wound: The field winding is wired in parallel with the armature; it operates at high voltage and low current.
      - Compound wound: There are two field windings, one wired in series with the armature, and one in parallel with the entire motor drive voltage.
    - Separately excited (SEPEX): The magnetic field is independent of the motor drive because a separate supply powers it.
- **Brushed slip ring:**
  - Homopolar motor and rail gun: These are true DC motors.
- **Brushless:**
  - Electronic commutator motor, brushless DC motor (ECM BLDC): When people say BLDC they mean this motor, although there is also an externally commutated BLDC motor type; magnets in the rotor generate a magnetic field; windings in the armature generate a rotating magnetic field; sensors detect the position of the rotor and turn on the windings in the armature that result in the magnetic field rotating just ahead of the rotor, to keep the motor turning.

B.3.4.2 AC Motors

AC motors are driven with a sinusoidal AC waveform, single-phase, three-phase, or more phases. The frequency may be fixed (typically the 50/60 Hz from the AC line) or variable. They can be brushed (slip ring, not a commutator) or brushless, and synchronous or quasi-synchronous.

AC motors have these characteristics:

- The torque is roughly proportional to the motor current.
- The motor voltage is roughly proportional to the motor speed.
- The speed is either directly proportional to the drive frequency (synchronous motors), slightly different from it by an amount called the slip (quasi-synchronous motors), or unrelated (fully asynchronous motors).

There is an even greater variety of AC motors:

- **Asynchronous:**

- Repulsion motor: AC is applied to the stator windings; the rotor also has wind-ings, connected to a commutator; two brushes are shorted together, to activate a rotor winding at a time; changing the position of the shorting brushes changes the speed and direction; formerly used in trains.

- Universal motor: Typically AC powered, but it can also work at DC and any other voltage drive; the power is applied to the rotor through a commutator, so that the correct winding is powered for a given angle of the rotor; because the current flows in the same direction in both the armature and the field windings, regard-less of the polarity of the applied voltage, the torque is always in the same direc-tion; therefore, this motor can operate at AC or DC, and cannot reverse direction.

- Slip:

  - Commutator:

    - Repulsion start induction run motor: When starting, same as above, for a much stronger starting torque compared to an induction motor; once the motor reaches speed, the brushes are placed in a position that makes it run like an induction motor.

  - Brushless:

    - Induction motor: The windings in the rotor are permanently shorted; there are no brushes.

    - Linear induction motor (LIM): Same as above, but linear; usually the rail acts as the rotor in an induction motor and the sliding part acts as the stator in an induction motor.

  - Slip ring:

    - Wound rotor induction: Same as an induction motor, except that there are variable resistors in series with the rotor winding to set its characteristics; since the resistors are stationary, they are connected to the rotor through slip rings.

- Synchronous:

  - Brushless:

    - Without magnets:

      - Hysteresis: The rotor in this motor is uniform (it has no poles) copper and iron; the motor uses the copper to start in slip mode, just like an induction motor; it then magnetizes the rotor's iron and switches to operating like an internal permanent magnet motor, running synchronously.

      - Reluctance/self-excited synchronous: These motors rely on the shape of the rotor to complete the magnetic path of the field generated by the stator; it comes in two types:

        - Synchronous reluctance:

        - Switched reluctance, also known as variable-reluctance (SRM).

    - With magnets:

      - Permanent magnet alternating current (PMAC).

      - Permanent magnet synchronous motor (PMSM)

      - Brushless alternating current (BLAC).

      - Directly excited synchronous motor (DESM).

      - Internal permanent magnet (IPM): Known by five different names, this brushless motor has magnets in the rotor and a winding in the stator; once it starts, it runs synchronously to the frequency of the drive.

- BLAC motors are very similar to BLDC motors, except that they are designed for a sinusoidal waveform; while some have used BLDC controllers (trapezoidal output) to drive induction or BLAC motors (sinusoidal), the efficiency of the motor suffers and the motor runs a bit more roughly.

- Slip ring:
  - DC-excited synchronous motor: same as the PMAC motor, except that instead of magnets, the rotor has an electromagnet; DC powers its winding through a slip ring.

B.3.4.3   Trapezoidal Drive Motors

Externally commutated BLDC[7] motors are driven with a trapezoidal waveform. They are brushless and synchronous. The frequency of the drive sets the motor speed.

There is nothing that is DC in BLDC motors: these are truly AC motors, so we should call them BLAC. However, the name BLAC is used for slightly different motors that require a sinusoidal drive (instead of trapezoidal drive). While some have used inverters with BLDC motors (trapezoidal), the efficiency of the motor suffers, and the motor runs a bit more roughly.

BLDC motors use an electronic commutator; it detects the position of the rotor and controls solid-state switches that power the correct winding for the present position. They are more reliable than motors with an electromechanical commutator but cost more. A magnet in the rotor of a BLDC motor generates the magnetic field.

B.3.4.4   Rectangular Drive Motors

Stepper motors are driven with a set of switched DC waveforms, either unipolar or bipolar. They are brushless and synchronous. The frequency of the drive sets the motor speed. They do not rotate smoothly; instead, they snap into place. As they are synchronous, there is no need for a position encoder for feedback for their speed; however, an encoder of some form may be used to detect the position or to confirm it. There are effectively two motors that use rectangular drive:

- Without magnets:
  - Variable reluctance stepper: Uses the same concept as the reluctance AC motor, but uses rectangular drive instead of AC.

- With magnets:
  - Permanent magnet stepper: May use a unipolar or a bipolar drive.
  - Hybrid stepper: Combines the operation of both of the permanent magnet stepper and the variable reluctance stepper motors.

B.4   MOTOR DRIVERS

Motor drivers generate the voltage (and possibly waveform) required to drive the motor at the desired speed and torque. They are matched to the type of motor (see Section B.3.3.4): using the wrong waveform with a motor may result in inefficient operation (motor runs hot), louder operation, lower torque, and synchronous motors not starting.

Motor drivers generate a varying DC voltage on their output; most do so with a technique called pulse width modulation (PWM). The output DC level is proportional to the duty cycle of this high-frequency PWM (Figure B.4(a)); to generate a waveform, they modulate the duty cycle (Figure B.4(b)).

This PWM wave is often not filtered, relying on the inductance of the motor itself to smooth the PWM voltage wave into a continuous current.

Motor drivers have different names depending on their function:

- DC motor controller: To drive a DC motor, including an electronically commutated BLDC motor.
- Sepex motor controller: Same as above, but specifically for a separately excited motor;
- Inverter: To drive an AC motor from a DC voltage;
- Variable frequency drive (VFD): To drive an AC motor from the AC line power;
- Stepper motor driver: To drive a stepper motor with rectangular waves;
- BLDC motor driver: To drive an externally commutated BLDC motor.

B.4.1   DC Motor Controllers

These motor controllers have one DC output (2 wires) with a variable DC voltage (Figure B.2(a)).

A DC motor controller is a step-down DC-DC converter. When driving the motor in the first quadrant (see Section 3.3.2) (positive speed and torque):

- It increases the current (relative to the battery current) to the current required by the DC motor to generate the desired torque.
- It reduces the voltage of the traction battery to the voltage required by the DC motor at a given speed and torque.

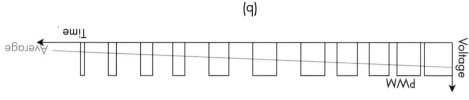

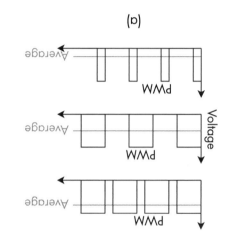

FIGURE B.4
PWM: (a) average DC level, and (b) creating a waveform by varying the duty cycle.

Most DC motor controllers are one-quadrant; they can drive only in one direction and they cannot brake. With the addition of a set of reversing contactors, they can drive in either direction but cannot brake. Two-quadrant motor drivers are not common. Several are four-quadrant; they can drive or brake in either direction, without the need for reversing contactors (see Section 3.3.2).

DC motor controllers may be

- **PM controller:** Drives the two terminals of the DC motor.
- **SepEx controller:** Also provides a separate excitation for a motor with additional terminals for the field winding.

B.4.1.1 PM Motor Drivers

Permanent magnet DC motor controllers work with any DC motor other than Sepex. DC motor controllers for small EVs are available from many vendors (Table B.4).

B.4.1.2 DC Sepex Motor Controllers

These motor controllers work with any DC motor. They have two DC outputs (three or four wires):

- A 2-wire, high-power output with a variable DC voltage for the rotor;
- A 2-wire, lower-power output, with a (usually fixed) DC voltage, for the armature.

Sepex motor controllers for small EVs are available from a few vendors (Table B.5).

B.4.2 AC Motor Drivers

An AC motor driver generates a variable frequency, AC, single-phase or three-phase to drive an AC motor (Figure B.2(c)).

Table B.4
DC Motor Controller
Manufacturers

Manufacturer	Location	Voltage [V]	Motor Current [A]	Quadrants*
4QD†	United Kingdom	10–56	20–320	1,2,4
ASI‡	Canada	12–96	30–100	1,2,4
Alltrax	United States	12–72	125–650	1
Belktronix	United States	120–144	800–200	1
Curtis	United States	24–120	45–700	1
Curtiss-Wright	United Kingdom	24–80	80–650	1
Kelly	United States	12–156	200–1,800	1,2,4
NetGain Controls	United States	160–360	1,000–1,400	1
Piktronik	Slovenia	24–48	260	1
Roboteq	United States	30–96	20–500	1
Sevcon	United Kingdom	24–80	75–600	1,2,4
Zapi	United States	24–120	70–1,000	1,4

*One quadrant = one direction, no regen; two quadrants = one direction with regen; and four quadrants = two directions with regen.
†Not isolated.
‡ Some not isolated.

TABLE B.5
DC Sepex Motor Controller Manufacturers

Manufacturer	Location	Voltage [V]	Motor Current [A]
Alltrax*	USA	24~48	300~600
Curtis	USA	24~48	200~700
Curtiss-Wright	UK	24~80	80~650
Kelly	USA	12~144	200~800
Navitas	Canada	24~48	500
Sevcon	UK	24~48	180~600
Zapi	USA	24~72	110~500

*Not isolated.

Inherently, AC motor inverters operate in four quadrants:

- *Reversible:* They can drive a motor in either direction.
- *Bidirectional:* They allow regenerative braking.

AC motor drivers may be

- AC motor inverters, with a DC input (for vehicles);
- Variable frequency drives (VFD), with AC line input (for industrial applications).

B.4.2.1 AC Motor Inverters

Most of these AC motor inverters (Table B.6) are for high voltage applications (>100V), such as passenger vehicles, heavy-duty, and industrial. Most of these AC motor inverters are isolated, and they typically have a CAN bus interface.

Originally, the term "inverter" referred to devices to convert DC to a fixed frequency (e.g., 50 or 60 Hz) and a fixed voltage (e.g., 110, 220 Vac). Later, the name was also applied to AC motor drivers, even though their voltage and frequency vary considerably.

B.4.2.2 Variable Frequency Drives

These motor drivers are the same as AC motor inverters, except that their input is AC power at line frequency (Table B.7).

TABLE B.6
AC Motor Inverter
Manufacturers

Manufacturer	Location	Supply Voltage	Motor Current
AC Propulsion*	USA	336~360 Vdc	580A
Curtis	USA	24~96 Vdc	150A~650A
Curtiss-Wright	UK	24~80 Vdc	24A
HEC Drives (2)	Netherlands	100~820 Vdc	280A~900A
Jon Elis	Lithuania	144 Vdc	750A
Kelly (1)	USA	12~180 Vdc	50A~1,000A
Navitas (1)	Canada	24~48 Vdc	80~225A
TM4 (2)	Canada	200~400 Vdc	300A~1,000A
Piktronik	Slovenia	24~400 Vdc	50A~400A
Roboteq (1)	USA	30~96 Vdc	20A~500A
Rinehart Motion	USA	160~720 Vdc	150A~300A
Sevcon	UK	24~48 Vdc	180A~600A
Tritium	Australia	160~450 Vdc	100A~300A
UQM†	USA	250~430 Vdc	69A~464A
Zapi	USA	24~72 Vdc	110A~500A

*Includes motor.
†Trapezoidal drive for BLDC motor.

Table B.7
Variable Frequency Drive
Manufacturers

Manufacturer	Location	Supply Voltage	Motor Current
Baldor	United States	Line voltage	Various
Eaton	United States	Line voltage	Various
Siemens	Germany	Line voltage	Various

B.4.3 BLDC Motor Drivers

A BLDC controller generates three bipolar, trapezoidal waves, as required by an externally commuted BLDC motor (Figure B.2(d)). Feedback from the motor provides a reference used to advance to the next phase. Therefore, the motor sets the speed (based on torque), not the driver. The top voltage may be variable.

A high-quality BLDC controller generates trapezoidal waves; beware that the typical BLDC controller approximates a trapezoidal wave with a bipolar switched DC (negative, 0V, positive, 0V negative...).

B.4.4 Stepper Motor Drivers

A stepper motor driver produces a unipolar rectangular wave for a synchronous motor (Figure B.2(e)). The motor speed is proportional to the frequency. The top voltage is fixed. There is no need for feedback. This is a rather simple driver and therefore low cost. Even though a driver itself is unipolar, it can drive a bipolar motor by connecting the two ends of a winding to two outputs of this driver (Figure B.2(f)).

B.5 HIGH POWER CHARGERS

Table B.8, showing high-power chargers, is based on a list that I maintain online [1]. Companies that offer chargers appear and disappear quite rapidly. Brusa has unfortunately stopped selling its great chargers.

B.6 HIGH-VOLTAGE CONVERTERS

Chargers, inverters, and invergers convert between AC and DC in one direction, the other, both, or within one (see Volume 1, Section 1.8.1).

B.6.1 Inverter Circuit Topology

This may be getting into more electronics than you care for, but some insight into how an inverter for high-voltage batteries works is required to understand why the battery voltage must exceed the peak AC voltage. Moment by moment, the inverter steps down the battery voltage to create the sine wave of the AC line voltage.

Let's start with an inverter for single-phase AC. The system uses a split battery (see Volume 1, Section 6.5) and an inverter with two transistors (represented as two switches, SW1 and SW2) in a topology that is called a half bridge (Figure B.5(a)). This is a very simplified circuit, but it gives you the general idea.

The switches are turned off and on very rapidly (~10 kHz) at a variable duty cycle PWM[8], to generate the desired sinusoidal waveform (Figure B.5(b)). To generate the positive half of the sine wave, SW1 is turned on, initially at a low duty cycle (to generate just a few volts), then at a constantly increasing duty cycle (to generate increasingly higher voltages) and finally, at the top of the sine wave, the switch is turned

on most of the time. During this time, SW2 is off. To generate the negative half of the sine wave, SW1 is turned off, and SW2 is turned on at a duty cycle proportional to the desired voltage. The inductor in the circuit smooths out the output: the output is a smooth sine wave, not a PWM.

Another way of driving a single-phase AC line is to use a single battery and four switches (in what's called a full bridge) (Figure B.5(c)). The switches "flip" the battery every half cycle, so that the same battery can be used to create both the positive and the negative half cycles of the sine wave.

Note that the battery voltage is half of the total voltage in the previous circuit: doubling the transistors in the inverter allowed the use of a lower voltage and, therefore, cheaper battery.

This is a noisy solution because the battery's voltage relative to earth ground is not fixed; rather, it bounces up and down, emitting electromagnetic interference. During the positive half-wave, SW2 is closed, and B- is at earth ground (through the neutral wire). During the negative half-wave, SW1 is closed, and B+ is at earth ground. Consequently, the battery "jumps up and down" at 50 or 60 Hz.

For split-phase AC (e.g., 2 × 240Vac in the United States), two half-bridge circuits are powered by a split battery[9] (Figure B.6(a)). One half-bridge (SW1 and 2) creates one sine wave, while the other one (SW3 and 4) creates its mirror image. For three-phase A, the inverter uses three half-bridge circuits, one for each phase (Figure B.6(b)); for three-phase Y, the battery is split, and the center is connected to the neutral (Figure B.6(c)).

---

9. Having a positive and a negative battery simplifies the design of the inverter considerably because it generates a positive half and a negative half simultaneously. The positive battery powers the positive half of the AC sine wave, and the negative one powers the negative half.

Table B.8
High-Power Charger
Manufacturers

Manufacturer	Location	3Φ	Power [kW]	Output [Vdc]	CAN
Bassi	Canada	✓	0.5~50	12~130	✓
BesiGo	China		3~20	48~360	
Current ways	United States		3, 6.6	42~450	✓
Curtis	United States		0.6~1.5	12~96	
Chennic	China		1.1	12~144	
DeltaQ	Canada		0.65~1.5	24~96	(✓)
EDN Group	Italy	✓	0.1~3	12~120	
ElCon	China	✓	1.5~5	24~417	✓
Iota	United States		0.1~1.1	13~54	
Jinan e-shine	China		0.6~4	12~360	
Jon Elis	Lithuania	✓	18	144	✓
Kingpan Industrial Co	China		0.5~5	14~352	
Lester*	United States		0.6	24~48	
Manzanita Micro †	United States		5.8~11.5	25~400	
Meanwell	China		0.03~1	12,24,48	
Quick Charge*	United States		1.2~4.8	12~120	
Sinexcel	China	✓	12.5	200~700	
SPE Electronica industriale	Italy		0.1~16	24~96	✓
Zivan	Italy		0.6~7	12~312	(✓)

*Not high frequency.
†Not isolated.

FIGURE B.5
Simplified inverter
circuits and resulting
waveforms: (a) single-
phase with split battery,
(b) proper variation of
duty cycle results in the
desired waveform, and
(c) single-phase flipping
a single battery.

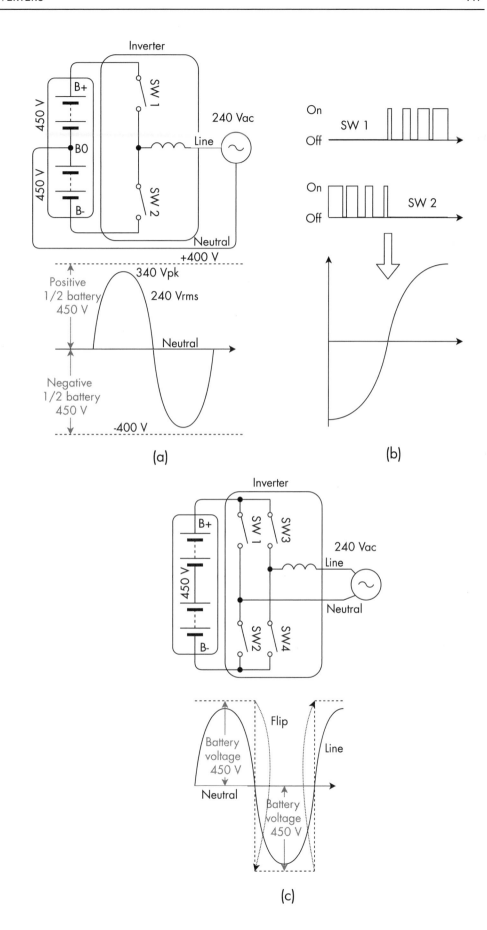

(a)

(b)

(c)

FIGURE B.6
Multi-output inverter
circuits: (a) split phase
with split battery, (b)
three-phase Y with single
battery, and (c) three-
phase Y with split battery.

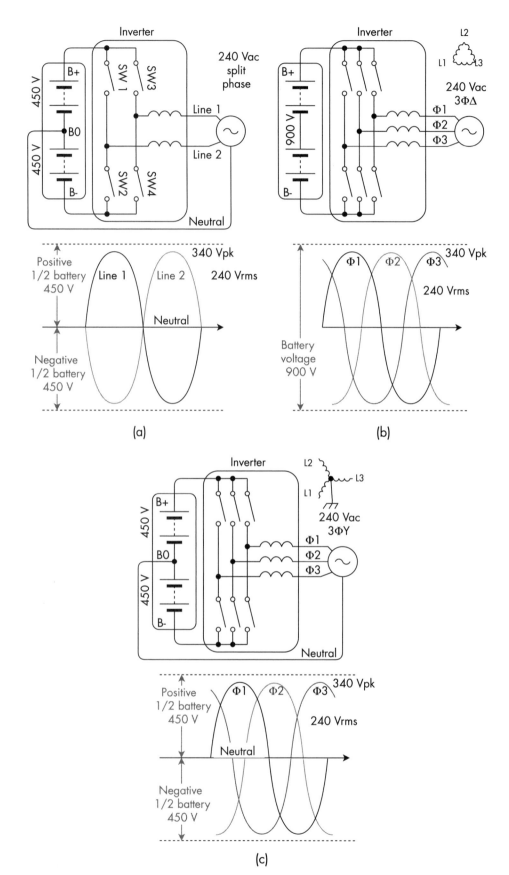

All of the circuits discussed above are inherently bidirectional:

- Power can flow from the battery, through the switches as they chop the battery voltage into a sine wave, through the inductor to smooth the current, into the AC circuit.
- Power can flow from the AC circuit, through the inductor to store energy, through the switches as they chop the sine wave and distribute chunks (sometimes to B+, sometimes to B-), and into the battery.

The basic power circuit for an inverger is identical to the power circuit for an inverter. The difference is in the way the control circuit and the software use this power circuit. The control in an inverter prevents power from flowing back to the battery; in an inverger, it doesn't.

These circuits can only reduce the DC bus voltage down to the instantaneous AC voltage. Therefore, a 450V battery can only produce 240 Vac. An inverger for higher AC voltages would need a higher voltage battery; yet, the industry prefers to use 450V batteries. This is resolved by placing a bidirectional DC-DC converter between the battery and the bridge to raise the DC bus voltage (Figure B.7).

By using a nonisolated DC-DC converter to raise the DC bus voltage, the inverger is still a transformer-less device.

This discussion assumed that the inverter drives an AC line; for an inverter that drives an AC motor, the difference is that that output voltage and frequency are variable.

B.6.2 SMA Sunny Island Invergers

I stated that, under certain circumstances, the SMA Sunny Island charged the battery despite the BMS telling it not to (see Section 2.5.2.5). Here are some details, and a rebuttal from SMA. This was tested with model 6048-US.

Figure B.7
DC–DC converter between the battery and the bridge.

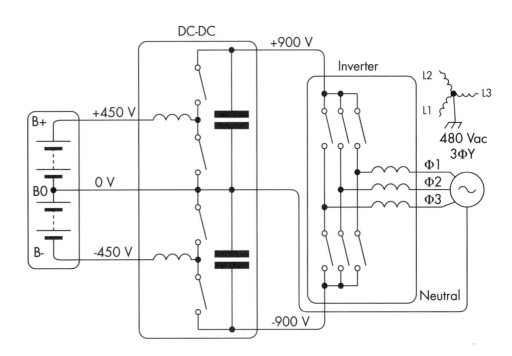

We found two instances when the Sunny Island charges the battery despite the BMS telling it not to, by setting the charging current limit (CCL) to 0A:

- Five minutes after enabling the Sunny Boys, the Sunny Island charges the battery at full current for about 30 seconds.
- If powered by the grid, the Sunny Island trickle charges the battery at about 2A (which may be fine for lead–acid, but not for Li–ion).

SMA's response, verbatim: "The statement in the passage was based off of your experience with one system using a non-approved Li-Ion battery. Of course the battery and BMS being used will have an effect on whether or not the system operates correctly, which is why we have the approved list."

That list includes batteries from these manufacturers: ADS-TEC, Akasol, Aquion Energy, Axitec, BMZ, BYD, GNB, Hoppecke, IBC, Leclanché, LG Chem, Mercedes-Benz Energy, Sony, SSL Energie, and Tesvolt.

B.7   DC-DC CONVERTERS

Table B.9 lists some manufacturers of DC-DC converters.

B.8   HYBRID SYNERGY DRIVE

The best-known example of a series/parallel hybrid is Toyota's trademarked Hybrid Synergy Drive, first generation (Figure B.8). This technology is used in the Toyota Prius, Camry, and Highlander, and in the Ford Escape.

In practice, this is a parallel hybrid because the only useful function in series mode is to charge the battery when the vehicle is stopped. When the car is moving, this drivetrain operates mostly as a parallel hybrid, since the engine can propel the car by itself and the motors can move the wheels by themselves. Table B.10 summarizes the operating modes.

TABLE B.9
DC-DC Converters

Manufacturer	Voltage [V]	Power [W]
Iota	12~48	Various
Vicor	~48	<400

FIGURE B.8
Hybrid Synergy
Drive topology.

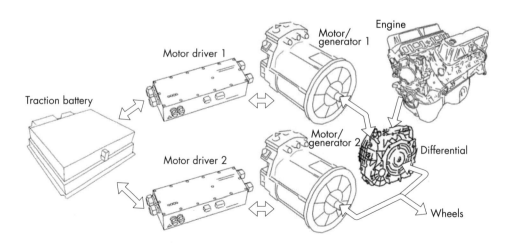

Mode			Battery	Figure Number	Engine	Battery	MG1	MG2	Wheels
Series	Stopped		Not full	B.9(a)	On	Charge	Gen	Stall	Stop
	Drive		Not empty	B.9(b)	On	Either	Stall	Motor	Drive
Parallel	EV	Drive	Not empty	B.10(a)	Off	Discharge	Motor	Motor	Drive
		Brake	Not full	B.10(b)	Off	Charge	Gen	Gen	Brake
			Full	B.10(c)	Brake	Neither	Motor	Gen	Brake
	HEV	Assists		B.11(a)	On	Discharge	Motor	Motor	Drive
		Neither		B.11(b)	On	Charge	Stall	Motor	Drive
		Charges		B.11(c)	On	Charge	Gen	Idle	Drive

TABLE B.10   Hybrid Synergy Drive Operating Modes

The SoC of the battery plays a strategic role. A high SoC increases the likelihood that the car uses the EV mode. This is exploited by companies that retrofit an HEV to a PHEV, to force the vehicle to minimize fuel consumption[10].

The series mode is only useful for charging the battery while stopped (Figure B.9(a)) or crawling (Figure B.9(b)). It is pretty useless at high speed because the output of the differential is turning fast yet does not contribute to propelling the car. The engine would have to run at twice the speed to compensate for this, which is not a good idea.

In the parallel EV mode, the engine is off. When driving, the battery powers both motors and both propel the car (Figure B.10(a)). When braking, both motors turn

FIGURE B.9
Hybrid Synergy Drive topology, series HEV mode: (a) stopped, and (b) crawling.

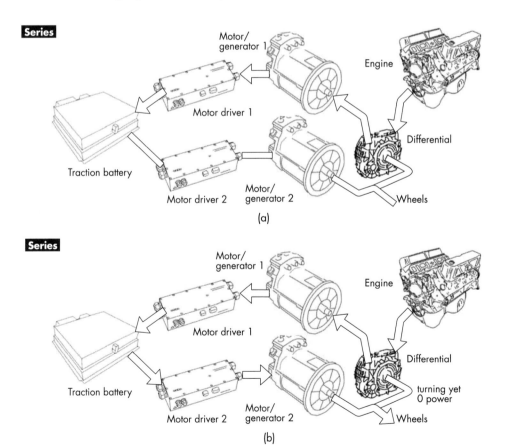

---

10. The Orion BMS is uniquely tailored for this conversion.

FIGURE B.10
Hybrid Synergy Drive
topology, parallel EV
mode: (a) drive, (b)
brake, and (c) brake
with full battery.

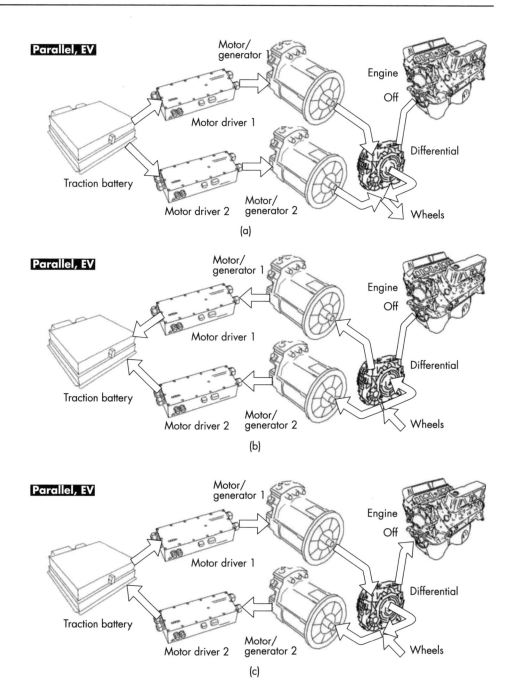

into generators, charging the battery (Figure B.10(b)). If the battery is full, power circulates in the system, wasting energy by heating the motors and the drivers (Figure B.10(c)). Some of the power spins the engine to add engine braking. The mechanical brakes handle the rest of the braking.

In the parallel HEV mode, the engine is on. The battery may contribute to the propulsion (Figure B.11(a)), or not (Figure B.11(b)), or be recharged by extra power from the engine (Figure B.11(c)).

B.9   ENERGY STORAGE OTHER THAN BATTERIES

Li–ion batteries may never provide the entirety of the storage required by the utilities because they do not scale easily. Instead, other ESS technologies may continue to

FIGURE B.11
Hybrid Synergy Drive
topology, parallel HEV
mode: (a) drive with
electric and ICE power,
(b) drive with ICE power,
and (c) drive and recharge.

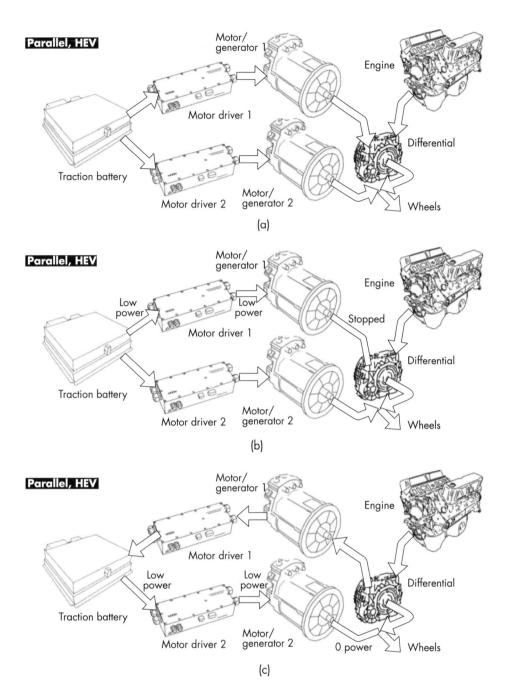

provide better value and prove to be more effective at a large scale. Pumped hydro, although inefficient, is effective. However, few locations can support it. The efficiency of storing compressed air in caves is terrible.

I find gravity power to be the most practical proposal for grid-scale storage. A deep well is filled with water and topped with a massive piston. Energy is stored by pumping water in the well to lift the piston and retrieved by letting the piston drop, pushing water out and spinning generators [2]. This solution can be implemented just about anywhere and is as efficient as pumped hydro.

B.10   TOROIDAL HYPERLOOP TEST TUBE

Instead of Space X's linear test tube, what is needed to test the technologies to be used in an actual vactrain is a circular test tube, ~2 km in diameter (Figure B.12(a)). A pod could accelerate and decelerate in it at the desired, comfortable rate of 0.5 G and travel 500 km.

As the pod speeds up, centrifugal forces rotate it (its bottom climbs the outer wall of the tube), eliminating any sideways forces. Simultaneously, the track inside the tube reshapes itself: initially, the track is horizontal and at the bottom (Figure

FIGURE B.12
Proposed round test tube: (a) top view, (b) cross section, pod at rest, and (c) cross section, pod at top speed.

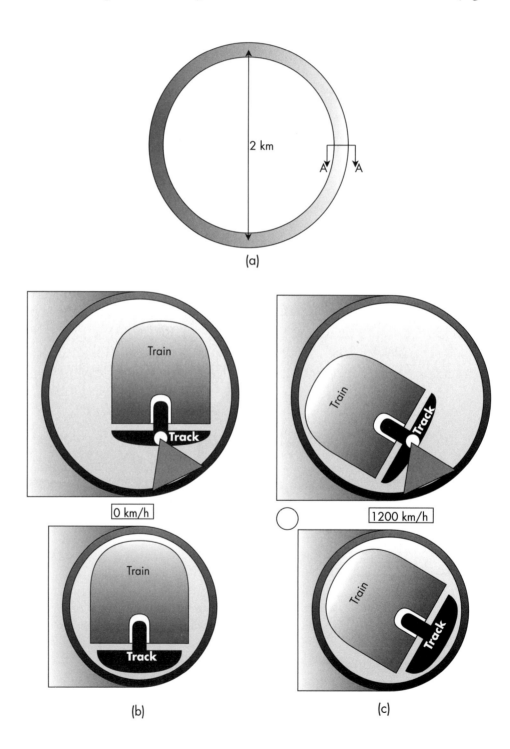

B.12(b)). As the train speed increases, the track either climbs up the outer wall, or pivots in place (Figure B.12(c)), so that it remains beneath the pod. The track would need to be made of resilient materials or somehow telescope to accommodate the change in shape.

B.11 THOUGHTS ON LEGAL MATTERS

While I am not a lawyer (INAL), I have had some experience in this business. Allow me to make some suggestions regarding forming contracts and seeking legal counsel. The following is not legal advice; rather, consider it as food for thought when retaining the services of a lawyer[11].

B.11.1 Prepare Contracts

If your company provides a battery to another company, you need a written contract that specifies exactly what you will deliver, when, with what specifications, and with what warranty. Consider how the customer's actions will affect your schedule, such as limiting your responsibility in case of its delays. Include expiration dates so that you're not stuck with having to suspend your performance or to provide support for indefinite periods. Specify the name of the individual working for the customer to whom you are expected to report, to discuss change orders, and to resolve conflicting directives. Be specific about what constitutes misuse, include a logger to detect misuse, and explain that misuse will void your warranty.

For purposes of preparing such a contract, retain a business lawyer, or a transactional lawyer, who is versed in the Uniform Commercial Code[12] and Products Liability litigation or the corresponding code in your country. Have them draft Terms and Conditions that apply to all your contracts, specifying that you and the customer are bound by them once you have accepted the customer's order. At a minimum, your Terms and Conditions should include the following:

- A precisely defined, limited warranty, drafted to limit your exposure to claims for breach of express or implied warranties;
- A specification of your remedies if the customer should default;
- A notice to the customer that your company is not responsible if the product is not installed and used per instructions and intended use, or if the user is not technically competent to use it;
- A provision stating that your company is not liable for any consequential damages; and, in any case, your company's total cumulative liability is limited to the price paid by the customer;
- Notice that all contract modifications must be in writing and signed by both parties;
- A specification that disputes must be resolved through binding arbitration, and that the forum for any such proceeding or a proceeding brought in court, is to be the jurisdiction where your company has its principal address.

If your company buys a battery from another company, prepare a contract that specifies exactly what you expect and by which time. Include expiration dates so that you're not stuck with a vendor who is too slow or unresponsive. Include penalties for

---

11. Thank you to Professor L. Scott Gould, Esq., for help in this section.
12. The United States uses the Uniform Commercial Code. Other countries use a different set of rules.

late delivery. Insist on a pro-rated warranty, with compensation inversely proportional to the battery's performance and the time in use.

B.11.2   Avoid Getting Sued

Do your very best to design and manufacture a good product. Do not cut corners. Cover your ass: always keep good records of internal communications that relate to engineering and business decisions, and especially of communications with customers. If the communication was verbal, send an email afterward: "To recap our recent conversation, I understand that we agreed to … If you disagree with this recapitulation, please explain your differences as soon as possible. Otherwise, I will assume we can proceed as I've outlined." Be responsive and support the customer even beyond what it is entitled to.

B.11.3   If About to Be Sued

If your customer notifies you that it intends to sue you, consult your lawyer before you say anything. In my experience, you may get different advice, depending on the lawyer's specialty:

- A litigation lawyer may tell you: "Do not speak to the customer at all; have all communications go through me."
- A business or transactional lawyer may tell you: "Put all your effort into helping the customer to try to defuse the situation, while not saying anything that can be used against you."

Routing all your communications through your lawyer costs you every time your customer wants to buy a piece of wire and then again every time your ship that wire. The advantage is that it keeps you from inadvertently giving the plaintiff material that can be used against you.

Continuing to work directly with the customer may either help (by avoiding a suit) or may backfire (by just postponing the inevitable, while the customer is milking you for material that can be used against you). Use your knowledge of your customer and their gripe against your company to decide which scenario is more likely. My experience (directly or through clients) is limited to the latter: trying to diffuse the situation. In each case, the customer did not follow through with the threat, or parties came to a financial resolution.

The following strategy has helped my customers and me:

- Appoint a single individual to talk on behalf of your company; instruct all other employees to respond that they are not authorized to speak to the customer, and to defer to the appointed individual.
- Do not stop communicating with the customer; rather, be as responsive as possible, presenting a positive attitude: "you're my partner, not my enemy."
- Politely point to the Terms and Conditions in the quote they placed an order on, which limit your liability and may specify arbitration.
- Offer free services and product well beyond your responsibility: engineering services, custom design, and repair or replacement of out-of-warranty product.
- Make yourself indispensable to the customer, while politely implying that all the support will vanish the moment a suit is actually filed.

- Hire an independent expert to look into the matter: a forensic expert in case of an accident, a software consultant in case of a malfunction.
- If you honestly believe that your company is blameless, help the customer see it that way, by presenting written evidence to that effect; if you honestly believe that another party is to blame, offer to side with the customer against that party; if you do believe your company is partially or fully at fault, have your lawyer help your company to do the right thing in a way that does not bring undue exposure.
- Offer to buy back your product at a reasonable price, even if it's old and out of warranty.
- Do not sell the customer anything else because every dollar it pays you increases your potential loss by $10; help the customer find an alternative supplier.

The goal is to have the customer step back from the threat. In the end, the cost of all the extra hand-holding will be well worth it, compared to the legal costs.

B.11.4 If Actually Sued

If your company is actually sued:

- Stop all direct communications with the customer while you set up a legal team.
- If your company has business insurance with full coverage, notify your insurance company; that company's lawyers will step in.
- Engage the services of a business litigation lawyer, in addition to your business lawyer.
- Gather all relevant documentation and present it to all of your lawyers.
- The insurance company may quickly decide that the circumstances are such that it is better to pay up than to fight; the insurance company will pay $1 million or $2 million to the plaintiff, and you can go on with your life, although paying a much higher premium from then on.
- Or the insurance company may decide that you have a case; it will help you with your defense because it does not want to pay for damages. Have your lawyers work closely with the insurance company lawyers.
- You may still be required to do depositions and provide materials for a suit against other parties; be prepared for a grueling deposition; get trained on how to answer questions in a deposition (e.g., tell the truth, be sure to understand the question before answering; stop and think before you answer; be sure of your answers, and if you are not sure or don't know, say so).
- Hopefully, this will end in a settlement, which your insurance company will pay.

REFERENCES

[1] liionbms.com/php/charger_options.php.
[2] www.gravitypower.net/, www.powermag.com/let-gravity-store-the-energy/.

APPENDIX C

RESOURCES

C.1 ONLINE RESOURCES

These are resources on the World Wide Web.

C.1.1 Web Sites, Blogs

General:

- Battery University (Isidor Buchmann, Cadex): batteryuniversity.com/;
- Electropaedia (Barrie Lawson, Woodbank Communications): www.mpoweruk.com/bms.htm;
- All About Batteries;

Small batteries:

- Li-ion BMS (Davide Andrea, Elithion): liionbms.com/php/index.php;
- Lygte Info (V Henrik K. Jensen): lygte-info.dk.
- Parametric battery search: batteries.parametric.com

C.1.2 Blogs

General:
- Battery Blog (Randy Smith, Energy Storage Instruments): batteryblog.ca/.

Small batteries:

- Battery Bro: batterybro.com/blogs/18650-wholesale-battery-reviews;
- Battery education: www.batteryeducation.com/.

C.1.3 Forums

General:

- Reddit: old.reddit.com/r/batteries/;
- Quora: www.quora.com/topic/Batteries.

Small batteries:

- RC Universe: www.rcuniverse.com/forum/batteries-chargers-84/;
- Reddit: old.reddit.com/r/18650masterrace.

Large low voltage:

- Solar panel talk: www.solarpaneltalk.com/forum/off-grid-solar/batteries-energy-storage/general-batteries, www.solarpaneltalk.com/forum/off-grid-solar/batteries-energy-storage/lithium-ion;
- Outback users: outbackpower.com/forum/.

Traction:

- DIY Electric car: www.diyelectriccar.com/forums/;
- V is for voltage: endless-sphere.com/forums/.
- Endless Sphere: visforvoltage.org/;

Racing:

- FSAE: reddit.com/r/FSAE/, ieee-collabratec.ieee.org.

C.1.4   Newsletters

Cells:

- Shmuel De-Leon, www.sdle.co.il/services/newsletter/ — news about the cell and battery industry; warning: relentless marketing; once they have your email address, kiss your mailbox goodbye.

Batteries:

- Battery Power, www.batterypoweronline.com/tag/battery-newsletter/;
- Electrical Energy Storage, www.electrical-energy-storage.events/en/newsletter/newsletter.html.

Traction:

- EVDL, www.evdl.org/ — EV conversion enthusiasts.

C.2   PRINT RESOURCES

These resources are on paper.

C.2.1   Books

These books are listed in alphabetical order by first author's last name.

C.2.1.1   Cells

This book is related to cells (Table C.1).

C.2.1.2   Batteries

These books are related to batteries (Table C.2).

C.2.1.3   Li-Ion Battery Management Systems

These books are related to BMSs (Table C.3).

Table C.1
Book Related to Cells

Title	Author(s)	Publisher	Description	Tone
Lithium-Ion Batteries: Science and Technologies	Masaki Yoshio	Springer	Chemistry inside Li-ion cells	Academic

Title	Author (s)	Publisher	Description	Tone
Batteries for Electric Vehicles: Materials and Electrochemistry	Helena Berg	Cambridge University Press	Detailed description of Li-ion cells and their degradation, with a passing look at traction batteries and BMS	
Batteries in a Portable World: A Handbook on Rechargeable Batteries for Non-Engineers	Isidor Buchmann	Self-published	Long-time expert in this field; review of all types of small batteries: safety, charging, caring, decay and failure, testing, applications	Very practical
The Complete Battery Book	Richard A. Perez	McGraw-Hill	Explains how batteries store and transfer energy, describes the structure of different kinds of batteries and discusses inverters, energy management, and new technology	Academic
Design and Analysis of Large Lithium-Ion Battery Systems	Shiram Santhanagopalan, A. A. Pesaran, Gi-Heon Kim, J. Neubauer, K. Smith	Artech House	Sharing scientists' experience with thermal analysis of Li-ion cells used in automotive applications	Academic, somewhat practical
Linden's Handbook of Batteries	Thomas Reddy	McGraw-Hill	Chemistry, construction and characteristics of primary and secondary cells; also fuel cells and capacitors	Academic as well as practical
The TAB Battery Book: An In-Depth Guide to Construction, Design, and Use	Michael Root	Self-published	Review of all battery technologies, history and science	Practical
DIY Lithium Batteries: How to Build Your Own Battery Packs	Micah Toll	Self-published	Guide in putting a Li-ion battery together, Dumpster diving, charging, with actual case studies	Very practical, although many errors, skeptical of BMSs
Lithium-Ion Battery Failures: A Systems Perspective	Ashish Arora, Sneha Arun Lele	Artech House	Analysis of what goes wrong with small batteries and how to avoid it	Practical
Modern Battery Engineering: A comprehensive introduction	Edited by Birke Peter Kai	World Scientific Pub. Co. Inc.	Anthology of a patchwork of subjects related to Li-ion cells, battery design, some way-out-there concepts, and recycling	Mixed

TABLE C.2   Books Related to Batteries

C.2.1.4   Applications

These books are related to battery applications (Table C.4).

C.2.2   Magazines

- Batteries International: www.batteriesinternational.com/;
- Battery Power: www.batterypoweronline.com;
- Batteries and Energy Storage Technology (BEST): www.bestmag.co.uk/;
- Charged EVs, Electric Vehicles Magazine: chargedevs.com/;
- Energy Storage Journal: www.energystoragejournal.com/.

Title	Author(s)	Publisher	Description	Tone
Battery Management Systems for Large Lithium Ion Battery Packs	Davide Andrea	Artech House	General information on Li-ion BMSs, and hardware design guide; Chapter 5 details many designs for the analog front end for a BMS for a large, high-voltage battery pack; other than this, the book covers many of the subjects also covered in this book	✓ Practical
Battery Power Management for Portable Devices	Yevgen Barsukov, Jinrong Qian	Artech House	Specific design info for small Li-ion batteries using off-the-shelf, complete BMS ICs; from the gurus at Texas Instruments	✓ Practical
Battery Management Systems: Design by Modelling	H.J. Bergveld, W.S. Kruijt	Springer Science and Business Media	BMSs for small Li-ion batteries for cell phones	Academic, not very useful
Battery Management Systems, Volume I: Battery Modeling Battery Management Systems, Volume II: Equivalent-Circuit Methods	Gregory L. Plett	Artech House	An in-depth book on sophisticated algorithms to determine the SoC of Li-ion cells; a professor with real-world experience guides you through the various aspects of battery and BMS design, with an in-depth and practical analysis of algorithms for state of charge evaluation; Volume I discusses models based on physical phenomena in the Li-ion cells, while Volume II discusses how to use those models in the discrete-time realization algorithm for evaluation of the state of the cells	✓ Academic, but also practical
Battery Systems Engineering	Christopher D. Rahn, Chao-Yang Wang	Wiley	An in-depth book on sophisticated algorithms to determine the SoC of Li-ion cells; probably more useful as a college textbook than as a tool for practical BMS design	Academic
Lithium-Ion Battery: The Power of Electric Vehicles: Basics, Design, Charging Technology & Battery Management Systems	Subodh Sarkar	Self-published	Focused on many aspects of BMS design for electric vehicles; it also discusses other topics related to EVs	Practical
A Systems Approach to Lithium-Ion Battery Management	Phillip Weicker	Artech House	In-depth academic book on sophisticated algorithms to determine the SoC of Li-ion cells; detailed coverage on some of the more complex software and hardware challenges in designing a BMS for a large battery	✓ Practical but also academic

TABLE C.3   Books Related to BMSs

C.2.3 Market Reports

These reports typically list companies in a particular industry and may forecast the growth in that industry. They can be quite expensive: $500 to $5,000. In a few cases, that price may be justified for marketing people and investors. A summary or a snippet is usually available, which may give you an idea of what information is included.

Frankly, my limited experience with market reports is disappointing. Researchers contact me often to request information to include in such reports. They don't pay

Title	Author(s)	Publisher	Description	Tone
Battery Book One: Lead Acid Traction Batteries	Ken Marsh	Curtis Instruments	Look at lead-acid battery-powered vehicles	Practical, outdated
The Handbook of Lithium-Ion Battery Pack Design: Chemistry, Components, Types and Terminology	John T. Warner	Elsevier Science	Traction battery design for electric vehicles, focuses of Li-ion, mentions BMS, thermal, mechanical, recycling	Practical
Modern Electric, Hybrid Electric, and Fuel Cell Vehicles	Mehrdad Ehsani, Yimin Gao, Stefano Longo, Kambiz Ebrahimi	CRC Press	Complete coverage on electric vehicles, not just the battery	Practical
Lithium-Ion Battery: The Power of Electric Vehicles: Basics, Design, Charging Technology & Battery Management Systems	Subodh Sarkar	Self-published	Electric vehicles, Li-ion cells, BMS characteristics, motors, charging, case studies (Nissan Leaf, PHEV)	Practical
The Electric Car: Development and Future of Battery, Hybrid and Fuel-Cell Cars (Energy Engineering)	Mike H. Westbrook	The Institution of Engineering and Technology	Overview of the history and technology that goes into electric vehicles; small coverage of batteries.	Practical
Off Grid Solar: A Handbook for Photovoltaics with Lead-Acid or Lithium-Ion Batteries	Joseph P O'Connor	CreateSpace Independent Publishing Platform	Covers the entire subject of solar installations; batteries are covered in just a few pages	Practical
Solar Energy: The Physics and Engineering of Photovoltaic Conversion, Technologies and Systems	Olindo Isabella	UIT Cambridge Ltd	Covers just the PV panels: no batteries	Practical

TABLE C.4   Books Related to Battery Applications

their sources for their time while providing data; yet they expect to sell the report at full price to their sources.

None of those researches who contacted me seem to know much about the subject. Once I spent 1 hour educating a market researcher on the difference between cells, batteries, and battery management systems, between primary and secondary cells, and other fundamental concepts. He was grateful, but he unable to offer a copy of the final report.

Most reports that intend to list BMS companies reveal a fundamental ignorance of what a BMS is. Lists of BMS companies include companies that don't sell BMSs. One report includes building management systems in a BMS list. In another report [1] (that sells for $4,950), few of the companies in the list of 10 "BMS companies" actually sell BMSs:

- One does programmable logic controllers: Johnson Matthey.
- Three make integrated circuits: Linear Technology, NXP Semi, and Texas Instruments.
- One made complete batteries: Valence, which is now out of business.
- One is a general consulting firm: Vecture.
- Only four of these companies actually sell BMSs: Elithion, Lithium Balance, Nuvation, and Ventec.

This report forecasts that the market value will be $11.17 billion by 2025; yet the yearly revenue from BMS sales for the four BMS companies listed in the report cannot be more than $3 million.

Here are some sources of market reports:

- Shmuel De-Leon, www.sdle.co.il/services/market-research-reports/;
- MarketWatch, www.marketwatch.com/;
- Research and Markets, www.researchandmarkets.com;
- Grand View Research, www.grandviewresearch.com/.

C.3   TRADE SHOWS

Europe:

- Automotive Battery Management Systems, Germany;
- BTE Battery Tech, Germany;
- EES, Electrical Energy Storage, Germany;
- ITEC, Germany;
- Weiterbildung, Netzgekoppelte PV/Batteriespeicher-Anlagen, Germany.

Asia:

- Asia Guangzhou Battery Sourcing Fair, China;
- Battery China, China;
- Battery Osaka, Japan;
- CIBF China International Battery Fair, China;
- ESS Expo, South Korea;
- Energy Storage Summit, Japan;
- InterBattery Seoul, South Korea.

North America:

- AABC, Advanced Automotive Battery Conference, USA;
- Battcon, USA;
- Battery Congress, USA;
- Power Mart Expo, USA;
- The Battery Show, USA.

C.4   BATTERY ASSOCIATIONS

- BAJ Battery Association of Japan http://www.baj.or.jp/e;
- Battery council https://batterycouncil.org/;
- China Association of Power Sources http://www.ciaps.org.cn/;
- EASE European Association for Storage of Energy http://www.ease-storage.eu/;
- EPBA European Portable Battery Association http://www.epbaeurope.net;
- EUROBAT Association of European Automotive and Industrial Battery Manufacturers http://www.eurobat.org/;
- Recharge https://www.rechargebatteries.org/association/;
- KBIA Korean Battery Industry Association http://www.k-bia.or.kr/;
- PRBA The Rechargeable Battery Association http://www.prba.org;

- ZIV, Zweirad-Industrie-Verband, http://www.ziv-zweirad.de/;
- ZVEI, Zentralverband Elektrotechnik- und Elektronikindustrie e.V. http://www.zvei.org.

C.5 SERVICES

These companies provide services to the battery industry.

C.5.1 Training

High-voltage training:

- e-Hazard: www.e-hazard.com/;
- MTCS, Online High Voltage Training Courses & Certification, www.mtcsuk.com/mtcs-online/high-voltage.

C.5.2 Testing

Battery testing:

- Intertek: https://www.intertek.com/;
- Mobile Power Solutions: https://mobilepowersolutions.com/;
- Shmuel De-Leon: www.sdle.co.il;
- UL: https://ctech.ul.com/en/industries/battery-testing/.

REFERENCE

[1] Battery Management System Market Analysis by Battery Type (Lithium-Ion Based, Lead-Acid Based, Nickel Based, Flow Batteries), by Topology (Centralized, Distributed, Modular), by Application, and Segment Forecasts, 2018–2025, www.grandviewresearch.com/industry-analysis/battery-management-system-bms-market/toc.

GLOSSARY

These definitions reflect how the following terms are understood in the battery industry, rather than by the general public.

AC adapter   Converts AC to DC, typically to power a consumer product

AC generator   1. Converts mechanical energy to AC power; 2. converts fuel to AC power

Agricultural battery application   Electrical storage for farming

Alternating current   Voltage that alternates from positive to negative

Alternator   Converts mechanical power to AC first, then to DC, in a vehicle

Amp   Unit of measure of current

Arrangement   See Cell arrangement

Automatic transfer switch   A transfer switch that switches automatically when one power source disappears

Automotive battery application   Electrical storage for vehicles, particularly passenger vehicles

Auxiliary load   A load that may be turned off in case of grid failure

Auxiliary power unit   Fuel-powered generator in a vehicle, independent of the engine

Back-feed   The act of sending power back to the grid

Balanced   In a series string, all cells are at the same state of charge at some point

Balancing   The act of restoring balance in a series string

Base transceiver station   Telecommunication equipment at a base station such as next to a cell phone tower

Battery   A collection of cells, BMS, and other components; a battery has one current, one state of charge, one protector switch

Battery array   Two or more complete batteries (each with its own protector switch) connected together

Battery electric vehicle   Electric vehicle with a battery, a charger, and no engine

Battery energy storage system   Energy storage system that uses a battery

Battery management system   Device or system that manages and protects a battery

Battery management unit   BMS for a large battery; it has no protector switch, relying instead on the system to obey its current limits

Block   See Parallel block

Breakdown voltage   Maximum voltage across two conductors; an arc occurs if exceeded

Breaker   See Circuit breaker

Buffer battery   Battery that is kept at about 50% state of charge and used at a high rate; contrast with power battery

Bus   1. A communication link; 2. a high-power conductor to interconnect batteries and other devices

Business battery application   Electrical storage for a large office building

Bypass balancing   Balancing by removing extra charge from the most charged cells in a series string and wasting in heat

Bypass switch   Transfer switch in which one of the power sources is a battery

C   Unit of measure of specific current, more accurately, 1/hour

C-rate   See Specific current

Cable   Bundle of wires in a single tube; contrast with wire, harness

Calendar life   How long a cell retains sufficient state of health as it degrades while not in use

CAN bus   Digital serial data bus used in automotive and industrial applications

Capacitance   Measure of size of a capacitor, expressed in farads (F)

Capacitor   Electronic component that stores a charge; it has lower energy density yet higher power density than a battery

Capacity   How much charge a cell or battery can store, expressed in ampere-hours (Ah)

Cell   Fundamental electrochemical storage unit inside a battery

Cell arrangement   The way two or more cells are interconnected

Cell chemistry   The material used for one of the electrodes in a cell

Cell format   The physical shape of the cell enclosure

Charge   1. Measurement of a quantity of electrons, expressed in ampere-hours (Ah); 2. the act of storing a charge into a cell or battery

Charge acceptance   Analog to state of charge, but for a battery array in which a battery may be disconnected, expressed in ampere-hours (Ah)

Charge current limit   Maximum allowed charging current, communicated from the BMU to the external system

Charge depleting mode   How an energy battery or a power battery is used

Charge sustaining mode   How a buffer battery is used

Charge transfer balancing   Balancing by moving charge between cells

Charger   Current-limited power supply, specifically designed to charge a battery

Charging curves   Graphs that describe the behavior of a cell during charging

Charging station   Stationary equipment into which an electric vehicle is plugged to recharge or into which a vessel is plugged to power itself while docked

Circuit breaker   Switch that opens automatically in case of over-current, is reclosed manually

Combiner   Selector switch to select power from one of two batteries or both in parallel

Connector   Allows connecting or disconnecting a wire, a cable, or a printed circuit board (PCB)

Contactor   Like a relay, but larger

Coulomb counting   Technique to evaluate the state of charge indirectly

Coulombic efficiency   The portion of the charge placed into a cell during charge that can then be retrieved during discharge, expressed in 5; practically 100% for Li-ion

Critical load   Load that must remain powered in case of grid failure

Current   A flow of charge through a conductor, expressed in amperes (A)

Current limits   The charge current limit and the discharge current limit

Current shunt   Device that measures current

Cycle   The act of charging and discharging the battery once, fully

Cycle life   How long a cell retains a sufficient state of health as it degrades while being charged and discharged

DC-DC converter   Converts a DC voltage to a different DC voltage; often isolates the two voltages, but not necessarily

Depth of discharge   How much charge a battery or cell delivered, expressed in Ah or percentage

Direct current   Voltage that remains continuously positive or negative

Discharge   The act of releasing a charge from a cell or battery

Discharge availability   Analog to depth of discharge, but for a battery array in which a battery may be disconnected, expressed in Ah

Discharge current limit   Maximum allowed discharging current, communicated from the BMU to the external system

Discharging curves   Graphs that describe the behavior of a cell during discharging

Distributed BMS   A BMS that uses a printed circuit board assembly mounted on each cell to sense its state

Dual-port battery   Battery with an input for charging and a separate output for discharging

Dual-switch, single-port battery   Battery with two contactors or MOSFETs, one for charging, one for discharging

Electric power   Power in the form of electricity

Electric vehicle   Vehicle propelled by electric power

Electric vehicle supply equipment   See Charging station

Electronic generator   Like an AC genset, but uses a battery instead of burning fuel, an uninterruptible power supply (UPS)

Emery   A neologism meaning unit of measure of ratio of total charge in the lifetime of a battery over the capacity of a battery

Energy   Given by a morning coffee, expressed in watt-hours (W-h)

Energy battery   Battery that is fully charged and then discharged at a low rate; contrast with power battery

Energy cells   Cells optimized for maximum capacity

Energy density   How much energy a cell or battery of a given volume can store

Energy efficiency   How much of the energy generated internally by a battery is seen by a load

Energy storage system   Any form of energy storage, not just electric

Engine   Converts fuel to mechanical power; contrast with motor

Fault protector   Redundant BMS, tripped when cells are operated beyond a wider range than the main BMS

Frequency   Rate of variation in voltage, such as in an AC voltage, expressed in hertz (Hz)

Fuel gauge   Informal name for state of charge evaluation

Fuse   Link that opens automatically in case of over-current, not resettable

Galvanic isolation   Isolation capable of withstanding a high voltage across it

Ganged batteries   A battery array in which the protector switches are either all on or all off, simultaneously

Generator   1. Converts mechanical power to electrical power. 2. See Genset

Genset   A set that includes an engine and a generator to convert fuel to AC power

Grid   Network providing electric power from the power company

Grid-interactive inverter   Inverter able to back-feed or to power the local loads in case of grid failure

Grid-tied   Device or system connected to the grid

Gross balancing   Fast balancing performed manually when the series string is badly imbalanced

Ground fault   See Isolation loss

Ground fault test   See Isolation test

Hall effect sensor   Device that measures current

Hard short   Short circuit with zero resistance; contrast with soft short

Harness   Loose bundle of wires, prepared for rapid installation inside a product; contrast with wire and cable

Hertz   Unit of measure of frequency

High voltage   Variously defined as above 48V, or above about 500, depending on the context

House power   Local, low-power electrical system, not for traction

High-voltage stationary battery   Large, fixed battery, high voltage, may be connected to the grid

Hybrid electric vehicle   Vehicle with more than one power source, one of which may be electric; specifically, vehicle with an engine and a motor, which may or may not have a battery; contrast with plug-in hybrid

Hysteresis   Effect that prevents the cell voltage from relaxing completely to the open circuit voltage

Imbalance   A measure of how much the state of charge differs among cells in a series string

Impedance   Same concept as resistance, but for AC

Industrial battery application   Electrical storage for heavy-duty manufacturing

Inrush current   Pulse of excessive current when connecting a battery directly to a device with a large input capacitance; continued excessive current when connecting a battery directly to another battery

Inverger   A neologism meaning combination charger and inverter, operates in either direction

Inverter   Converts DC to AC, either at a fixed line frequency to power AC loads or at a variable frequency to drive a motor

IR drop   The change in voltage when a cell or battery powers a load; the difference between the open-circuit voltage and the terminal voltage

Isolation   Lack of a path for current between the battery and earth ground or chassis

Isolation loss   Unintentional path for current to earth ground or chassis in an otherwise isolated battery

Isolation test   Used to detect an isolation loss

Kalman filter   Technique to evaluate the state of charge indirectly

Large cylindrical cell   Large, round, hard cell. See Cell format

Large prismatic cell   Large, rectangular cell with plastic case. See Cell format

Large, stationary, low-voltage battery   A 12-V to 24-V battery at a fixed location, such as for solar storage

Li-ion cell   Stores charge by moving lithium ions between electrodes, forms the basis of a Li-ion battery

Load   A device that uses power

Low voltage   A voltage up to about 40V or 48V

Maintenance balancing   Slow balancing by the BMS

Marine battery application   Electrical storage for a vessel

Master/slave BMS   BMS subdivided into modules

Maximum power point   Operating point where the power source provides as much power to the load as possible

Maximum power time   A neologism meaning the duration of a discharge performed at the maximum power point; characteristic of performance for cells and batteries for use in high-power applications (lower is better)

Micro hybrid   Small hybrid electric vehicle whose engine stops when the vehicle is stopped

Microgrid   Same as the regular grid but much smaller and local

Mid balancing   Balancing a series string at around 50% state of charge

Modular battery   A single battery subdivided into modules

MOSFET   A solid-state transistor used as an electrically controller switch, such as a protection switch

Motor   Converts electrical power to mechanical power; contrast with engine

Motor controller   Driver for a DC motor

Motor driver   Converts DC power to the voltage required to drive a particular motor

Off-grid   Device or system disconnected from the grid

On-grid   Device or system connected to the grid

Open-circuit voltage   The voltage of a cell or battery after a long period at rest at no current

Parallel block   Two or more cells connected in parallel. See Cell arrangement

Parallel-first   Two or more parallel blocks connected in series to achieve the desired capacity. See Cell arrangement

Pedestal   See Charging station

Photo-voltaic   See Solar panels

Plug-in hybrid   Electric vehicle with a motor, a battery, an engine, and a power cord for charging

Pouch cell   Flat, soft cell; see Cell format

Power   Rate of conversion of energy into work, expressed in watts (W)

Power bank   Small storage device with AC input and DC output; contrast with uninterruptible power supply

Power battery   Battery that is fully charged and then discharged at a high rate; contrast with energy battery and buffer battery

Power cells   Cells optimized for minimum resistance, maximum ability to deliver power

Power density   How much power a cell or battery of a given volume can store

Power efficiency   How much of the power generated internally by a battery or cell is seen by a load

Power source   Device or system that provides electrical power, AC or DC

Power supply   Device that converts AC to DC; usually not current-limited; contrast with charger

Precharge   Technique to avoid inrush current

Printed circuit board assembly   Panel on which electronic components are mounted

Protector BMS   Small BMS that includes a protector switch

Protector switch   Electrically controlled switch to disconnect the battery if required to protect the cells

Pulse width modulation   Rapid turn-on and turn-off of a voltage to reduce its average to the desired level

Quadrants   Characteristic of a motor or motor driver, consisting of four permutations of forward and reverse, drive and brake

Radar chart   Graph that compares multiple parameters of cell technologies at a glance

Ragone plot   Graph that compares the power and energy of cell technologies at a glance

Recreational vehicle   Vehicle with living space

Redistribution   High-power technology to keep all cells at the same state of charge at all times

Relaxation   The process that slowly brings the cell voltage toward the open-circuit voltage

Relay   An electrically controlled switch; compare to MOSFET and contactor

Residential application   Electrical storage for a house

Resistance   Characteristic of a device that limits current; specifically, inside a cell, the series resistance that limits the current and causes the voltage to sag when powering a load

Resistor   Electronic component that converts electrical power to heat

RS-232   Older, digital serial data link

RS-485   Digital serial data bus used in industrial applications

Safe operating area   The range inside which a device may be operated without damage

Safety disconnect   Manual switch to open the battery circuit as it is being serviced

Self-discharge   Effect that causes a cell's state of charge to slowly drop over time

Series string   Two or more cells connected in series to achieve the desired voltage. See Cell arrangement

Series-first   Two or more series strings connected in parallel. See Cell arrangement

Shore power   See Charging station

Short circuit   Unintentional connection that draws excessive current

Shunt   See Current shunt

Single-switch battery   Battery whose protector switch uses a single contactor

Small battery   A low-voltage battery typically used in consumer products

Small cylindrical cell   Small, round, hard cell. See Cell format

Small prismatic cell   Small, rectangular cell with metal case. See Cell format

SMB   Digital serial data links used in small batteries

Soft short   Short circuit with a high resistance; contrast with hard short

Solar array   Several interconnected solar panels

Solar charge controller   Charger powered by solar panels

Solar panel   Converts sunshine to electrical power

Source   See Power source

Specific current   Current relative to the capacity of the cell or battery, expressed in C or 1/h; also known as C-rate

Specific energy   How much energy a cell or battery of a given mass can store

Specific power   How much power a cell or battery of a given mass can store

Specification sheet   Describes the characteristics of a given cell

Split battery   Two batteries connected in series with a center tap

Starter battery   See Starter lighting ignition battery

State of charge   A measure of how much charge is available for discharge in a cell or battery, expressed in percentage

State of health   Poorly defined measure of how well a battery is doing

Stater lighting ignition battery   Standard car battery, typically 12V

String   See Series string

Super-capacitor   See Ultra-capacitor

Switch   Electric component that lets current through when closed and does not when open

Tap wires   Wires that connect a wired BMS to its cells

Telecom application   Electrical storage for a base transceiver station equipment

Terminal voltage   Actual cell voltage, such as when powering a load

Thermal runaway   Unstoppable process that destroys a cell, releasing smoke and fire

Thermistor   Electronic component that senses temperature

Top balancing   Balancing a series string at 100% state of charge

Total specific charge transferred   Ratio of the charge transferred over the lifetime of a battery, over the battery capacity, expressed in emery

Traction battery   Battery that is used in the propulsion of a vehicle

Traction motor   Motor that propels a vehicle

Transfer switch   Manual switch that selects one of two power sources

Transformer   Electric component that converts an AC voltage to another AC voltage, isolates the two voltages

Ultra-capacitor   Capacitor with high energy density compared to a standard capacitor, though still much lower than a battery

Uninterruptible power supply   AC-powered battery storage device with an AC output

Unmanned aerial vehicle   Flying device with no people, drone

USB   Digital serial data link used in consumer computers

Variable frequency drive   Converts AC to variable frequency AC to drive a motor

Vehicle   A device used to transport people or freight

Vehicle control unit   Computer that controls a vehicle

Vessel   A marine vehicle such as a boat, yacht, or ship

Volt   Unit of measure of voltage

Voltage   Measure of the potential difference between two conductors, expressed in volts (V)

Voltage sag   The reduction in the terminal voltage of a cell or battery while powering a load

Voltage translation   A technique to evaluate the state of charge indirectly

Watt   Unit of measure of power (W)

Watt-hour   Unit of measure of energy (W-h)

Weak cell   Cell suffering from low capacity or high resistance

Wind generator   Generates electric power from wind

Wire   Single conductor; contrast with cable and harness

Wired BMS   BMS connected to its cells through wires

ACRONYMS

AC Alternating current
APU Auxiliary power unit
BESS Battery energy storage system
BEV Battery electric vehicle
BMS Battery management system
BMU Battery management unit
BTS Base transceiver station
CA Charge acceptance
CAN Control area network
CB Circuit breaker
CC Constant current
CCCV Constant current/constant voltage
CCL Charge current limit
CD Charge depleting mode
CS Charge sustaining mode
CV Constant voltage
DA Discharge availability
DC Direct current
DCL Discharge current limit
DoD Depth of discharge
EAPU Electric auxiliary power unit
ESS Energy storage system
EV Electric vehicle
EVSE Electric vehicle supply equipment
GIGO Garbage in, garbage out
HEV Hybrid electric vehicle
HV High voltage
IR Current times resistance
LCO $LiCoO_2$
LFP $LiFePO_4$, $LiFeYPo_4$
Li-ion Lithium ion
LMO $LiMnO_2$, $LiMn_2O_4$
LNO $LiNiO_2$
LTO Li_2TiO_3
LV Low voltage
MHEV Micro hybrid electric vehicle

MOSFET   Metal oxide semiconductor field effect transistor

MPP   Maximum power point

MPT   Maximum power time (neologism)

NCA   LiNiCoAlO2

NMC   LiNiMnCoO2

OCV   Open circuit voltage

PCB   Printed circuit board, protector circuit board

PCM   Protector circuit module, phase change material

PHEV   Plug-in hybrid electric vehicle

PV   Photo voltaic

PWM   Pulse width modulation

RV   Recreational vehicle

SLI   Starter lighting ignition

SMB   Smart battery system

SOA   Safe operating area

SoB   State of balance

SoC   State of charge

SoE   State of energy

SoH   State of health

SoI   State of imbalance

SoP   State of power

SoV   State of voltage

TSCT   Total specific charge transferred

UAV   Unmanned aerial vehicle (drone)

UPS   Uninterruptible power supply

USB   Universal serial bus

VCU   Vehicle control unit

VFD   Variable frequency drive

INTERNATIONAL DICTIONARY

A standard dictionary may not offer appropriate translation for some of the battery-related terms used in this book. This book was originally written in English; to aid non-English speakers, I would like to offer a limited dictionary of battery-related terms in Spanish[1], German and simplified Chinese[2]. In later editions, with your help, I'd like to add a few more languages.

A single list alphabetized by English would be of limited use when translating into English. Therefore, I thought it would be easier to divide this dictionary into shorter sections of various topics.

English	Español	Deutsch	中文
Measures	Medidas	Maßnahmen	测量
Capacitance (capacitor)	Capacitancia	Kapazität	电容
Capacity (battery)	Capacidad	Kapazität	容量
Current	Corriente	Strom	电流
Efficiency	Eficiencia	Effizienz	效率
Energy	Energía	Energie	能量
Impedance	Impedancia	Impedanz	阻抗
Inrush current	Irrupción de corriente	Einschaltstrom	浪涌电流
Maximum power time (MPT)	Tiempo a máxima potencia	Maximale Leistungszeit	最大电力时间
Open circuit voltage	Voltaje de circuito abierto	Leerlaufspannung	开路电压
Power (physics)	Potencia	Leistung	电力
Resistance	Resistencia	Widerstand	电阻
State of charge	Estado de Carga	Ladezustand	充电状态
State of health	Estado de Salud	Gesundheitszustand	健康状况
Voltage	Voltaje, tensión	Spannung	电压
Electrical Components	Componentes eléctricos	Elektrische Komponenten	电气元件
Bus bar	Barra colectora	Sammelschiene	母线
Cable (multiple conductors)	Cable multiconductor	Mehrleiterkabel	电缆
Capacitor	Condensador	Kondensator	电容器类
Contactor	Contactor	Schütz	接触器
Fuse	Fusible	Sicherung	保险丝
Protection switch	Interruptor de protección	Protektorschalter	保护开关
Relay	Relé	Relais	继电器
Resistor	Resistor	Widerstand	电阻器
Ring terminal	Terminal de anillo, de ojillo	Ringanschluss	环形端子

---

1. Thank you to Gabriel Villaseñor in Mexico.
2. Although the Chinese language has a word for "cell" (细胞), it mostly refers to biology. Unfortunately, when talking about electric storage, the word for "battery" (电池) is used for both batteries and cells.

English	Español	Deutsch	中文
Switch	Interruptor	Schalter	开关
Tap wire (cell voltage sense)	Derivaciones de batería, cable de toma	Abzweigdrahtes	
Wire (1 conductor)	Cable unipolar, alambre	Kabel	电线
Devices	Dispositivos	Geräte	设备
Assembly	Montaje, ensamblaje	Montage	
Battery	Batería	Akkumulator, Akku, Batterie	电池
Bus	Bus	Bus	
Capacitive load	Carga capacitiva	Kapazitive Last	容性负载
Cell	Celda	Zelle	電池
Charger	Cargador	Ladegerät	充电器
Dual bus	Bus dual	Duale Bus	
Inverger	Inverdor	Ladungsrichter	
Inverter	Inversor	Wechselrichter	逆变器
Lead-acid battery	Batería de plomo-ácido	Blei-Säure-Batterie, Bleiakku	铅酸蓄电池
Li-ion cell	Celda de iones de litio	Li-Ionen-Zelle	锂离子电池
Li-ion battery	Batería de iones de litio	Li-Ionen Batterie	锂离子电池
Load	Carga	Last, Ladung	负载
Module	Módulo	Baugruppe	模组
Single-bus	Bus sencillo	Einzelbus	
Cell types	Tipos de celdas	Zelltypen	电池类型
Format	Formato	Format	
Large cylindrical cell	Celda cilíndrica grande	Große Zylinderzelle	
Large prismatic cell	Celdas prismáticas grande	Grosse prismatische Zelle	大号方形电池
Pouch cells	Celdas de bolsa	Pouch-Zelle	聚合物电池
Protected cell	Celda protegida	Geschützte Zelle	
Small cylindrical cell	Celda cilíndrica pequeña	Kleine Zylinderzelle	圆柱形小电池
Small prismatic cell	Celda prismática pequeña	Kleine prismatische Zelle	小方形电池
Cell characteristics	Características de las celdas	Zellmerkmale	特性
Capacity fade	Desvanecimiento de capacidad	Kapazitätsverlust	
Degradation	Degradación	Verschlechterung	降解作用
Hysteresis	Histéresis	Hysterese	磁滞现象
Mismatched cells	Celdas no coincidentes	Nicht übereinstimmende Zellen	
Self-discharge	Autodescarga	Selbstentladung	自放电
Spec sheet	Hoja de especificaciones	Datenblatt	规格表
Cycle life	Vida cíclica	Zykluslebensdauer	循环寿命
Cell arrangement	Arreglo de celdas	Zellenanordnung	排列
Battery bank	Banco de batería	Batteriebank	电池组
In parallel	En paralelo	Parallel	并联
In series	En series	In Reihe	串联
Parallel blocks	Bloques en paralelo	Parallschaltung	
Parallel-first arrangement	Arreglo paralelo-primero	Parallel-Erste Anordnung	并联第一排列
Series string	Cadena en serie	Reihenschaltung	
Series-first arrangement	Arreglo serie-primero	Serienerste Anordnung	串联第一排列
Stack of cells	Pila de celdas	Zellenstapel	
Strings in parallel	Cadenas en paralelo	Parallele	
Battery types	Tipos de batería	Batterietypen	电池类型
Battery	Batería	Akku, Batterie	电池
Battery array	Matriz de baterías (arreglo)	Akku-Array, Batterie-Array	电池组
Buffer battery	Batería amortiguadora	Pufferbatterie	缓冲电池

English	Español	Deutsch	中文
Energy battery	Batería de energía	Energiebatterie	能量电池
Ganged batteries	Baterías en brigada	Ganged Batterien	联动电池
Modular battery	Batería modular	Modulare Batterie	模块化电池
Power battery	Batería de potencia	Leistungsakku	动力电池
Split battery	Batería partida	Geteilter Akku	分体电池
Battery functions	Funciones de batería	Akkufunktionen	电池功能
Charge (noun), Charging	Carga	Ladung	充电
Charge (verb)	Cargar	Laden	充电中
Discharge (noun), discharging	Descarga	Entladung	放电
Discharge (verb)	Descargar	Entladen	放电
Storage	Almacenamiento	Speicher	存储
Battery balance	Balance de batería	Akkubalance	电池平衡
Balance (noun)	Balance	Ausgleich, Balance	平衡
Balance (verb)	Balancer	Ausgleichen, Balancieren	平衡
Balanced	Balanceada	Ausgeglichen	衡的
Balancer	Balanceador	Ausgleich Maschine	
Balancing	Balanceo	Ausgleichen	
Charge transfer balancing	Balanceo por transferencia de carga	Ladungswechsel Ausgleich	电荷转移平衡
Bypass balancing	Balanceo de desvío	Brückenausgleich	旁路平衡
Imbalance	Desbalance	Ungleichgewicht	失衡
Mid balanced	Balanceada en el centro	Mitte Ausgeglichen	中间平衡
Top balanced	Balanceada en la cima	Top Ausgeglichen	顶部平衡
Unbalanced	Desbalanceada	Unausgeglichen	不平衡
Battery design	Diseño de la batería	Batterie Entwurf	
Breakdown voltage	Tensión de ruptura	Durchbruchspannung	击穿电压
Clearance distance	Distancias de despeje	Luftstrecke	电气间隙
Connection	Conexión	Anschluß, Verbindung	连接
Creepage distance	Distancias de arrastramiento (?)	Kriechmindeststrecke	爬电距离
Grounding	Puesta a tierra	Erdung	接地线
Isolation (electrical)	Aislamiento	Galvanische Trennung	电气隔离
Noise immunity	Inmunidad al ruido	Störfestigkeit	抗噪声
Regulatory standards	Normas regulatorias	Regulatorische Normen	监管标准
Reliability	Confiabilidad	Zuverlässigkeit	可靠性
BMS types	Tipos de BMS	BMS-Typen	
Analog protector	Protector analógico	Analogschutz	类比保护板
Array master	Maestro de matriz	Array-Master	阵列主机
Banked topology	Topología en bancos	Banktopologie	银行拓扑
Battery management system	Sistema de Gestión de Batería	Batterie-Management-System	电池管理系统
Battery management unit (BMU)	Unidad de Gestión de Batería	Batterie-Management-Einheit	电池管理单元
Digital protector	Protector digital	Digitaler Protektor, Schutz	数字保护板
Distributed topology	Topología distribuida	Verteilte Topologie	分布式拓扑
Master/slave topology	Topología maestro/esclavo	Master / Slave-Topologie	主/从拓扑
Wired BMS	BMS cableado	Verkabelte BMS	有线 BMS
Wireless BMS	BMS inalámbrico	Drahtloses BMS	无线 BMS
BMS functions	Funciones de BMS	BMS-Funktionen	
Communication link	Enlace de comunicación	Kommunikationsverbindung	
Configuration	Configuración	Konfiguration	组态
Coulomb counting	Conteo de Coulombs	Coulomb-Zählung	库仑计数

English	Español	Deutsch	中文
Current limits	Límites de corriente	Strombegrenzung	
Detection	Detección	Erkennung	检测
Fault protector	Protector de fallas	Fehlerschutz	
Input	Entrada	Eingang	输入
Measurement	Medición, Medida	Messung	测量
Open drain output	Salida con drenaje abierto	Open drain Ausgang	开漏输出
Output	Salida	Ausgang	输出
Port	Puerto	Port	接口
Power (supply)	Alimentación	Versorgung	供应
Power supply	Fuente de alimentación	Netzteil	电源
Precharge	Precarga	Voraufladen	预充电
Protection	Protección	Schutz	保护
Sensing	Detección	Erfassungs	
Shutdown (noun)	Apagado	Herunterfahrens	关掉
Shut-down (verb)	Apagar	Heruntergefahren	
Turn off (verb)	Apagar	Ausschalten	关掉
Turn on (verb)	Encender	Einschalten	打开
Warnings and faults	Advertencias y fallas	Warnungen und Fehler	
Thermal management	Gestión térmica	Wärmemanagement	
Airflow	Flujo de aire	Luftstrom	空气流动
Cooling	Refrigeración, enfriamiento	Kühlung	冷却
Heat (noun)	Calor	Hitze	热力
Heat (verb)	Calentar	Heizen	发热
Heat transfer	Transferencia de calor	Wärmetransfer	传播热量
Heating	Calefacción	Heizung	加热
Insulation (thermal)	Aislamiento	Wärmedämmung	保温
Liquid cooling	Refrigeración líquida	Flüssigkeitskühlung	液体冷却
Temperature	Temperatura	Temperatur	温度
Dysfunctions	Disfunciones	Funktionsstörungen	
Ground fault	Falla a tierra	Grundfehler	接地故障
Short circuit	Cortocircuito	Kurzschluss	短路
Troubleshooting	Diagnóstico	Fehlerbehebung	故障排除
Applications	Aplicaciones	Anwendungen	
Energy storage system	Sistema de Almacenamiento de Energía	Energiespeichersystem	储能系统
Grid	Red eléctrica	Netz	电网
Grid back-feed	Retroalimentación a la red eléctrica	Netzrückführung	电网回馈
High voltage	Alto voltaje, alta tensión	Hochspannung	高压
Low voltage	Baja tensión	Niederspannung	低电压
Maximum power point	Punto de máxima potencia	Maximaler Leistungspunkt	最大功率点
Microgrid	Microrred	Mikronetz	微电网
Stationary battery	Batería estacionaria	Stationäre Batterie	固定电池
Traction battery	Batería de tracción	Traktionsbatterie	牵引电池

INDEX

Lithium-Ion Battery Failures in Consumer Electronics, Ashish Arora,
Sneha Arun Lele, Noshirwan Medora, and Shukri Souri

Microgrid Design and Operation: Toward Smart Energy in Cities, Federico Delfino, Renato
Procopio, Mansueto Rossi, Stefano Bracco, Massimo Brignone, and Michela Robba

Plug-in Electric Vehicle Grid Integration, Islam Safak Bayram and Ali Tajer

Power Grid Resiliency for Adverse Conditions, Nicholas Abi-Samra

Power Line Communications in Practice, Xavier Carcelle

Power System State Estimation, Mukhtar Ahmad

Renewable Energy Technologies and Resources, Nader Anani

A Systems Approach to Lithium-Ion Battery Management, Phil Weicker

Signal Processing for RF Circuit Impairment Mitigation in Wireless Communications, Xinping
Huang, Zhiwen Zhu, and Henry Leung

The Smart Grid as An Application Development Platform, George Koutitas and Stan McClellan

Smart Grid Redefined: Transformation of the Electric Utility, Mani Vadari

Synergies for Sustainable Energy, Elvin Yüzügüllü

Telecommunication Networks for the Smart Grid, Alberto Sendin,
Miguel A. Sanchez-Fornie, Iñigo Berganza, Javier Simon, and Iker Urrutia

For further information on these and other Artech House titles, including previously considered
out-of-print books now available through our In-Print-Forever® (IPF®) program, contact:

Artech House	Artech House
685 Canton Street	16 Sussex Street
Norwood, MA 02062	London SW1V 4RW UK
Phone: 781-769-9750	Phone: +44 (0)20 7596-8750
Fax: 781-769-6334	Fax: +44 (0)20 7630-0166
e-mail: artech@artechhouse.com	e-mail: artech-uk@artechhouse.com

Find us on the World Wide Web at: www.artechhouse.com